FIELD CROP PESTS IN THE NEAR EAST

MONOGRAPHIAE BIOLOGICAE

EDITOR

W. W. WEISBACH

Den Haag

VOLUMEN X

SPRINGER-SCIENCE+BUSINESS MEDIA, B.V. 1962

FIELD CROP PESTS IN THE NEAR EAST

BY

E. RIVNAY

Head of the Department of Entomology
The National and University Institute of Agriculture, Rehovot.

SPRINGER-SCIENCE+BUSINESS MEDIA, B.V. 1962

ISBN 978-94-017-1546-1 ISBN 978-94-017-1544-7 (eBook)
DOI 10.1007/978-94-017-1544-7

Originally published by Uitgeverij Dr. W. Junk, Den Haag in 1962
Softcover reprint of the hardcover 1st edition 1962

Zuid-Nederlandsche Drukkerij N.V. – 's-Hertogenbosch

CONTENTS

PREFACE

In this book data pertaining to the biology, ecology, and control methods of field crop pests are presented. In the term "field crops" there are included grain crops, vegetables, fodder and industrial crops. These data may be of use to the student of entomology, the naturalist, as well as to the farmer in his endeavour to protect his crops from insect pests.

The author has limited himself to the entomology of field crops in the Near East, because of the lack of any treatise of this type. Judging from titles alone, there are a number of books in several languages, that apparently deal with insects from all over the world. In reality, however, these books have a much narrower scope. An American book deals with insects of the United States; an Australian with pests of that country, and a French book with French and North African entomology. Many entomologists from Spain, France and Italy have written on the insects of the western section of the Mediterranean area. This book intends to fill the gap on pests of the eastern section.

During the past thirty years, the subject of our study has never been stable and unchanging; almost every year has presented problems that demanded study; a new pest, a familiar one that changed its habits, a new crop with all the problems of pest control it entailed, or a new insecticide — all claimed the writer's attention.

In 1948, in a pamphlet on insect pests of field crops in Palestine, the author stressed the need for a book on this subject. He wrote then: "... During that period (1920—1945), much uncultivated land was tilled, swamps were drained, hills were cleared of rocks and stones, and the settling of the Negev was begun. Here the problem of insect pests arose. These creatures inhabited the uncultivated land and fed on its plants, and when these plants disappeared, the insects migrated to cultivated fields. Sometimes the cultivated land is still surrounded by untilled areas which are the source of thousands of insects...".

Since these words were written, the tillage of uncultivated land has continued and is going on even now. More settlements have been established in the Negev. The Hulah swamp region has been drained and the Beisan Valley is flourishing with settlements. The systems of irrigation pipe lines and channels are expanding. This is taking place not only in Israel, — the same thing is happening in other countries on the eastern shore of the Mediterranean. Vigorous efforts are being made to till more land, to drain more swamps, and to increase the number of fields under irrigation.

As could be foreseen, the natural balance in insect populations has been upset. With more fields under irrigation, the crop rotations in the summer increased, and thus better conditions were offered for the development of certain insects. These insects, which were formerly more or less harmless, became serious pests.

It is therefore understandable that entomological problems are still prevalent, and have become even more important than previously.

From 1945 onwards air transportation and commerce between countries increased tremendously. This made possible the migration of insects to countries in which they were not indigenous. Such invasions caused new insect pest control problems.

Last but not least, the synthetic organic contact insecticides have been developed during this period. Their influence on practical entomology was greater than expected. Our approach to entomological problems has changed completely. It is true that these insecticides have solved problems that formerly could not be solved, but on the other hand, the insecticides have raised questions that are sometimes more serious than those solved. In any case, these insecticides changed practical entomology to a great extent.

The author often makes use of examples taken from conditions in Israel. This is not only a result of closer acquaintance. Israel serves as a bridge between the northern and southern countries of this area. Conditions prevalent in both European and African Near Eastern countries exist also in that little state. In addition to being part of Asia, Israel has its own specific aspects too. Furthermore, in this country, more than in any other country in the Mediterranean region, agriculture has progressed tremendously. The first steps in draining swamps, reclaiming uncultivated land, and increasing the water supply for irrigation in this area were taken in this country. The entomological results of these changes can be seen in Israel more than in any other country.

The author has been concerned with pests of field crops for the past thirty years. Many of the insects that are discussed in this book have been reared by him or by his co-workers in the laboratory. Some unpublished results are also embodied in this work. To this have been added the findings gathered by other scientists from both nearby and distant countries. With less familiar pests the literature on the subject was studied and the information condensed as presented in this book.

The literature consulted is given at the end of each chapter. However, a great many of these references were not read in their entirety, nor in their original edition but as abstract or reviews. This was particularly so with papers which appeared in languages unfamiliar to the writer. In this respect the writer is greatly indebted to the "Review of Applied Entomology". Without it the work would have been far less complete. Also some Text Books and other compilations were consulted as follows: METCALF, FLINT & METCALF – a General Text Book of Entomology, MENOZZI – Beet Pests in Italy, SILVESTRI, – a text book in applied Entomology, BODENHEIMER, Pests and Animal lifes of Palestine, SHWEIG – Vegetable pests, BALACHOWSKY & MESNIL, Injurious pests of Plants, and PEARSON, Cotton pests and Central Africa – The last two were consulted to a great extent.

Many of the drawings are original, many others were borrowed from other publications and due acknowledgement is made in each case – Sincere thanks are extended herewith to all concerned.

Many chapters in the book were read by some colleagues, the chapters concerning their particular line of work. Thanks are thus due to Mr. I. HARPAZ, Mrs. V. MELAMED-MAJAR, Mr. E. SWIRSKI and Miss S. YATHOM. Mr. A. BRINDLE read the whole manuscript before it was sent to the Printers. Sincere appreciation is extended herewith.

The responsibility for errors which may have crept into a book of such diverse subjects, rests solely upon the writer.

I. INTRODUCTION

Agriculture in the Near East

The climatic dissimilarities and contrasts in the countries of the Near East, and the differences between the peoples inhabiting them, have caused distinct differences in the agriculture of certain countries in the Near East. In some of these countries, Europeans have settled, either bringing with them European cultural methods or introducing changes in some of the customary agricultural practices. The French in the central part of North Africa, the English in Egypt and Iraq, and the Germans and the Jews on the eastern coast of the Mediterranean Sea all brought with them methods to improve local agriculture.

This has resulted, in some countries, in old and new agricultural methods existing side by side. Even in Israel, which introduced both mechanization and new methods in all fields of agriculture, contrasts can be found in neighbouring fields. In one field the practice is deep plowing with a tractor and harvesting with a combine which produces sacks of wheat almost ready for milling. In an adjacent field it may be shallow plowing with a "nail" plow pulled by a camel or donkey, harvesting with a sickle or merely by plucking out the plants from the soil, and stacking of grain to dry and be threshed with a threshing sledge during the summer.

Distinction should be made therefore between the old and the new. The local agriculture before the Europeans modified it will be discussed first.

Agriculture Typical to the Near East

In spite of differences between individual countries, there are, nevertheless, typical patterns throughout the entire region. In accordance with the climate, there are two growing seasons; each season has its typical winter and summer crops.

In the winter, cereals are grown for grain. Seeding takes place from November to December in Israel and Jordan, in January in North Africa; harvesting occurs from May to June. Summer crops are sown from March to May, and are harvested from July to August with some harvesting also in September.

Accordingly, a primitive crop rotation exists which, except for slight changes, is typical for all these countries.

It is as follows:
First year — Winter Crops Second Year — Summer Crops
First year — Winter Crops or Second Year — fallow or summer fallow
First year — Winter Crops or Second Year — watermelon

The usual winter crops are: cereals — wheat, barley, and to a small extent, oats. The wheat is mainly of the hard wheat varieties and in every country there are local strains. The following are examples of accepted varieties in a few countries.
In Syria — Churani, Salmoni;
In Israel and Jordan — Churani, Nursi, Gulgula and Abu Fashi;
In Tunisia — Machmudi, Amira, and Agili.
There are also local varieties of barley in every country.

Besides cereals, some legumes are sown, but only to a small extent. These are: chick peas, *Cicer arietinum;* lentils, *Lens esculenta;* "Julbania", *Lathyrus sativus; Trigonella foenum graecum;* "Ful", *Vicia faba;* "Turmus", *Lupinus termis;* and the clover, *Trifolium alexandrinum.*

The climate in the southern countries of the Near East is not suitable for winter cereals, and the crops are therefore unable to compete with those of the European countries. The success of the crops in certain countries around the Mediterranean Sea is dependent upon the quantity of rain that falls during the season; the climatic conditions at both ends of the season are the deciding factors. Sometimes the rains start early, causing the seeds to germinate, and then cease for such a long period that the tender plants dry up. Sometimes the rains fall late and thus retard the crop; or sometimes they cease too early so that there is not enough moisture in the soil to form a healthy and fertile plant. In more arid regions or in sandy soil, barley, which ripens quicker, is sown in place of wheat.

Since cereals are the major crop and the farmer's livelihood depends upon it, it is sometimes found that he sows wheat year after year. This method reduces the fertility of the soil; and as the second year's yield is poor, this method is not advantageous.

Summer crops necessitate persistent plowing in order to retain the moisture which has accumulated in the soil during the rainy season. This is particularly important for certain crops such as sesame and watermelons.

The summer crops are *Sorghum vulgare* and *Sesamum indicum.* Their success depends, to a great extent, upon the amount of rain during the previous rainy period, the character of the soil tillage, and the number of plowings.

Vegetables typical to countries of the Near East in the summer are eggplant, okra, tomato, cucumber, squash, "fakus", and "lubia". Winter vegetables are onion, lettuce, carrot, radish, cabbage, and cauliflower. Industrial crops are cotton and sesame.

Agriculture in Egypt

In the land of the Nile, where agriculture is based on natural water resources, Alexandrian clover is one of the important crops and ranks high along with wheat and cotton. According to the methods accepted since the beginning of this century, the Egyptians grow their crops in a three year rotation as follows:

First Year	Second Year	Third Year
Wheat	Clover	Cotton

Since the question of water is no limiting factor — the climate is warm and the soil is both rich and fertilized by the water of the Nile — it is possible to grow two crops in the same year and the industrious farmer can sow five or six crops in a three year crop rotation as follows:

	First Year	Second Year	Third Year
Winter Crop	wheat	clover	clover
Summer Crop	corn	corn or fallow	cotton

In poor soils barley is grown instead of wheat, and rice instead of corn. In places where sugarcane is grown, there is a four year rotation as follows:

First Year	Second Year	Third Year	Fourth Year
sugarcane	sugarcane	winter: clover summer: corn	winter: wheat summer: fallow

Modern Agriculture in Israel

The settlers of European origin in countries of the Mediterranean area have introduced and are still introducing improvements in local agriculture. In order to understand the character of these improvements and their motives, Israel will serve as an example. In this country efforts are being made:

a. to improve the fertility of the soil by beneficial crop rotations;

b. to increase the yields by thorough cultivation and by adopting better crop varieties;

c. to lower the costs of production through mechanization;

d. to enlarge the area of arable land by draining swamps and reclaiming wasteland;

e. to extend the land under irrigation by developing water resources and a network of irrigation pipelines;

f. to variegate farming by the introduction of crops which have not been grown before in the country;

Examples will be given on these points:

A. Crop rotations — The usual crop rotation on the new farms is a three year cycle:

First Year	Seond Year	Third Year
Winter grain	Summer crop	Fodder or Winter grain

Some farms practice a four year rotation. The following are examples:

First Year	Second Year	Third Year	Fourth Year
Winter cereals	Summer crops	Winter cereals	Summer fallow
or Winter cereals	Summer crops	Winter cereals	green manure
or Winter cereals	Summer crops	Winter cereals	hay
or Winter cereals	legumes for grain	summer crops	legumes for hay

The first three rotations are alike but for the difference in the fourth year; in all of them there are two years of winter cereals. In the fourth method there are two years of legumes, one for hay, and the other for grain.

Some agronomists advise a six year rotation for the dry fields of the northern Negev as follows:

Winter cereals — summer crops — legumes for hay or manure — winter cereals — Sudan grass for pasture — legumes for hay or grain.

The purpose of all these rotations is to retain the fertility of the soil.

B. Increased Production — The varieties of hybrid corn and sorghum, which were introduced into the country, gave higher yields per area. Also the intensive cultivation of corn, both of the local and foreign varieties, increased yields from 70 kg to 190 kg per dunam*.

* 1 dunam = 1000 m^2 = approx. ¼ acre.

C. Mechanization — Field crop farming in Israel is based completely on mechanization; i.e. on plowing and tilling with the aid of tractors, and harvesting with combines and other machines. Crops which cannot be cultivated by machines have been excluded from the crop rotation. The tall and hard sorghum, the so-called "durra" which cannot be harvested by machine, has been excluded and in its place low sorghum is sown. For the same reason, the growing of sesame and various legumes such as lentils and chick peas, requiring manual labour in harvesting, has been limited.

D. Reclamation of Areas — If we take into account the activity of the Jewish settlers from the start of the century, most of the Jezreel Valley can be considered as a property which was made available only by the drainage of the swamps which covered large areas of this valley. The swamps of the Northern Sharon and other areas were also drained. The peak of this project will be reached with the drainage of the Hulah lake and surrounding swamps, whose waters will flow to remote arid areas. By installing extended water pipe lines, farming areas have been regained in the Beisan Valley and the northern Negev where rain is scanty. Some farming regions have been reclaimed by bringing water from far away wells and by terracing slopes. This process is continuing and reclaimed regions will increase with the completion of the Jordan — Negev pipeline.

E. The Enlargement of Land under Irrigation by Exploitation of the Soil Water Table — The springs and streams in this country are not only utilized for irrigation at distant sites, but also for the increment of nearer areas under irrigation. A crowded network of irrigation pipes is being laid. Portable pipes, which can be shifted from one place to another, have aided in this development.

F. The Import and Development of New Crops — The increase of land under irrigation resulted in a demand for the introduction of new crops which previously were not grown at all or were neglected. Today large areas of alfalfa provide the raw material for drying and milling plants; peanuts have become an important item of export; cotton fields satisfy a large proportion of the local demand for cotton; sugar beet fields will gradually provide for the entire sugar need of the country; and crops of oil seeds, safflower, sunflower and sesame supply the local oil presses. This situation did not exist 20 years ago.

The Causes of Fluctuation of Insect Populations

In the countries of the Mediterranean area, there are insects of various zoogeographical elements, and which are typical of different climates. There are insects which occur all year round, others which are prevalent in the summer, and still others which are found in the winter only. The winter in the Mediterranean area is mild and temperatures do not reach extremely low levels, and therefore some insects are active during the winter. The fluctuations of insect populations are outstanding from season to season and year to year. In the following paragraphs the factors which are responsible for these changes in populations will be discussed.

A. The Problem of Food

Food, available throughout the year in abundance, creates the conditions necessary for the increase of the number of animals which feed upon it. If there were no other limiting factors, these animals would multiply to such an extent that they would destroy their food or host-plants. When cereals, e.g. wheat, are sown year after year without any crop rotation, the cereal leaf miner is capable of completely destroying the plant upon which it feeds.

In this connection, it should be pointed out that man himself often upsets the natural balance between the plant and the insects feeding upon it. In Israel, for example, great changes in the density of the population of various insects took place because of new practices in agriculture. A few examples will be considered here. Before the increase of land under irrigation, sorghum was the only plant of summer cereals which was grown in this country. Only one generation of *Sesamia cretica* developed on this crop, the peak of growth of which was in May-June. Today, with the introduction of irrigation and the extension of irrigated land, various varieties of corn, sorghum, grasses etc. are being cultivated throughout the summer. These summer crops cover large areas from May to October, and create an unlimited supply of food. The insects which feed upon them can now raise additional generations in summer; or in other words, a sharp increase in population has taken place. These new conditions gave rise to insects whose presence was nearly unnoticeable previously. In the former period, for example, the sorghum maggot *Atherigona excisa* could not find suitable conditions for its development. This species feeds on young plants and thus, when the sorghum was in a suitable state for this fly, the pest was not yet active. Today, young plants are available throughout the summer, so that this pest could spread greatly. Similar to the example of sorghum, cucurbit crops should be mentioned; the cropping cycles follow each other through the entire summer. These conditions have permitted the increase of *Rhaphidopala foveicolis*, *Baris granulipennis* and others. In the above, the term "food" refers not only to the plant as a whole, but often rather to a specific part, or to a detailed state of maturity which is suitable to the particular pest under discussion. For instance, *Toxoptera aurantii* feeds on immature citrus leaves and the rise in its population cannot occur but in a period when soft new leaves are plentiful on the trees.

B. The Influence of Temperature

In countries of the Mediterranean area, the summer temperature may rise to such a degree as to have a harmful influence on the insects, and thus the course of building up of populations is interrupted. *Epilachna chrysomelina*, for example, ceases to reproduce at a temperature of 32—33 °C, and at a temperature of 28—32 °C the oviposition decreases a great deal (MELAMED, 1956). Females of *Agrotis ypsilon* which were kept during their larval stage at a temperature above 26 °C, laid less eggs than when the larvae were kept at 22 °C, or did not lay at all (RIVNAY, 1961). The number of adults of the Mediterranean fly trapped in a week in August 1956 when the temperature rose to 35 °C showed a decrease of 90% in comparison with that of the week before the heat wave.

Too low a temperature also causes a decrease in the population of insects, both directly and indirectly. Cold weather in the Mediterranean area only rarely causes the sudden death of insects in large numbers. Sometimes, when the temperature drops to 0 °C or below, dead Mediterranean fruit flies can be found in citrus fruits.

In a cold, long and rainy winter, the death-rate of over-wintering females of *Vespa orientalis* is very high, and *Prodenia* larvae do not survive to pupation.

However, low temperatures influence the population mainly by slowing the rate of insect development. For example, the development of *Prodenia* in the winter takes more than three months, instead of less than one month in the summer. The development of the Mediterranean fruit fly in winter takes more than ten weeks instead of three weeks in summer. Low temperatures may prevent reproduction even if the insect is active. The threshold of reproduction of some insects is above their threshold of development. Examples for this among insects of the Mediterranean area are as follows:

Name of Insect	Threshold of Development (°C)	Threshold of Reproduction (°C)	Optimum Temperature (°C)
Capnodis carbonaria	19	26	28—31
Ceratitis capitata	9—11	13—15	23—27
Epilachna chrysomelina	14	22—23	25—28
Hypera variabilis	9.4	10	17—20
Liogryllus bimaculatus	15	19	28—34

It is often found that a low temperature, which permits reproduction and development, reduces the activity of the hatching larva to zero. Thus larvae may die of starvation immediately upon hatching from the egg, because of inability to move about in search for food: larvae of *Coccinella 7—punctata* hatched from eggs at a temperature of 13°C, but remained near the egg shells unable to move, and finally died. In the winter, many dead neonate larvae can be found under the scales of the female of *Chrysomphalus aonidum*. Because of the cold weather they were unable to crawl out and to look for a suitable place to feed and thus died of starvation. This is an important factor in the course of building up the scale population in citrus during the citrus season.

C. The Influence of Humidity

In the countries around the Mediterranean relative humidity often drops to low levels; it is sometimes very dry for a number of hours during the day. This lack of moisture influences the death rate of insects to a great extent, mainly in those stages of development in which the insect does not take up food, namely in the egg and pupal stages. Eggs of *Epilachna chrysomelina* almost all die at humidities below 40% (MELAMED, 1956). Dormant larvae of cereal leaf miner usually die at humidities of less than 50% (RIVNAY, 1956). Also pupae of *Heliothrips haemoerhoidalis* die when the humidity is less than 50% (RIVNAY, 1935). Eggs which are in the soil are especially sensitive in this regard; eggs of locusts, crickets, *Marseulia*, *Rhaphidopalpa*, *Sitona* and others need a humid environment and even a thin film of contact moisture in order to develop. As long as these conditions do not exist in the soil, they do not hatch. On the other hand, eggs of other insects are resistant to low humidity and, on the contrary, are sensitive to high humidity, as for example the eggs of *Capnodis*, from which the insects may hatch at a humidity of less than 10% but are injured when it is above 85% (RIVNAY, 1944a).

The moisture in the soil may influence other biological phenomena of the insect. The awakening from summer dormancy of many insects depends on the amount of water supplied by the rains. The presence or absence of different pests in certain years, depends on the amount and timing of rain in those years. The appearance of *Phyllopertha* and the cereal leaf miner *S. temperatella* depends upon the time and quantity of rain (RIVNAY, 1956).

D. Diapause

The life history of certain insects, the food of which is absent during a definite part of the year, must be fitted to the annual cycle of the food host. During the period in which the plant is absent, the insects which depend entirely upon this plant become dormant. This may be either a summer or a winter dormancy, in accordance with the cycle of the host.

It should be pointed out that in the term "diapause" the extreme slowing down of development due to lower temperatures is excluded.

In the chapters dealing with *Heliothis*, *Leptinotarsa*, *Platyedra* and others the reader will find detailed discussions on the factors influencing diapause. Here only a few examples will be mentioned:

1. The cereal leaf miner attacks winter cereals. The larva awakens from its dormancy at the beginning of the winter, when its host is at the start of its growth. It completes its development after a few weeks. The pupa, adult and egg develop in a month's time, but the young larvae of the ensuing generation do not continue to develop, but go into diapause which lasts until the beginning of the following winter (RIVNAY, 1956).

2. *Phyllopertha nazarena* feeds on the roots of cereals during the winter, and in the summer, when its tender and fresh food is lacking, it is in a state of diapause (larval stage) at a 40—70 cm depth in the soil. The insect awakens from its dormancy the following winter (RIVNAY, 1944b).

3. *Hylemyia antiqua* emerges in November from the pupa which was dormant during the whole summer. The pupae of the generation which these flies raise, become in part dormant and in part continue to develop. The same thing occurs in the following generations (YATHOM, 1960).

When the diapause of these insects in the Mediterranean area is compared with that of insects in northern countries, a fundamental difference is seen. The dormancy of certain insects in the Mediterranean region is in the summer, while in the north it is in the winter. Certain Scarabaeids related to *Phyllopertha nazarena* in Europe and America are active in the summer and dormant in the winter. *Hylemyia antiqua* in Holland, Germany, England etc. raises the same number of generations as in Israel, its dormancy in the above mentioned northern countries being in the winter, while in Israel this occurs in the summer. In other words, these insects are dormant when their hosts, cereals or onions, are not available. On the other hand, other insects, such as *Baris granulipennis* or *Rhaphidopalpa* are dormant in Israel in the winter because their host, melons, are not available then.

There is, of course, another difference which may occur, namely an additional generation in the Mediterranean area in comparison with northern countries.

Lixus junci, for example, raises in Italy only one generation, which matures in July and goes into winter dormancy until the following spring. In Israel, one generation matures in May and in part continues to develop and reproduce, and only the insects of the second generation, which mature in July, go into dormancy until the following winter. In other words, in Italy the insect is univoltine while in Israel it is bivoltine (RIVNAY & MELAMED, 1956).

E. Insect Migrations

One of the causes of the widespread outbreak of a certain insect in an area may be due to their migration, or their invasion from other countries. There are insects, which under certain conditions develop a tendency to migrate. The result of this tendency is to leave unfavourable environmental conditions and to settle in places with more favourable conditions.

In the chapters on locusts, aphids and noctuid moths longer discussions are presented regarding the migration habits of these insects. Here the types of migration are enumerated and examples of each are brought.:

a. Inter-Host Migration. Within the same area an insect may pass over from one plant to another. The time of migration often depends upon the physiological state of both the plant left by the insect and of the plant towards which it is attracted and migrates. This fact is well known especially in the case of aphids, which frequent many hosts in accordance with the latter's growth and the season. Typical is the migration of *Forda* sp. and *Geoica* sp. which feed in the summer on leaves of *Pistacia,* and in the winter on the roots of cereals. Among other insects, mention should be made of the Mediterranean fruit fly which is polyphagous and attacks many hosts indiscriminately. The pomegranate butterfly *Virachola livia* attacks in one season the pods of the *Acacia* and the fruit of *Punica* in another. The purpose of these migrations is the search for proper food and is indirectly a result of climatic conditions.

b. Intraterritorial Migration. Certain insects, such as *Coccinella 7—punctata* spend the winter in the warm valleys and the summer in cooler regions of the same country. This is true with some butterflies; for example *Pontia daplidice* spends from March to May in the Jordan Valley, and from June to November in the hills. The causes of migrations like these are climatic.

c. Inter-Regional Migration. Various insects, such as the locust, mentioned above, migrate to distant countries. In addition to locusts, there is a list of insects of other orders that migrate long distances. In this field of research the unknown is greater than the known. However, it is possible to point out a number of pests whose local population is renewed through invasions from other countries; *Pyrameis cardui* migrates from the south as does the sphingid *Deilephila lineata livornica.* Among the Noctuidae, *Laphygma exigua* and *Heliothis armigera* are migrants, more details of which are in the chapter on Noctuidae.

LITERATURE

V. Melamed, 1956. *Ktavim,* **7,** *83–95.*

E. Rivnay, 1935. *Bull. Soc. Roy. Ent. Egypte, 119-124.*

E. Rivnay, 1944a. *Bull. ent. Res.,* **35,** *235-242.*

E. RIVNAY, 1944b. *Bull. Soc. Ent. Fouad, 101-108.*
E. RIVNAY, 1956. *Ktavim*, **7**, *1-27.*
E. RIVNAY, 1961. Contribution to the Phenology and Ecology of *Agrotis ypsilon*. MS.
E. RIVNAY & V. MELAMED, 1956. *Ktavim*, **7**, *63-82.*
S. I. YATHOM, 1960. *Agr. Res. Sta. Special Bull.* **25**, *1-169* (in Hebrew).

Agrotechnical Methods of Insect Control

The development of insect pests may be prevented by agrotechnical procedures. One may thereby prevent the build up of an insect population, which may be otherwise hard to control. The life histories of pests of cultivated plants are often adapted to the season and to the development of their host plants. Often even only a slight deviation from the ordinary farm practice may exclude insect pests and the exploitation of one of their weaknesses may prevent the damage they cause. In order to succeed with such practices the farmer must have a thorough knowledge of the life history and habits of the pests. Certain agrotechnical methods of insect control may be beneficial to the plant from other points of view. In other words, it is the cheapest and best method of fighting pests.

A. Thorough Plowing

Many of the insects whitch attack field crops spend some part of their life in the soil. With insects which are not subterranean this period is a resting stage; the insect is either in the form of a pupa, egg or larva. The insects select a suitable place in the soil according to their requirements of temperature and soil moisture and remain there. A change in these conditions may be fatal to them and being immobile the insects die in large numbers. This can be brought about by plowing: the turning over of the soil brings the pest to the surface where unfavorable humidity and temperature conditions destroy them. Thorough and frequent plowing is the best and most desirable method for destroying the cereal leaf miner *Syringopais temperatella* and the leaf beetle *Marseulia dilativentris*. These insects exist at depths that can be reached by the plow. However, this method is not suitable for those insects that are at depths beyond the plow's reach, as in the case of *Phyllopertha nazarena* which spends the summer in dormancy at a depth below 40—50 cm.

B. Field Sanitation

Plant remnants, such as stalks, remain in the fields after harvesting. These often contain the pupae or larvae of pests. Pupae of the Hessian fly may be found near the roots in wheat; the larvae of *Pyrausta* and *Sesamia* spend their resting period in corn stalks and the larvae of the grain saw-fly *Cephus* in wheat stubble.

Such crop residues in the fields provide a source of infestation for the crop which grows during the following season. Most entomologists therefore advocate turning over of the soil as soon after harvest as possible. Many of the larvae in the resting stage perish in the process of plant decay. However, this is not the solution to pest control in every case. Insect pests may migrate long distances and may reinfest a field in spite of healthy cultural control methods.

C. Weeds or Volunteer Crops

Insects may lay their eggs in a field before the crop has sprouted. They have been attracted there by the weeds growing in the field. The larvae which develop feed on the weeds and then on the cultivated crop which has sprouted in the meantime.

On the borders of fields, on road sides and in ditches, weeds often grow extensively. Insect infestation usually occurs in these places *Agrotis ypsilon* usually begins its ravages in fields adjacent to roads and ditches.

Volunteer cotton of the year before has usually served as host to the pink boll worm and to the spiny boll worm. A great improvement took place after sanitary conditions in the cotton fields were corrected.

D. Crop Rotation

In an uncultivated field where several plants grow together, the limited food supply causes a natural balance of the insect population. The elimination of this vegetation and its replacement by one particular crop in that field destroys this natural balance. The pests, to which this particular crop serves as host, become more numerous because of the abundance of the food supply. Furthermore, if the same crop is grown year after year on the same land, as is often the practice in the Near East, its pests are liable to build up populations to dangerous proportions. Thus, if wheat is sown in a field where wheat was grown the previous year, it is as a rule more infested with the leaf miner *S. temperatella*. Consecutive cultivations of a crop on the same or neighbouring fields should therefore be avoided as much as possible.

Crop rotation may bring about benefits other than the prevention of pests. For example, a legume causes the accumulation of nitrogen in the soil which can then be available to wheat grown the following year. Such wheat is better adapted to withstand the maggots of *Chortophila flavibasis*.

Again, crop rotation involves practices which are of benefit not only to the crop under cultivation, but also to the crop which follows. A case in point is wheat grown after sesame or watermelons. For the latter crops the soil is ploughed several times during the spring, by which the insects hidden in the soil are destroyed. Wheat and barley when grown after these crops were not infested with the cereal leaf miner, while adjacent fields were heavily infested with this pest. It should be remembered, however, that such a method may be employed only against insects which do not fly or migrate.

E. Resistant Crops

Usually insects are not limited to one plant but feed on various plants. Even monophagous insects are not restricted to just one plant; they feed upon a number of species of the same family. There seems to be, however, a degree of preference, as regards the host plant. *Epilachna chrysomelina* which feeds on various Cucurbitaceae prefers the watermelon, while *Rhaphidopalpa* selects the melon. *Sesamia* is more attracted to sorghum, while *Pyrausta* prefers corn. This preference is due without doubt to differences in certain plant characteristics. Such differences have influenced the development of varieties resistant to certain insects in the past and may do so in the future.

Corn varieties in America have coarse leaves which prevent damage from *Blissus leucopterus*. The corn growers expect to develop a variety of corn whose silk is so small that it will not attract *Heliothis zea* which oviposits in the silk.

F. Irrigation

Many insects are attracted to fields after irrigation. In many cases it is hard to control pests such as *Prodenia* and *Laphygma* because of their repeated oviposition in the same field. This should be remembered at the timing of control applications and irrigation.

G. Sowing Density

Some soil insects feed on and destroy the seed in the soil before it sprouts. Some farmers increase the number of seeds per area so that a sufficient number shall remain for germination. After developing, the superfluous plants are thinned out. In countries where wire worms destroy wheat seeds it is customary to increase the number of seeds to such an extent that, even though the pest destroys many seeds, the wheat develops well. In Israel seeds of sorghum and corn are attacked by Tenebrionidae such as *Hionthis*, Millipedes such as *Strongylosoma syriacus*, and others. Dense seeding results in full rows and excess sprouts are thinned out subsequently.

H. Planting Time

A correlation usually exists between the life history of the pest and that of its usual host in the same country. In Israel, for example, the larvae of the cereal leaf miner awaken from their dormancy when there are young wheat and barley plants into which they can penetrate and feed upon.

The farmer can sometimes impede this harmony and prevent the damage by the pest. In the U.S., for example, it is customary to delay sowing in the fields that are known to be attacked by the Hessian fly, so that the insect oviposits before the seeds sprout. The pest meanwhile dies out and when the crop comes up there are no laying females in the field.

It is not possible to make use of this method of insect control with non-irrigated crops in Israel. Here, seeding and growth depend upon the winter rains and seeds are sown by tractor during a short period. Nevertheless, it is known in countries of the Near East that wheat or barley sown between at the end of December or the beginning of January is free from the cereal leaf miner, while that sown in November is attacked by it. In neighbouring clover fields of different ages it often happens that the earlier sown plants are completely consumed by the pests *Laphygma* and *Prodenia* while those sown later are not attacked at all. The opposite case also occurs: early sowing may produce healthy plants which can withstand the attack. The damage depends on when the insect appears. The experienced farmer must remember which of his fields are attacked and when the attack occurs. He can then exploit all his resources against pests.

I. Healthy Plants

A healthy plant is likely to survive both insects and diseases. Neighbouring fields often give a striking example of this. In 1941 fields in the region of Tel-Adashim and Mizra were attacked by the wheat maggot *Chortophila*. In fields with healthy plants, with stems hardened early because of rapid growth, the damage was slight. Even where the main stem was attacked, new stems were able to develop.

Quarantine Measures

Closer international contact, the greater exchange of merchandise between countries and the improving facilities of communications have brought about an increased dispersal of insects from one country to another, or even from one continent to another. Many of the invading insects are notorious pests. As one of the earliest cases, mention should be made of the Hessian fly which was introduced from Europe to North America as early as 1780. On the other hand, the *Phylloxera* was introduced from America into Europe a short period after steamboats began to cross the Atlantic. The Mediterranean fruit fly spread to many countries with the aid of the steamboats travelling from Europe to India along the western shores of Africa.

Air communication gave a boost to the dispersal of insects, because with the shortening of the duration of the journey it became easier for insects to survive the trip.

It is well-known that pests have become more troublesome in their newly adopted country than they have been in their country of origin. In the United States, for instance, large sums were and still are spent to control the European Corn borer and the Japanese beetle, pests which cause far less damage in their countries of origin. Great efforts have been exerted in Florida to eradicate the notorious Mediterranean fruit fly.

In Israel, too, many pests were introduced accidentally from foreign countries. During the past twenty years alone about a dozen pests have established themselves in this country, some of which cause great trouble and involve expenditures in their control, while some are not under control yet. To mention a few: The russet mite *Vasates destructor* on tomatoes; the rust mite *Phyllocoptruta oleivora* on citrus; the gall thrips *Gynaikothrips ficorum* on *Ficus nitida;* the Eucalyptus borer *Foracantha semipunctata* on *Eucalyptus* and the Fig borer *Batocera rubromaculata* on figs.

In view of this, it is no wonder that many countries have established regulations with the purpose of preventing the incidental import of undesired insects. Such regulations prevent or limit the import of cargo which is liable to carry with it certain pests. Often the regulations require that a health certificate from the country of origin accompany the merchandise. Other regulations demand that prior to its entry into the importing country the goods should be fumigated or treated otherwise against certain pests. Some regulations may be specific against certain pests, while others may limit the import from some specific countries only. Thus Italy limits the import of fresh fruit only from countries where fruit flies of the genera *Dacus* and *Anastrepha* prevail, i.e. South American countries, India, Australia and South Africa. On the other hand, certain countries may be exempted from restrictions. Thus Cyprus allows the import of tomato seedlings from Israel only.

Many countries arrange special plants for fumigation or other treatments of certain goods. However, regulations often do not take into account the circumstances by which insects are transferred by means other than by man, so that often restrictions may be of no avail. Wind, birds and the like may transfer insect pests long distances and introduce them into a new country, defying its existing regulations — a good example is the Mediterranean fruit fly in Florida. Furthermore, not in all cases are the restrictions justified. Some countries have regulations against certain pests which will not survive its conditions even if introduced there. For instance, there are reasons to believe that the Colorado potato beetle will not be able to exist in the warm arid countries of the Near East. Still many of these countries have restrictions against this pest.

II. POLYPHAGOUS INSECTS

ORTHOPTERA

Locusts and Grasshoppers

Damage to agriculture by Orthoptera is caused chiefly by members of the Grasshopper family Locustidae (formerly Acrididae). Grasshoppers of different kinds are permanent residents in the Near East, and at times, when conditions are especially favourable for their development, they cause damage. In certain years some species attack various field crops and often do serious harm. Occasionally migratory locusts coming from remote places invade the countries in the Near East.

Relationship between Grasshoppers and Locusts

The riddle of the identity and origin of the migratory locust, which has perplexed mankind for many generations, was solved at the beginning of this century by the well-known entomologist UVAROV. He discovered that the locust is nothing but a phase of the usual grasshopper. Under certain conditions, which will be discussed later, some species of grasshoppers are able to give rise to new generations which differ in body and in habits from their parents. The same species appear therefore in two phases — a solitary phase, the grasshopper, and a migratory phase, the locust. The differences in colour and structure of body between the two phases are sometimes so great that it is difficult to discern any relationship between them. As regards differences in behaviour – the migratory phase is inclined to congregate, form swarms and migrate long distances from its birthplace, whilst the solitary phase remains where it is and lives singly. In recent years, however, indirect evidence has accumulated to the effect that the solitary phase, too, may migrate further than was formerly thought (GUICHARD, 1955).

The desert locust *(Schistocerca gregaria)* is the most common species in the Near East. Its origin, it appears, is in the Sudan and surrounding areas. There the locust lives in the solitary phase, and is known by the scientific name of *Sch. gregaria* ph. *flaviventris* BARN. In the countries of its migration it is known as *Sch. gregaria* ph. *gregaria* FORSK.

The Desert Locust — Schistocerca gregaria Forsk

Distribution. It appears that this locust is found in the solitary phase on the steppes of the Sudan, in the southern part of Tripolitania (GUICHARD, 1955) and in eastern Eritrea (JANNONE, 1953).

Centres of the solitary phase are present also in West Africa (Senegal) in the West, and Pakistan and neighbouring territories in the East.

In the migratory phase it is present over a large area of the continents of Africa and Asia. From the maps reproduced by the Anti-Locust Research Centre, it is seen that the area visited by migratory swarms extends over the whole of North Africa, the southern boundary being approximately along the 10th latitude North; in East Africa the area includes Ethiopia, Somalia and Kenya as far as the 5th latitude South in that country.

In Asia it includes north-west India, Pakistan and all other north-west Asian countries, the northern limit being a line drawn along the southern shores of the Caspian Sea. Occasionally swarms may penetrate further north into Asia Minor to the east and Portugal to the west of its area of distribution.

Type of Injury. Damage by the locust is caused at all stages of its active development, but not to an equal degree. The pink adults which have not reached sexual maturity eat very little; the yellow adults eat much more, but most of their time is spent in mating and laying eggs. The biggest eaters are the ravenous hoppers which take in more and more food progressively as they grow.

Locust larvae devour all herbaceous plants. They chew every leaf and soft portion of trees and damage the bark too. There is no better description of the damage they may cause than that given in the book of Joel, chapter II, verse 3 "The land is as the Garden of Eden before them, and behind them a desolate wilderness".

Description of the migratory phase. The adult. Body length from the head to the tip of the abdomen is 58-72 mm, and from the head to the wing-tips 62—78 mm; the length of the antennae is 15 mm. The anterior part of the prothorax is constricted. Two transverse furrows divide it into 3 zones — the anterior being the widest and the second the narrowest. The posterior margin of the prothorax is wide and forms a wide rounded angle. It is bisected by a slight longitudinal ridge. Its surface is spotted, the anterior part being smooth.

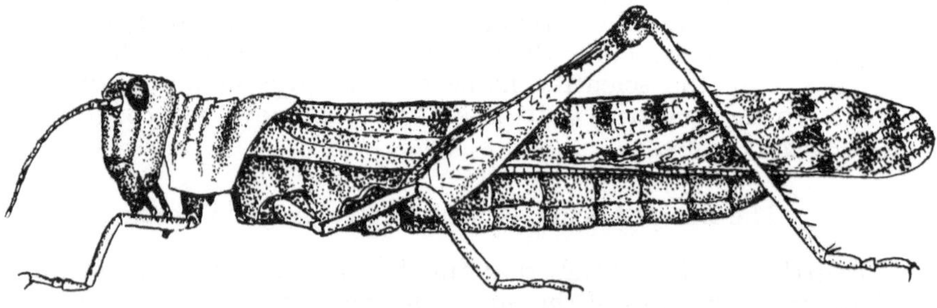

Fig. 1. *Schistocerca gregaria* – adult.

The colour of the adult immediately after the last moult is pinkish-brown and its abdomen is dark brown. The tegmina are spotted and near the tip the spots may run together to form a wide irregular band. The hind wings are transparent and more pink towards the base. The wing-tips are brownish. The hind tibiae are light yellowish-red becoming reddish at the tips. Their spines are light and the tips black.

When the adult of the migratory phase is fully mature, its body is completely yellow.

The colour of the adult in the solitary phase is a very light greyish yellow during all its stages. Along the middle of the pronotum there is a light-coloured stripe, with two additional bands, one on each side of the prothorax. A dark brown area can be distinguished between the stripes. The surface of the prothorax is reticulate.

The egg is yellow or yellowish-red in colour; its length is 8 mm and its width 12 mm. The larva on hatching from the egg is 8—10 mm long and looks like the adult except that it is wingless. In proportion to the length of its body it has a big head — up to 4 mm long. The hind legs are also bigger in proportion to the body than in the larger larva. Wing-pads appear in the fourth larval instar in the shape of small triangular scales. In the fifth instar the pads become larger — up to 15 mm in length. The larvae are black and speckled with yellow patches and lines (the colour of the larva of the solitary phase is grey-green).

The Biology and Ecology of the Desert Locust

The biology of the solitary grasshopper in the Sudan has hardly been investigated, while that of the migratory locust has been studied more fully in the countries of its invasion.

The larva moults five times before maturity, and it has therefore 5 larval instars. The period between the last moult and egg-laying is the pre-oviposition period.

The Pre-oviposition Period

Under favourable conditions of temperature 3 weeks pass between the last moult and egg-laying. The male is ready to mate about 3 weeks after its emergence. At lower temperatures, such as prevail during the spring in Israel, the pre-oviposition period lasts about a month, while about 6—8 weeks are necessary at the lower temperatures common in winter; at still lower temperatures 3—4 months are required.

The length of the pre-oviposition period may be influenced by factors other than temperature, such as density of population. Mrs. Norris (1954) found that when females of *Sch. gregaria* were reared in isolation, oviposition was delayed. But when females were together with males, this period was shorter and was correlated to the maturing period of the male (about 10 days at a favourable temperature). Over 28 days were required before females without males began to lay. Males also have a reciprocal influence on their own maturation; the presence of mature males accelerated the maturation of younger males, while young males and young females retarded the maturation of older males.

Ecological Conditions

Most writers have remarked that the most important requirement for oviposition is moisture in the soil. BODENHEIMER & FRANKEL (1929) state that moist, sandy, low-lying soils, such as occur in wadis, where moisture is stored in the upper layers, are preferable. POPOV (1958) says that moisture at a depth of 6 cm is sufficient, but that soils with a dry surface are better than completely saturated ones. Miss WALOFF (1946), in her survey of the breeding territories in East Africa, points out that each breeding period occurs in a rainy season. Areas with sparse vegetation are preferred to those of dense growth or to bare ground (POPOV, 1958).

Parthenogenesis

From recent studies it appears that females of *Sch. gregaria* may reproduce parthenogenetically (NORRIS, 1954, HAMILTON, 1955). Six generations of individuals were reared by HAMILTON from unfertilized eggs. Unlike certain Hymenoptera, only females developed from these eggs; they mated and oviposited like normal females, but, as mentioned above, their pre-oviposition period was longer, and mortality of eggs three times greater than that of fertilized eggs.

Oviposition

The female digs a hole in the ground about 8 cm deep, aided by the hooks on the tip of her abdomen. There she lays a mass of eggs covered by an envelope of froth, which dries quickly. This covering protects the eggs and facilitates the exit of the larvae. The number

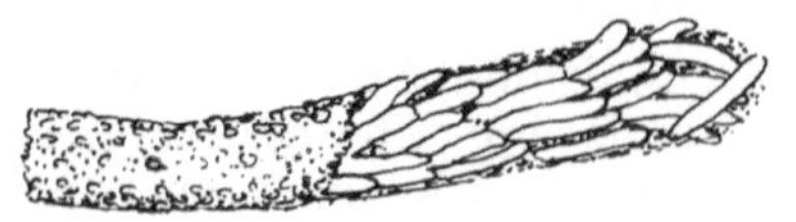

Fig. 2. Egg pod of *Schistocerca gregaria.*

of eggs in one egg-mass varies from 20 to 100; on an average there are 40—50; the female may lay more than one egg-mass. More than half of 116 females raised under laboratory conditions laid more than one egg-mass, as may be observed from the data of BALLARD (1931) (table I).

Table I.

Oviposition by the migratory phase (after BALLARD, 1931)

Eggs laid by 32 females		Egg-masses laid by 116 females	
No. of eggs laid	Percentage of females	No. of egg-masses laid	Percentage of females
70—100	21.8	1	39.2
101—150	31.4	2	31
151—200	25	3	14.7
201—250	12.5	4	7.7
251—300	9.3	5	5.2
		6	1.7

There are on an average 7—10 days between each egg-laying. In unfavourable conditions as long as a month may elapse.

The total number of eggs laid by one female is generally more than 100. In table I the number of eggs laid by 32 females under laboratory conditions are given.

Developmental period of the egg

BALLARD'S figures for the developmental period of the egg at different temperatures are given in table II.

Table II.

Developmental period of the egg of the desert locust (BALLARD, 1931)

Month	Mean Temp.	Period (in days)
March	15 °C	43
April	20 °C	30
May	25 °C	21

Eggs kept at a temperature below 18 °C failed to develop (BODENHEIMER, 1929). Thus, it seems from the observations of BALLARD that in March the eggs developed only during the day when the temperature was above 18 °C, and at night, apparently, no development at all took place.

This view is confirmed by the breeding experiments of SHULOV, 1952. Eggs kept at a constant temperature of 15 °C failed to hatch. The time of development of the egg at various degrees of temperature, as determined by this author, are given in Table III.

Table III.

Incubation of eggs of *Sch. gregaria* at various degrees of temperature (SHULOV 1952)

Temperature in °C	15	20	27	30	33	37
Days	—	45—50	14—15	12	11	—
Percentage hatched	0	25	70—90	70—90	10	0

Limiting Factors

Observations made by BALLARD on 1,055 eggs showed that only 83.5% of them hatched. BALLARD claimed that under ideal moisture conditions the percentage hatching may reach 100%. A more intensive study of the influence of temperature and moisture on the development of eggs was made by BODENHEIMER & FRANKEL (1929); their results are summed up in the following table IV.

From these data it is evident that the most favourable temperature is 28—32 °C. A temperature above 34 °C is detrimental to the eggs and the hatching percentage decreases. A humidity below 60% is also unfavourable for their development.

Table IV.

The influence of temperature and relative humidity on the development of eggs of the desert locust (BODENHEIMER & FRANKEL, 1929)

Temperature (°C)	% of eggs hatching	Relative humidity (%)	% of eggs hatching
18	0	below 60	below 20
22	30	80—90	60
28—32	60	100	100
36	15		

These conclusions on the effect of moisture upon the egg conflict somewhat with the findings of SHULOV, who made a more exact study of the moisture requirements of the eggs.

The following are his findings:

1. The egg of *Sch. gregaria* must absorb roughly as much as its own weight in water, before it can complete its development. This intake must be through direct contact with water and cannot take place through atmospheric humidity.

2. The absorption of water takes place during 3—5 days (at 27° C) after oviposition, which is at the end of the anatrepsis or beginning of the katatrepsis phase.

3. Lack of water intake does not interfere with the development of the embryo throughout its anatrepsic stage, but the katatrepsic stage cannot continue without intake of moisture.

4. An embryo in the anatrepsic phase may remain alive for two weeks at 20° C or probably longer at lower temperatures without moisture. If this period is prolonged, the embryo dies.

5. Under natural conditions prevailing during the winter in the Middle East, the interval between anatrepsis and katatrepsis due to lack of water is probably longer and the rate of development slower, which may account for the lengthy incubation period of 81 days as reported by BALLARD.

The Larva

BODENHEIMER kept larvae at a controlled temperature of 30° C and found that they became mature 43 days after hatching. Naturally at lower temperatures the development period is prolonged. He suggested that the threshold of development of the larva was identical with that of the egg, namely 18° C.

Fig. 3. *Schistocerca gregaria* – First instar. (Figure above natural size) after REGNIER.

Table V.

Number of days required to complete the larval period of the desert locust (BALLARD)

Instar / Period	1	2	3	4	5	Total
May—June	6.5	8	7.5	9	11	42
Sep. —Oct.	8.5	10.5	14	7	13	53
Nov.—Dec.	10	12	10	10	23	63

The larval period as obtained by BALLARD is given in table V. In this table, 5 instars are included. Under certain conditions larvae which develop into more instars may be produced. ALBRECHT (1955) states that when females of *Sch. gregaria* were kept in isolation, small hoppers hatched from those eggs which completed development after 7 instars, whilst eggs from crowded females produced hoppers which completed their development in 6 instars.

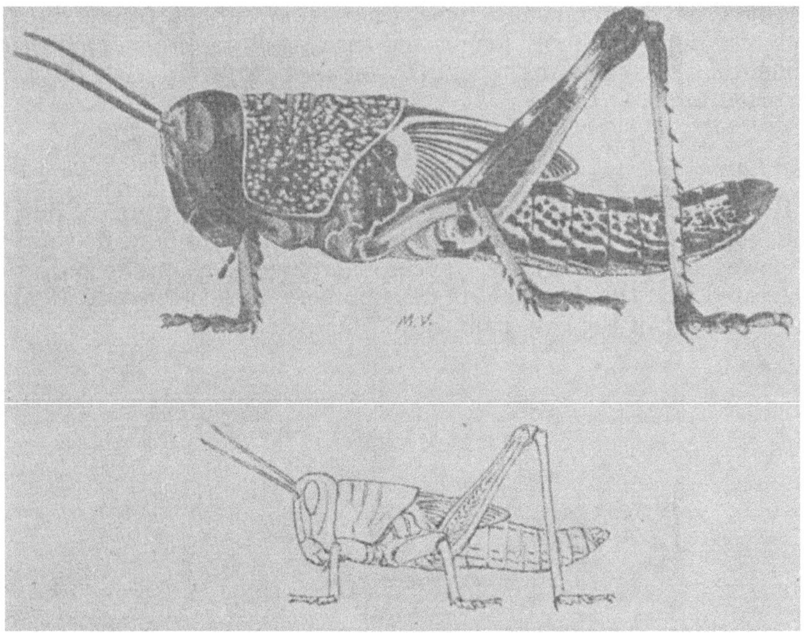

Fig. 4. *Schistocerca gregaria;* fifth instar, (Figure below natural size) – after REGNIER.

Larval Behaviour

The behaviour of the hoppers was studied by FRANKEL (BODENHEIMER & FRANKEL, 1929) in Palestine and his description is as follows: immediately after hatching the larvae gather into swarms. They advance by walking or hopping. On sloping ground they walk, whilst in level areas they either hop or walk. Their speed is more or less equal in either case and is about 6 metres per minute at a temperature of 30° C. At lower temperatures they progress more slowly. At night, the larvae climb on to shrubs and rest there. In the morning, when the temperature is higher than 20° C, they climb down and warm themselves in the sunshine, and only begin to hop when the temperature reaches 24° C.

A more detailed account of the behaviour of locust hoppers in

Africa is given by ELLIS & ASHAL (1957), who claim that their activity is in accordance with a definite pattern, as follows: At night they roost. Half an hour before sunrise they stream down and begin to march. This march continues for 2—2½ hours, when they collect in bands and warm themselves in the sun for ½—2 hours. Then they start moving again, hopping rather than marching. When the temperature rises too much, they roost.

In the afternoon marching is resumed. The marching alternates with feeding and roosting, each stage lasting 15—45 minutes.

Activity takes place only at temperatures above 17° C. Marching begins at 24° C, it is slower below 28° C and above 36° C.

Factors influencing formation of the migratory phase

It was established that the primary factor which stimulates the transformation from solitary to gregarious phase is aggregation. Some entomologists maintained that in nature this transition is gradual, involving a few generations. In the laboratory, however, it proved to be quite abrupt. Thus, CHAUVIN (1941) transformed solitary hoppers into gregarious, by crowding them. Furthermore, the same individual females, when reared in solitary conditions, produced solitary offspring, and when reared in crowded conditions produced gregarious offspring.

In order to have aggregations, there must be a prolific development of individuals. This is possible when the development of the eggs is successful. Locust eggs develop well in wet soil, for they need contact moisture, and at a temperature between 27—32° C. Otherwise the death rate is high.

From the above, it may be assumed that outbreaks of migratory swarms in the field should come about as a result of the chain of events, as follows: a high percentage of hatching of eggs due to humidity in the soil, a satisfactory development of hoppers due to abundance of food, and crowding of the bands which should stimulate a migratory phase. In the vast arid zones of Africa this is difficult to explain. However, UVAROV (1957) sums up the facts regarding the dynamics of locust outbreaks as follows:

Migratory locusts are attracted to areas of air flow convergence where it eventually rains (RAINEY). Females will lay in moist soil only (POPOV). Thus the rains ensured oviposition and a full hatching of the eggs; the moisture in the ground provides also a vegetation which serves as food for the ensuing hoppers. However, in these arid zones the rains are often scanty, short-lived, and followed by drought. The vegetation may then dry before the hoppers complete their development. They are forced to congregate into bands in the patches of green vegetation left in low wadies, where the soil has richer reserves of moisture; this crowdedness creates the migratory habits, and the ensuing adults will rise in swarms.

This migratory phase is quite desirable for the existence of the species in such unfavourable environments. By the time the adults mature the vegetation in those patches may have been consumed, thus they must wander in search of food. As they mature sexually they must also wander in search of oviposition sites in humid areas in the vast arid zone. The existence of the species depends therefore upon its migratory habit.

These flights in search of food are characteristic not to the migratory phase only but also, as mentioned above, to the solitary phase. UVAROV calls this nomadic behaviour which is also an essential feature for the upkeep of the species in the poor conditions. It should be borne in mind, on the other hand, that migration into remote countries may bring the species into unfavourable ecological conditions.

The following observations were made by GUNN (1948) regarding the behaviour of migratory swarms:

At night the locusts roost. After sunrise they descend to the ground and orientate themselves at right angles to the sunrays. The most consistent factor that controls their flight activity is temperature. The descent from trees takes place at a temperature of 15—19° (average 17° C) and flight at 17—23° (average 20° C). Swarms settle to the ground when the temperature falls to 20—23° C. The flight is with the wind if its velocity is higher than five ft/sec.

With regard to difference in shape and behaviour of the two phases, there is little doubt that their origin is in the physiological

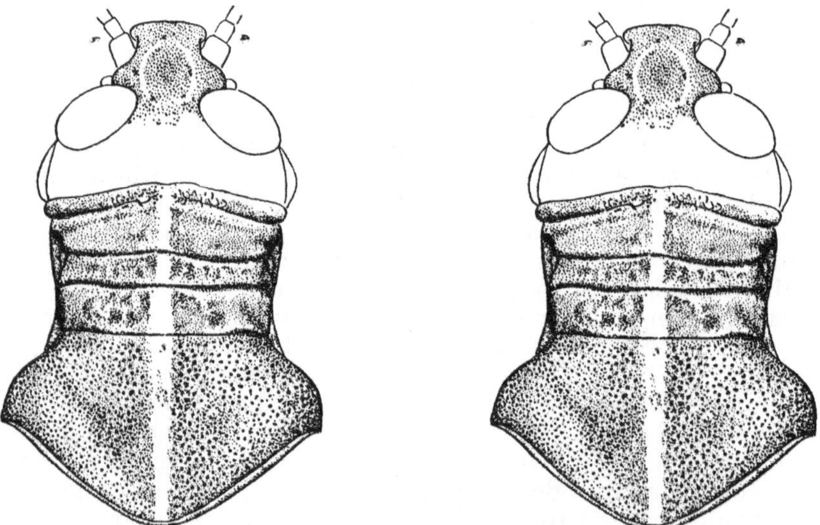

Fig. 5. *Schistocerca gregaria* – Pronotum of solitary (right) and gregarious phase (loft) after UVAROV. Courtesy Com. Inst. Ent.

changes that take place in the tissues of the body. Thus, the water content of the migratory phase is less than of the solitary one, while the content of uric acid and fat is relatively greater.

Although the solitary and migratory phases are quite distinct, the transitional phases often do not show distinct characteristics. The ratio of the length of the elytra to that of femur often serves to distinguish the phase. Lately DIRSH (1953) has introduced another distinguishing mark — namely the ratio of the length of the femurs to the width of the head (F/C), claiming that the elytra/femora ration (E/F) is not always reliable.

These characteristics, together with others are given in Table VI.

Table VI.

Summary of the differences in the phases of the desert locust

Characteristics		Solitary phase	Migratory phase
Colour of body	Larva	White-grey, turning grey-green	Black with yellow patches, and finally yellow with dark spots
	Adult	Grey-green	Reddish-pink, becomes yellow at sexual maturity
Structure of body	Surface of pronotum	Rough	Smooth
	Length of wings	Less than twice the length of femur	Twice the length of the femur
	Ratio femur/face (F/C)	3.75—3.85	3.15
Disposition		Reposed	Restless
Social behaviour		The larvae and the adults live singly	All the stadia are incline to congregate
Migratory properties		As a rule do not migrate; lay eggs at breeding place	Migrate distances; lay eggs in areas far from breeding place

Migration path in the Middle East

Swarms which invade the Middle East appear either in the autumn or early in the spring during February-March. Apparently these do not stem from the original source of outbreak. The summer- or "monsoon" generation in the Sudan matures in the autumn and may migrate in the direction of the Red Sea, among other places.

Similarly monsoon swarms from North-East Africa and Southern Arabia migrate north and north-east to these areas. Along its Sudanese and Arabian shores autumn rains cause favourable breeding conditions. The adults from these winter breedings mature during the winter, migrate north-wards too and invade the Middle Eastern countries. The generation which breeds in the Middle East matures in June, but thereafter adverse climatic conditons may prevent breeding. BODENHEIMER (1929) points out that the routes described above are in accordance with the prevailing winds in the respective season and country.

Later writers confirm the long-known fact that the locust migrates with the wind. Even more, RAINEY (1951) states that in whatever direction a swarm starts off, when it is in the air it changes with the wind. Having collected many meteorological data and recorded incidences of locust swarms, RAINEY finds a correlation between low barometric depressions and the occurrence of swarms. He further states that they are attracted to the lines of convergence of winds coming from opposite directions, and quotes a series of swarm incidents along the intertropical convergence zone.

In Iraq and Iran swarms may also appear, perhaps chiefly from outbreak areas in Pakistan or the adjacent territories.

Periodicity

It has been pointed out that in the Near East locust invasions are periodic events, occurring every few years. Eleven to thirteen-year cycles have been mentioned by some authors; thus in Israel, for instance, locust outbreaks were recorded for 1865, 1878, 1900—1902, 1915, 1928—1930, 1941—1944, 1950—1955, 1959—1960.

SHCHERBINOVSKIĬ (1952), too, states that in the USSR territories bordering Iran outbreaks occur every 11 years.

Miss WALOFF, who studied locust outbreaks for several years in the zone where the insect is always found, states that "there is no breeding cycle inherent to the species, as some authors claim". Of course they are not inherent, and the locust is always present in the central zones, but, as quoted, periodicity does occur in the peripheral zones which locusts invade occasionally. The chief factor in these invasions is the climate. This opinion is expressed by Miss WALOFF (1946) (and also verbally) as well as by SHCHERBINOVSKY, who ascribes this periodicity to the climatic changes caused by solar activity.

The Moroccan Locust — Dociostaurus maroccanus Thunb.

Another acridid found in the Near East is the Moroccan locust. This locust appears in the solitary phase in countries where the climate is less favourable for its development. In other countries, however, where climatic conditions are more favourable, it turns into a migratory locust and causes serious harm. In Cyprus, for example, severe damage has been reported on various occasions, while sporadic outbreaks occur in the Balkans also; in Iraq swarms frequently develop and cause destruction to crops.

In the initial stage of their development, the larvae feed on the leaves of *Poa balbosa* and similar grasses found at the sites of propagation. Later they eat anything they can and cause destruction of the vegetation.

Description. The male is 20—28 mm long and the female 28—30 mm. Their colour is yellow-grey, spotted with light or dark yellow. The posterior part of the thorax is darker than the anterior part. Yellowish-brown spots and stripes are present on the tegmina. There are 3 black bands on the hind femora and their tips are also black. The rear tibiae are reddish-white. The hind wings reach beyond the tip of the femora.

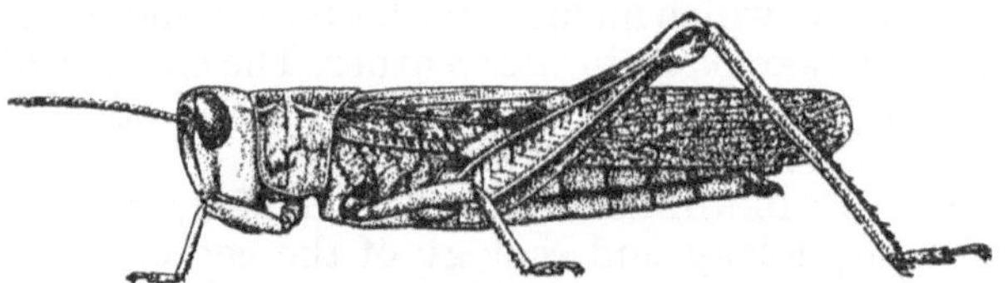

Fig. 6. *Dociostaurus maroccanus* – after Del Canizo. VI. International Congress of Ent. Madrid.

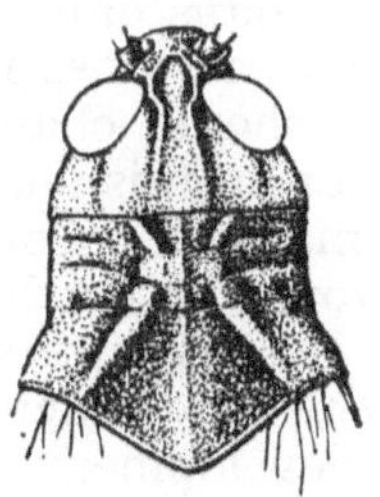

Fig. 7. *Dociostaurus maroccanus* – pronotum – after Del Canizo. VI. International Congress of Ent. Madrid.

The eggs are enclosed in a sac about 20—30 mm long, and 15 mm wide, closed with a plug of froth. This substance is also found within, together with the eggs, of which there are 20—50 per sac. The egg is of a clear yellow colour, 5 mm long and shaped like a grain of rye.

The newly-hatched larva is 5—6 mm long, brownish-black in colour, with black spots.

In the fifth instar the length is 18—20 mm, and the colour yellowish-red, with a white stripe along the body. The rear legs are reddish, and the wing pads extend to the middle of the abdomen.

Fig. 8. Egg pod of *D. maroccanus.*

Biology and Ecology of the Moroccan locust.

The egg. The period of incubation lasts 8—10 months. During this lengthy period, (which includes both summer and winter months) extreme changes take place in the weather. The changes of temperature in Iraq may be between 7° C in winter and 50° C in summer (and the rainfall from 50 to 500 mm).

According to Bodenheimer & Shulov (1951), who carried out a study on the physiology and ecology of the egg of *D. maroccanus*, the egg of this species goes through four phases in the course of its development: A. The initial period. During this period the embryo begins its development, which is soon interrupted; the interruption may last a few days only or for one to two months. B. Anatrepsis. The embryo resumes its development up to the end of this phase, when it should change its position within the egg-shell. C. Diapause. During this period the embryo is again at rest. In nature this period may last as long as five months, but may also be shorter under favourable conditions of contact moisture and temperature. D. The Katatrepsis. The embryo, as a result of imbibing contact water, and under favourable conditions of temperature, resumes its development up to the hatching stage.

The length of this period is affected by the temperature; Table VII presents the length of this stage at various degrees of temperature, as per Bodenheimer & Shulov.

Table VII.

Temperature	14.9	15.7	18.9	19.2	23.3	26.2	29.4	32
Katatrepsis period in days	62	51	21	28	17	13	11	12

The threshold of development was calculated accordingly and found to be 12.1° C, while the thermal constant 164 day – degrees. It became apparent from the experiments that a temperature of 32° C is the upper limit for favourable development. The authors found also that eggs moistened immediately after oviposition died.

The physiology and ecology of the egg, as outlined above, helps to explain the long incubation period, and also why years with below-average rainfall cause an increase of the population, and why excessive rains may retard the hatching. The various periods of egg hatching are explained, also. Thus, while in Iraq the eggs hatch in March (ROOKE, 1930), in Bulgaria they hatch at the end of April (CHORBADJIEFF, 1936).

The larva. The larval period lasts 35—55 days according to REGNIER (1931), 42 days according to ROOKE, and 33—41 days according to CHORBADJIEFF.

About 12 days from hatching, after the first moult, the larvae show a tendency to congregate and migrate. They arrange themselves in a wide but shallow formation, and in their advance consume all the vegetation they encounter.

The adult. The adults appear in May (at the beginning of the month in Iraq, and at its end in Bulgaria).

After the final moult, the maturing adults undergo a presexual period. This period, at the end of which the adults are able to mate, lasts 8—10 days with the male and 4 weeks with the female.

Egg-laying begins a fortnight after mating. Between egg-laying periods the female may copulate a number of times.

The Moroccan locust is different in colour at the presexual and the sexually mature stage. In the latter, white spots are present on its body and the patterns in other colours are less obvious.

With due regard to the above facts, the times of pairing and egg-laying in the various countries are given in the following table VIII:

Table VIII.

Times of pairing and egg-laying of the Moroccan locust

Country	Pairing	Egg-laying
Iraq	Mid-May	End of May—July
Morocco	End of May	Mid-June—July
Bulgaria	Mid-June	July

The females usually choose as egg-laying sites, desert areas where the soil is friable and loose. They are also wont to lay in the vicinity of woods.

The number of egg batches and number of eggs (according to various sources) are given in the following table IX.

Table IX.

Egg-laying capacity of the Moroccan locust

Country	Number of egg cases per female	Number of eggs per case
Iraq, Solitary	1	20
Iraq, Migratory	3—6	30
Morocco	4—6	20—50
Bulgaria	2—3	18—40

In a severe infestation, about 200 egg cases were found per square metre in Iraq and 300 in Bulgaria.

According to CHORBADJIEFF, the adults gather in swarms and migrate in "belts" more than a kilometre long. The height of their flight does not exceed 10 metres. Generally they fly between ten and eleven in the morning, when the temperature ranges from 22 to 26° C.

The Desert and Moroccan locusts compared. The differences between the two species are as follows:

1. The eggs of the desert locust of the migratory phase do not have an obligatory diapause. They develop without interruption as long as humidity and temperature are suitable for them. Eggs of the Moroccan locust, whether of the migratory or solitary phases, have an obligatory diapause lasting about 10 months, including those of winter.

2. The difference in the time they need for egg-development causes the two species to give rise to different numbers of generations. The desert locust in the migratory phase may have two to three generations per annum, while the Moroccan locust has only one.

3. The desert locust migrates thousands of kilometres, from arid areas to more rainy places. The Moroccan locust, however, migrates short distances only to areas with the same climatic nature. REGNIER (1931), who investigated the species in Morocco, states that it wanders no more than 50 kilometres in its lifetime. According to BODENHEIMER (1944), who studied the locust in Iraq, it travels at the most only 200 kilometres during its lifetime.

Breeding areas of the desert locust in one country are therefore a source of swarms to distant lands. Thus, from the steppes of Sudan swarms migrate to Arabia, and thence to Syria, and even to Turkey, whereas Moroccan locust swarms are from local breeding-areas.

4. Outbreaks of the desert locust are caused during years of rains above the average, while those of the Moroccan locust occur in years with subnormal rainfall.

Dociostaurus is called a "locust", no doubt because of its gregarious habits. However, in view of the differences between the two species outlined above, and in the light of the points raised by ANDREWARTHA (1945) about the distinction between locust and grasshopper, *D. moroccanus* should be undoubtedly considered a grasshopper.

The Mediterranean Grasshoppers — Calliptamus spp.

This genus, which comprises about a dozen species, is prevalent in the Mediterranean countries. Of all the species, *C. italicus* L. is the best known, and most widely distributed, and will therefore be considered as representative of the genus.

The Italian Grasshopper — Calliptamus italicus L.

This insect is found in all Mediterranean countries and also in Central Europe. In years when its population increases greatly it eats the vegetation around it. If the plants are young, they may be destroyed completely.

Description

The adult. The male is 16—25 mm long, and the female 20—35 mm. The colour varies from yellow to grey-brown, or dark brown. Sometimes the body is speckled with dark or grey spots. The prothorax is large, with its front margins narrower than the posterior ones, the latter broadening out at the sides of the body. The upper portion of the prothorax is smooth, with longitudinal ridges on either side. Three brown stripes are present on the prothorax. The tegmina may be longer or shorter than the body. The hind wings are transparent, with a pink or reddish base. The rear femora are short and wide, with a red or brown-purple inner side. At the tips of the femora are black spots.

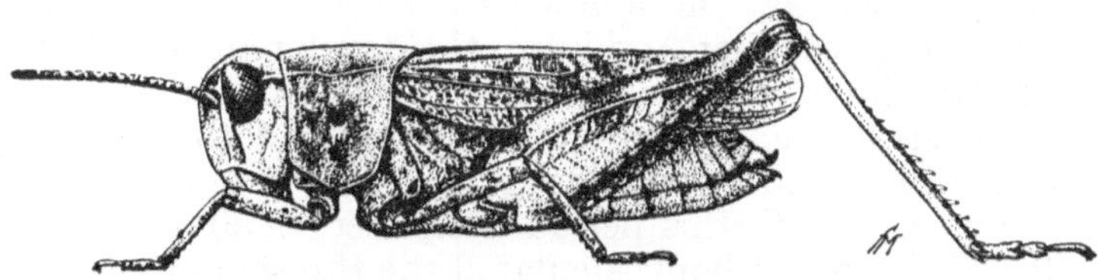

Fig. 9. *Calliptamus italicus* – after DEL CANIZO.
VI. International Congress of Ent. Madrid.

The egg is about 6 mm long. It is elongated, cylindrical, and somewhat curved. The egg pod is about 35 mm long and 6 mm wide. It is also cylindrical in shape, and curved. It has a hard thin shell and is closed by a frothy plug. The pod contains from 25—30 eggs.

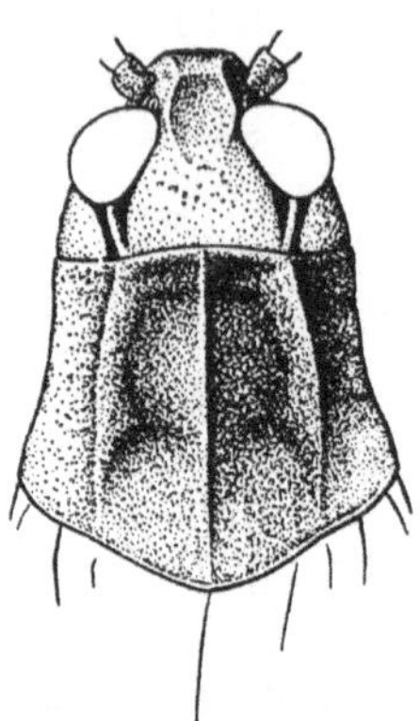

Fig. 10. *Calliptamus italicus* – Pronotum – after DEL CANIZO. VI. International Congress of Ent. Madrid.

The developing larva is about 6 mm in length. It is coloured grey-brown or black, its abdomen is yellow on the inner side, and it has 3 stripes on its prothorax. The fifth-instar larva is from 13—30 mm long, and has similar colouring to that of the adult.

Life History. Adults appear in the summer — in June in the southern parts of Europe, and in July in northern regions. In Israel, *C. palestinensis* adults also appear in June. A short time after maturity, 10—14 days after the last moult, egg-laying begins and lasts about 2 months. During this period, the female may lay 4—6 egg-pods. The eggs remain in diapause during autumn and winter, and do not hatch until the following spring. In Israel they hatch in April-May. If the winter and spring are cold they hatch only in May.

PENER & SHULOV (1960) made a detailed study of the development of the egg of *C. palestinensis*, the physiology of which is very similar to that of *C. italicus*. The following are the results of their study: As with the eggs of other Acrididae, the development period may be divided into three: A. The pre-diapause period. This includes the anatrepsis phase. Moisture is not essential for its development. At 20° C, this period lasts 50 days in moist soil and about 70 days in dry soil. At 27° C, this period is shorter — 30 days in wet soil and 40 in dry. The authors calculated the threshold of development accordingly and found it to be between 13—13.6° C.
B. The diapause. During this period the embryo is in apparent

diapause, whose length depends upon the water-supply. If contact water is not provided, the embryo dies, but it may remain alive in this stage for as long as 70 days at 20° or 27° C and resumes its development if moisture is provided. C. The post-dormant period. This includes the katatrepsis. This period starts only when contact moisture is provided, and is imbibed by the embryo. Its length of development depends upon the temperature. At a constant temperature of 24° C, most of the eggs do not develop, as they need, apparently, a colder temperature before entering the katatrepsis stage.

In a room in which the average daily temperatures ranged between 18—22° C and under adequate conditions of water supply, the total development period of the egg lasted 164—204 days; when less moisture was provided the period lasted over 250 days.

The larva feeds on weeds, *Artemisia* being the favoured food. After moulting five times, it matures within 5—6 weeks.

Adults are found in Israel from June to November and the population is at its peak in July.

Limiting Factors. The physiology of the egg is well-adapted to Mediterranean climatic conditions. Eggs oviposited in the dry summer start their development immediately and then they enter diapause until the rains come. If they occur in November and December, as is customary in this region, the eggs resume their development. If the rains are late there is mortality in the eggs. The later the rains the higher the mortality. Abundance of this species depends therefore upon the periodicity and quantity of the rains each year, or a few years in succession.

The European Locust — Locusta migratoria L.

This pest in its various subspecies, is familiar in Europe and other parts of the world as a destructive migratory locust, but in the Mediterranean countries only the solitary phase is known.

Description. The female is 37—60 mm long, and the male 30—35 mm. The general colour is green or light greeny-brown. The front wings are spotted. There are brown patches on the head and thorax. Two lateral lines are present on the pronotum, and on either side of the prothorax there is a spot. The tips of the tibiae and tarsi of the hind legs are reddish.

Along the prothorax there is a ridge, which appears convex from the side, higher near the head and lower posteriorly. The front edges of the notum form an angle, and the rear margins a rounded right-angle.

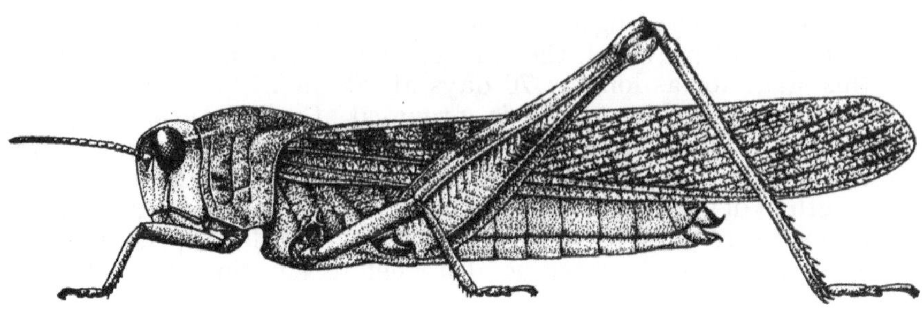

Fig. 11. *Locusta migratoria* – after DEL CANIZO.
VI. International Congress of Ent. Madrid.

Distribution. UVAROV (1921), who investigated this pest, came to the conclusion that this species is in fact the most abundant in the world, but that it appears in different forms and phases.

L. m. migratoria – the migratory, is the European locust found in Eastern Europe, Central Asia, and China. *L. m. danica* is the solitary phase of this species. This is found in Southern Europe, Western Asia and North Africa.

L. m. migratorioides, another form of this species, is found south of the Sahara. From outbreak areas around the river Niger it may migrate to East Africa and also southwards.

In Israel and in other countries around the Mediterranean, as mentioned above, only the solitary phase is found. Adults are generally seen in irrigated fields, or other damp places. Here and there forms may be found passing from one phase to another, but they do not develop into migratory swarms. It appears that conditions prevailing in these areas are not conducive to mass propagation, and swarms of migrants from other centres do not reach here.

The Egyptian Grasshopper — Anacridium aegyptiacum L.

This is a very large and conspicuous species. It, too, may be found in wet places and in irrigated fields.

Description. The female is 50—60 mm long, and the male 30—35 mm. The general colour is brownish-yellow or brownish-green. Three dark patches are present on the hind femora, whilst on the prothorax there is a comb notched horizontally in 3 places. A spine projects on the lower side of the prothorax near the front coxae.

Life Cycle. During the winter this grasshopper occurs in the presexual stage. Only when spring comes are they capable of propagation. In Israel, the females lay eggs in March and the larvae hatch in April. The peak of the larval population is in the months of June and July, and the peak of the adults in May to June. However, the adults may be found in Israel all the year round.

Oedipoda spp.

A number of species of this genus of grasshopper may cause damage in fields of summer crops.

The typical markings of this genus are: — 3 brown stripes on the tegmina and a black or dark-brown line near the margins of the hind wings.

The three most abundant species are:

a. Oe. aurea Uv. — Length of body from head to tip of abdomen 20—25 mm. Base of hind wings yellow. Found mainly on hillsides among bare and exposed rocks.

b. Oe. miniata Pall. — Body length as above (20—25 mm). Base of hind wing red. Often found together with the Italian grasshopper. Attacks field crops.

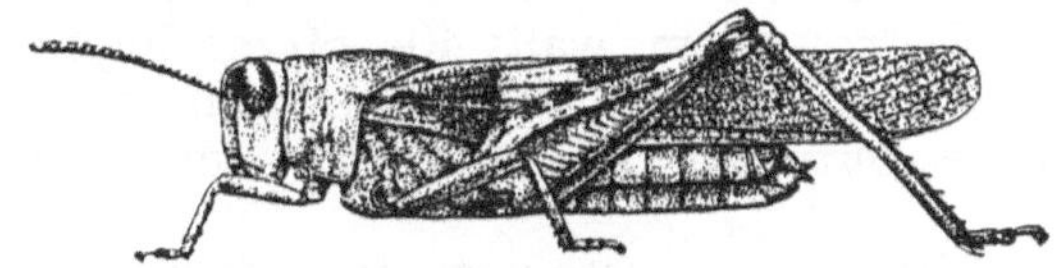

Fig. 12. *Oedipoda coerulescens* – after Del Canizo. VI. International Congress of Ent. Madrid.

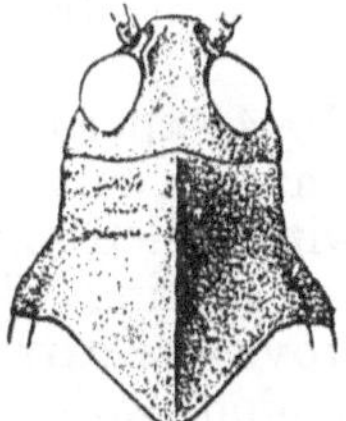

Fig. 13. *Oedipoda coerulescens* – Pronotum – after Del Canizo. VI. International Congress of Ent. Madrid.

c. Oe. coerulescens L. — Length of female's body 22—28 mm, hind wings bluish at the base. Prevalent from Southern Europe to Central Africa. Attacks various field crops.

Locust control

Plagues of locusts have been known ever since ancient times as one of the most severe punishments by which heaven could castigate mankind. However, as regards methods of control, great changes have taken place, especially during the last thirty years.

A complete picture of the improvement in methods of control of locust will be obtained through a study of locust invasions in the Middle East during the last half century.

In 1915, when it became known in a village that locusts had settled in its vicinity, the villagers would go out with sticks and tins in order to scare off the flying insects with their noise and to prevent egg-laying.

At about the same time the government brought out a law obliging each inhabitant to go out to the fields and gather up the eggs laid on the ground, each person having to gather a specified quota. We shall not here discuss why this law failed to be carried out effectively, but even had it succeeded it would have been impossible to clear all locust-eggs from the country. Here and there the eggs hatched and the emerging larvae ate all they found in their path. It is reported that one farmer set up walls of tin around his grove, and saved it.

During 1928—1930 the system of egg-eradication and the personal quota system were still in force. In order to stop the advance of the larvae and to destroy them, walls 40—50 cm high were built and channels dug alongside them. Hopper-swarms reaching these obstructions were burnt or destroyed by some other technical means.

In North-Africa it was customary to gather up larvae in large strips of cloth onto which they had hopped, and then to destroy them. At this period baits were used to a large extent. The composition of the bait was roughly as follows: — 5 kilos sodium arsenite, 10 kilos molasses and 20—50 litres of water to each 100 kilos of bran.

During the years 1941—1944 experiments were carried out in the destruction of the locusts at the outbreak areas. Thanks to the initiative of UVAROV and with Allied military aid, teams were sent to Arabia among other places. During this period it was also realised that molasses did not improve the attractiveness of the bait, and that the use of wet bran alone was adequate for this purpose.

After the Second World War, with the introduction of synthetic insecticides for agricultural use, a sweeping change took place in the approach to locust control.

In 1950, Benzene hexachloride was used against locusts and proved to be very effective in controlling this pest. A dust containing 5 % B.H.C. was adequate in controlling the hopper swarms. This material was used to control adults, and, in the form of a bait, to control both larvae and adults. Somewhat later another contact insecticide was discovered — aldrin — which was most effective against locusts, and was introduced in order to control them in the Mediterranean area.

Advances have also been made in methods of poison-distribution. Manure-spreaders have been used in place of hand-spreading; similarly, large areas have been dusted and sprayed by motorised equipment and aeroplanes have been employed against the locusts.

The following are extracts from the instructions on locust control issued in 1955 by the Plant Protection Department in Israel: "Any concentration of locusts must be reported immediately to the Head Office. Methods of control that each farm can undertake individually are: — In plantations — by dusting; make sure that the duster is in working order. Materials to be used are: — 4% aldrin dust, 4% dieldrin (5% B.H.C. should be added to this list), at the rate of 2½—3 kilos per dunam. In irrigated areas — by spraying; make sure that the sprayer is in working order. The materials to be used are: — 0.5% aldrex (aldrin emulsion), 0.4% dieldrex (dieldrin emulsion), 0.3% lindexane (lindane emulsion), 0.6% quammax (B.H.C. emulsion), 0.1% parathion (emulsion).

The rate of spray to the dunam is about 100 litres. At low volume, 15—20 litres should be given per dunam at a suitable concentration.

0.5—1% of the poison, according to its nature, is added to the bran in making the bait.

LITERATURE

H. G. ANDREWARTHA, 1945. *Bull. ent. Res.* **35,** *379-389.*
F. O. ALBRECHT, 1955. *J. Agr. Trop. Bot. appl.* **2,** 3-4, *109-192.*
E. BALLARD, 1931. *Tech. & Sci. Service, Ministry of Agr. Bull.* **110,** *1-149.*
F. S. BODENHEIMER, 1944. *Iraq. Minist. Agr. Bull.* **29,** *1-121.*
F. S. BODENHEIMER & G. FRANKEL, 1929. *Z. angew. Ent.* **XV,** *124.*
F. S. BODENHEIMER & A. SHULOV, 1951. *Bull. Res. Counc. Israel,* **1,** *59-75.*
R. CHAUVIN, 1941. *Ann. Soc. ent. France* **110,** 3, *133-272.*
P. CHORBADJIEFF, 1936. Ministry of Agriculture and Public domains, Sofia, **61,** *1-80.*
D. E. DAVIES, 1952. *Anti-Locust R.C. Mem.* **4,** 56 pp.
V. M. DIRSH, 1953. *Anti-Locust R.C. Bull.* **16,** 34 pp.
P. E. ELLIS & C. ASHALL, 1957. *Anti-Locust R.C. Bull.* **25,** 94 pp.
K. M. GUICHARD, 1955. *Anti-Locust R.C. Bull.* **21,** 35 pp.
D. L. GUNN et al., 1948. *Anti-Locust R.C. Bull.* **3,** 70 pp.
H. G. HAMILTON, 1955. *Proc. Roy. ent. Soc. Lond.* (A) **30,** 7-9, *103-114.*
G. JANNONE, 1953. *Boll. Lab. ent. agr. Portici* **12,** *189-248.*
M. J. NORRIS, 1954. *Anti-Locust R.C. Bull.* **18,** 44 pp.
M. P. PENER & A. SHULOV, 1960. *Bull. Res. Counc. Israel,* **9B,** *131-156.*
G. B. POPOV, 1958. *Anti-Locust R.C. Bull.* **31,** 70 pp.
R. C. RAINEY, 1951. *Nature, Lond.* **168,** *1057-1060.*
P. R. REGNIER, 1931. Dir. gén. de l'agric. Comm. et de Colon. **1.** Morocco.
H. G. D. ROOKE, 1930. *Dept. Agr. Iraq. Mem.* **13,** *1-13.*
N. S. SHCHERBINOVSKIĬ, 1952. Moscow GOS Isd. sel'shokh Lit. 416 pp.
A. SHULOV, 1952. *Bull. ent. Res.* **43,** *469-476.*
B. P. UVAROV, 1921. *Bull. ent. Res.* **12,** *135-163.*
B. P. UVAROV, 1928. Locusts and Grasshoppers. Imp. Bur. Ent. 346 pp.
B. P. URAROV, 1957. Arid Zone Research. **VIII,** *164–198.*
Z. WALOFF, 1946. *Anti-Locust Mem.* No. **1,** 74 pp.

Crickets — Gryllidae

The crickets differ from the grasshoppers in having a short body, longer antennae as a rule, and tegmina whose margins bend over the sides of the abdomen. The ovipositor is long and spear-shaped.

There are three distinct groups: 1. Crickets — one of which is discussed below; 2. Tree crickets, which do not concern us in this subject, and 3. the Mole crickets — important pests of field crops.

The Two-Spotted Cricket — Liogryllus bimaculatus De G.

Although not conspicuous in Mediterranean countries, this insect may on certain occasions cause sufficient damage to trouble farmers. In tropical countries, however, it is a more frequent pest to field crops.

Description. The two-spotted cricket is a large, glittering and black insect. Its legs are not hairy. The ocelli in its head are arranged in the form of an isosceles triangle whose apex is very obtuse, almost rectangular. The two-spotted cricket has two yellow spots at the base of its tegmina and is 20—28 mm long. It can be distinguished from another species, resembling it, by the pronotum which is wider than its head and whose posterior border is wider than the anterior one. The tegmina are longer than the body, approximately 12 mm. The ovipositor of the female is about 17 mm long. The egg is long and yellow and has the shape of a banana. The neonate larva is dark brown and 4 mm long.

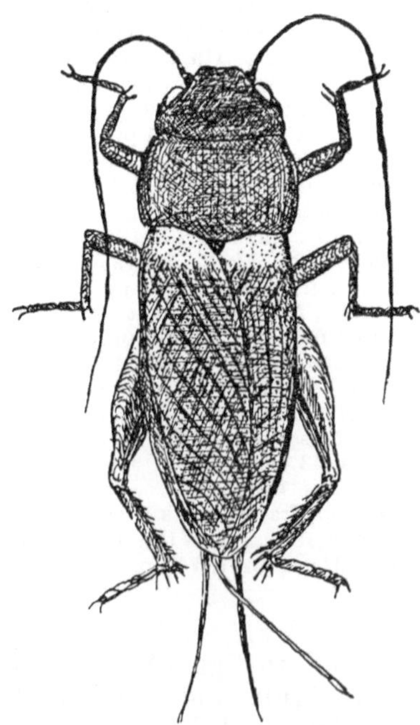

Fig. 14. *Liogryllus bimaculatus*

Distribution. Damage by this cricket has been recorded in India, Java, Tanganyika, Somalia, Erithrea, Rhodesia and Israel. It may be surmised therefore that the region of its distribution spreads over the tropical area of Asia and the eastern part of Africa, and northwards to the shores of the Mediterranean Sea and Western Asia, and in Southern Europe.

Type of Injury. This cricket, which is polyphagous and is not limited to special hosts, can cause damage in various ways. In Tanganyika, it was recorded that the pest ate cotton seeds before their germination. In other places, the cricket chewed and destroyed young seedlings or saplings of coffee plants, sugar cane, rice and cotton. In India the chewing of pea pods has been reported; in Israel it attacked ears of corn. The pest makes its way between the leaves covering the cob or bores holes through them, and feeds on the tender grain. In army camps in southern Israel and Jordan complaints were made of damage to silk cloth. This damage, it seems, came as a result of a hunger for animal protein.

Life History. The female lays its eggs in the ground in groups of ten. For this purpose she inserts the ovipositor into the soil; she can do it only when the soil is moist. The incubation period at a temperature of 26° C lasts about 12 days. Upon hatching, the larvae feed on various parts of plants and also on animal food. In the laboratory they were fed bread and potato or corn grain with the addition of fish meal. A detailed study on the biology of the species was carried out recently by RIVNAY & ZIV (1962) who raised the pest under various conditions. Rates of development at various degrees of temperature are given in table X.

Table X.

Length of development of *L. bimaculatus* at various degrees of temperature (RIVNAY & ZIV)

Development stage \ Temperature in °C	34	30	26	22	18
Preoviposition period	3—5	6—8	11—12	10—19	no ovip.
Incubation period	7—8	8—10	12—15	20—26	45—54
Larval development	30—39	32—55	55—76	129—142	—

Further investigation showed that the adults lived from 20—59 days. The average egg production from 12 females reared at a temperature of 30—34° C was 1379 — the maximum per female was 3240 eggs. The optimum temperature of the insect was 30—34° C; the threshold of development around 15° C, while the threshold of reproduction was 19° C.

Ecological Factors. Like the eggs of locusts, the eggs of this cricket need contact moisture in order to hatch. In laboratory breedings it was often necessary to moisten the soil to prevent the egg from shriveling. In the southern Mediterranean area the soil in the summer is too dry to permit the eggs to hatch, and in the rainy season the temperature is too low to allow a quick and normal larval development. Development therefore is possible only during the summer, in irrigated soils or along drainage canals where the soil is always damp.

In fact the large and heavy attack by this species witnessed by the writer was in the "Kabara" swampy area along the drainage canals. On that occasion only were there bands recalling the bands of locust hoppers. Otherwise the cricket in that region lives a solitary life.

The Mole Cricket

The mole cricket is one of the most notorious pests in the field, vegetable garden and in the nursery or flower bed. It is very polyphagous, though some plants are more favoured than others, as discussed below.

Description of the genus. The genus *Gryllotalpa* belongs to the cricket family, but in habits and morphology the species of this genus differ entirely from crickets.

Although a subterranean dweller, it is capable of flying. Special morphological adaptations enable it to live in this way. The prothorax is very large because it bears the forelegs and their respective muscles which serve in digging. The pronotum is covered with fine pubescence which reduces the friction during the digging process. The tegmina are short and do not cover the entire abdomen. The large transparent hind wings are folded like a fan; their tips are visible as two tails reaching beyond the end of the abdomen. (In the brachypterous types they reach the middle of the abdomen). The forelegs are short and strong; the tibia are flat and decorated with four hard teeth on the margin; the tarsal segments are also flat, and together with the flat tibia make a very efficient digging shear. The middle legs are short; the hind legs, especially the femora, are slightly larger but not as large as in the common cricket.

The egg is whitish yellow and becomes darker as it ages. It is elliptical, 2.5—3 mm long and 1—1.5 mm wide; before hatching, it is 4 mm long.

The Common Mole Cricket — Gryllotalpa gryllotalpa L.

The common mole cricket is 60—67 mm long and its colour is bright greyish-brown or dark brown. There are two types, the long-winged and the short-winged (brachypteron). In the former, the tegmina reach the fourth abdominal segment and the hind wing tails extend beyond the end of the abdomen. In the short-winged type, the tegmina reach only the second abdominal segment and the wing tails do not extend beyond the abdomen.

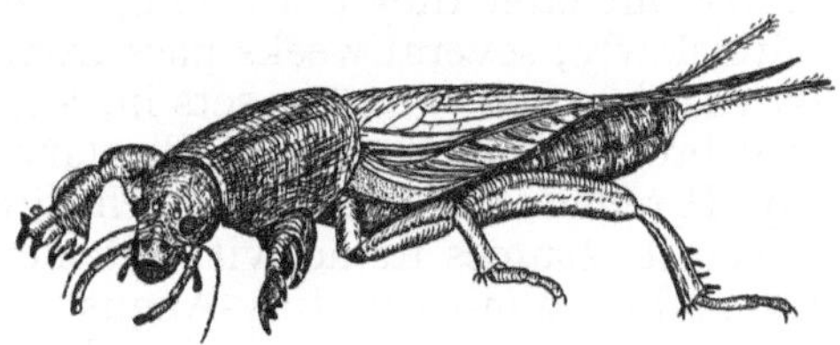

Fig. 15. *Gryllotalpa gryllotalpa.*

Distribution. This species is common in Europe, North Africa and Western Asia. Recently it has been introduced into New Jersey, U.S.A.

(Africa is inhabited by another species, the so-called African mole cricket *Gryllotalpa africana* PALISO DE BAUR. Its range of distribution is from Equatorial to North Africa, Sinai and Southern Israel. It occurs also in meridional Asia, Java, Formosa, China, Australia and Hawaii. It differs from the common mole cricket in its size, it being 24—31 mm long only, and by the shape of the anterior femora. The life history discussed below is that of the common mole cricket.)

Food and Hosts. Some entomologists believe that this insect is totally carnivorous and that its damage to plants is merely the result of its search for subterranean prey. Those who have witnessed its mode of life and method of injury cannot agree with this assumption. The pest is an omnivorous insect and feeds upon plants, but needs animal protein for its development.

Type of Injury. Injury is caused by the digging of tunnels in the ground, between rows of plants and in flower beds, by uprooting tender plants and by chewing the stems of young seedlings. In irrigated sandy soil the tunnels may be easily seen, but in newly plowed soil they are not so easily detected.

In potato tubers and fleshy vegetable roots such as carrots and beets, the mole cricket often gnaws cavities which render the product unmarketable. Germinating peanuts may also be eaten by this pest.

Life History. Because of this subterranean habits, the biology of the insect has not been studied sufficiently and many details about its life remain obscure.

The investigations of several students in Egypt and the USSR give us some idea of its habits.

In Russia a generation lasts two years, as follows: Mating takes place in May-June. Egg-laying begins immediately. The incubation period lasts about three weeks. In other words, the neonate larvae hatch late in June. They remain in the nest a few days, until they harden, and then disperse in search of food. At the beginning they remain in the vicinity but later they crawl longer distances. These larvae develop quite slowly; several weeks pass from one moult to the next. Thus, by the time cold weather sets in, and the development of the insect is brought to a standstill, the larva has reached its third instar only. It remains in this stage in the ground until the following spring when it resumes its activity. It moults then, and again in June-July, and once more in July-August.

At the end of the summer adults appear. By then, however, the temperature is low again and the cricket remains dormant throughout the winter. It awakens the following spring, mates and starts to reproduce anew. In the countries of the Near East, in the Caucasus, Egypt and Israel, as well as in Italy and Austria the rate of development is faster. The following details were observed by WILLCOCKS (1925) and by CASSAB (1937). Activity of the adults begins as early as March. Mating and oviposition take place in April. The incubation period lasts 7—10 days. The development of the larvae in all its instars does not last more than 4 months. Thus, the new adults appear as early as July-August and those of later hatchings in September-October. No information is available about a second generation during the same year. The adults pass a long preoviposition period throughout the winter months and begin to reproduce only in the spring, as mentioned above. The egg-laying period is also long. BODENHEIMER (1930) observed newly laid eggs in the summer. These must have been a late oviposition by a female of the first generation and should not be looked upon as a second generation.

Location. The adults are found as a rule in light, irrigated, sandy soils. The nests, however, are located in loose humous soil. This soil must be moist all year round, in particular during the egg-laying period. Consequently, nests of *Gryllotalpa* are found in non-cultivated humous lowlands near drainage canals or near wadis.

Courting and Mating. These activities take place from March to June. In courting, the male approaches the female while chirping. He circles round her with outstretched antennae, then tries to caress her antennae. The positive reactions to these approaches are expressed by the female by raising her body. The male hastens to crawl under her, directing the tip of his abdomen upwards to meet the tip of the female's abdomen. Mating takes place both in the tunnels and on the surface (CASSAB, 1936).

Construction of Nest. Soon after mating, the female begins to construct a nest. As a rule the nest is 10—15 cm deep. In cases where the soil is too moist the nest may be only 6 cm deep, while in dry soil it may be as deep as 35 cm. The female digs a tunnel, enlarging it at one end into a cell 3.5 cm high and 5 cm in diametre. Its walls are plastered from within. This cell is very often situated above the tunnel, connected to it by a short looped tunnel. One female can build more than one nest; as a rule she builds three.

Oviposition. Many eggs are laid in the nest in the course of several ovipositions. With each additional oviposition, however, the number of eggs decreases. Thus, the third oviposition has about half the number of eggs as the first.

Table XI presents the number of eggs laid by females of the two species of *Gryllotalpa* found in the Near East.

Table XI.

Number of eggs laid by two species of *Gryllotalpa* (after CASSAB, 1938)

	Gryllotalpa gryllotalpa			*Gryllotalpa africana*		
	minimum	maximum	average	minimum	maximum	average
1st laying	200	300	250	33	75	45
2d laying	140	200	160	25	55	29
3d laying	100	160	120	15	35	19

Cannibalistic Habits. Some writers point out that during the search of prey the individuals may attack their own kind. Also, the male may devour the female, or if the female is stronger, she may devour her mate. The male preys upon the offspring and these prey upon each other.

Attraction to Light. Individuals are attracted to light. Often they are seen around street lamps. (Trapping one summer in Egypt during September brought 49 gryllotalpas; 32 were *G. gryllotalpa*, and 17 were *G. africana*. Of these only 10% were females. (All were of the long-winged type.)

Control Methods. One system of control consists of agrotechnical methods, namely, cultivation of the soil to destroy the nests and flooding the area to drown eggs and larvae. From the discussion above it is clear why these methods do not control the pest satisfactorily. Another method is the placing of cups filled with water across the tunnel. The insects fall into the cups and drown.

A third method is the protection of seedlings with collars of reeds. This is used in Greece and by the Arabs in small vegetable gardens.

Another method, using a bait, consists of favorite food mixed with insecticide. This food may be bran, rice, corn, cotton seed cake, etc. The insecticide may be any stomach poison. In particular use is sodium fluosilicate. This is mixed at the rate of 5—10% and water is added to make a mash. About 5 kg/dunam are spread out towards evening. Some of the synthetic insecticides kill the insect just by contact. B.H.C., aldrin, and heptachlor give the best results. By spreading the insecticides along areas frequented by the pest or by working it into the soil 2—3 cm deep, satisfactory control can be obtained. However, because of the ill effects of these insecticides, general application must be limited. Only partial application, around plants for instance, should be allowed.

LITERATURE

A. Balachowsky & L. Mesnil, 1936. Insectes Nuisibles aux Plantes Cultivées, *1692-1702.*

F. S. Bodenheimer, 1930. Schädlingsfauna Palestinas, Berlin, 439 pp.

A. Cassab, 1934. *Bull. Soc. Roy. ent. d'Egypte, 421-426.*

A. Cassab, 1936. *ibidem, 23-25.*

A. Cassab, 1937. *ibidem, 82-87.*

A. Cassab, 1938. *Soc. Fouad 1er d'Ent., 397-399.*

E. Rivnay, 1948. Field Crop Pests in Palestine (in Hebrew). 72 pp.

E. Rivnay & M. Ziv, 1962, Contributions to the Biology of *Liogryllus bimaculatus* D.G. M.S.

F. Silvestri, 1939. Compendio di Ent. **1,** *80-81* and *87-90.*

F. C. Willcocks, 1925. The Insect and related Pests of Egypt, Agr. Soc. Tech. Section, Cairo.

The order Rhynchota which is composed of Heteroptera and Homoptera includes some of the most notorious pests as far as damage to agriculture is concerned. The sub-order Homoptera includes the scale insects, aphids, white flies, etc.

Pentatomidae

The sub-order Heteroptera includes also some very injurious pests. The family Pentatomidae, for instance, has the *Eurygaster* — the menace to wheat in the Middle East. Some members of this family are polyphagous and will be discussed below. The Pentatomidae or the so-called "stink bugs" are flat, broad insects, have five-segmented antennae, and a scutellum reaching the membranous part of the wings, or even covering a large part of the body. The eggs are cylindrical or barrel-shaped, with the typical lid cover. They are laid on the food plants in clusters, one egg standing near the other.

The Green Stink Bug — Nezara viridula L.

Description. The adult is 14—18 mm long and 9—11 mm wide; its colour is entirely green; the scutellum is long, tapering posteriorly and projecting into the membranes of the hemielytra; the head is small and pointed, and together with the prothorax forms an equilateral triangle with an arched base; the third segment of the antenna is longer than the second; the entire body is punctate.

The egg is cream-coloured, with the typical lid; its diameter is 0.75 mm and it is 1.2 mm high. As a rule the eggs are laid on the underside of the leaf.

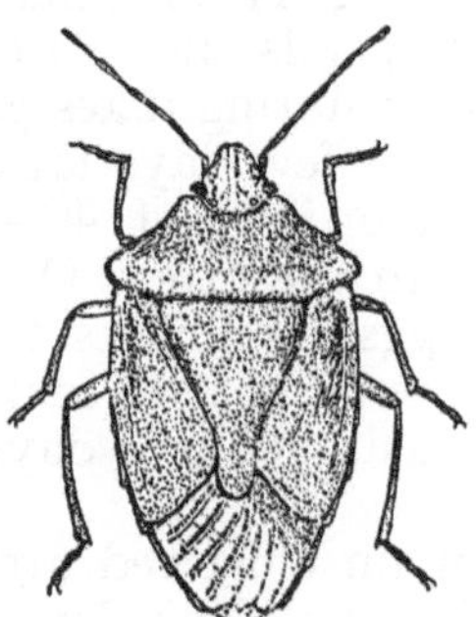

Fig. 16. *Nezara viridula.*

The neonate larva is 1.3 mm long, orange-coloured, and the head is red. Changes in size and colour occur as it grows. Details are given in table XII.

Distribution. The bug is distributed in countries having a warm climate. It is found in Europe, particularly in its southern countries; in Africa; in the southern states of the U.S.; in Western and South eastern Asia; and in Australia.

Hosts. The species is very polyphagous. As many as 115 species of plants belonging to 32 families were listed as its food plants. These include important crops such as corn, leguminous crops, cotton, tomato, and many others.

Table XII.

Development, size, and colour of instars of *N. viridula* (after KAMAL 1937)

Instar	Development in days at 26° C	Length in mm	Width in mm	Colour
end 1st	4.75	1.9		Dark red with bright spots
2d	10.12	2.2	1.5	Dark brown, ventrum has bright spots
3d	9.25	3.5	2.3	Dark brown, ventrum has bright spots
4th	11.37	5.5	5	Head and thorax green, abdomen green with red margin
5th	15.12	10	8	Body green, margin of abdomen and dorsal spots pink.

Type of Injury. The insect sucks the juice of the plant. When many feeding insects are gathered on one plant, it is liable to wither either partially or entirely. One insect may suck dry the premature seeds in the pods of crops such as alfalfa and *Ricinus*. One insect may also cause blemishes and malformations on young fruit such as tomatoes. It is an unpleasant experience to bite into a fruit containing the odour of a stink bug. The bug may also transmit diseases.

Life History. The adult spends the winter in dormancy, hidden in crevices, under rocks, etc. Mating takes place soon after awakening, and egg-laying begins a few days later. The eggs are laid in groups under the underside of the leaf. 40—70 eggs may be in one group, standing one next to the other. One female may lay 4—5 groups. The number of eggs diminishes in each subsequent oviposition. All told, one female may lay as many as 160 eggs.

The incubation period lasts about 6 days at 26° C and 4 days at 27—28° C.

The neonate larvae remain clustered around the egg shells for some time before they disperse. The development of each instar lasts several days, as indicated in table XII.

The entire development, including the incubation period, lasts 56 days according to KAMAL (1937). SILVESTRI (1939) quotes a period of 5—10 days for eggs and 25—35 days for the larval development, but does not mention the temperature. The pre-oviposition period may last 15—30 days.

Number of Generations. According to the above mentioned records the development of one generation takes about 2 months during the summer. If we assume that egg-laying begins in March, and that the bug enters diapause in November, then 4 generations can be raised within this period in Israel or Egypt. SHWEIG (1954) states this to be the case in Israel, while KAMAL observed only 3 generations in Egypt. Possibly the colder weather in the spring and at the end of the summer lengthens the period of development. JONES (according to SILVESTRI (1939)) speaks of 4 generations in Louisiana, while BOSELLI mentions 2 generations in Italy (SILVESTRI).

Limiting Factors. The bug exists in many countries without causing serious damage. There is reason to believe that one of the factors which keeps its population at a low level is the egg parasite *Microphanurus basalis* which lays its eggs in the egg of the bug. It is true that the parasite does not lay many eggs but its development is very fast; at a temperature of 26° C it completes its development in 10—12 days, while its host takes 56 days to develop (KAMAL).

When climatic factors limit the activity of the parasite, though not of the bug, a rise in the population of the pest may take place.

In countries where the bug exists without its parasite it may be quite troublesome. This was the case in Australia until the parasite was introduced from Egypt.

Control. In case of a severe infestation the bug may be killed with the aid of synthetic insecticides, such as BHC, DDT, dieldrin, toxaphene, and phosphoric insecticides.

In Australia and New Zealand good control was obtained with 0.05—0.1% DDT emulsion a.i. and also with a dust of BHC at the rate of 115 g gamma isomere per dunam. In Florida 10% dust of sabadilla was more efficient than DDT or other synthetic insecticides.

In cases where alfalfa and clover grown for seed are attacked, while in blossom, sabadilla (or even toxaphene) is preferred because it is less harmful to bees.

Dolycoris baccarum L.

Various crops are inhabited during the summer months by a smaller bug of this family, namely *Dolycoris baccarum* L.

Description. The body is 10—12 mm long and 5.5—6 mm wide; it is pale brown or walnut, with a greenish or purplish tinge; the colour is mottled, with small darker dots; the tip of the scutellum is straw-yellow; the abdominal tergites are black and yellow, ventrum and legs are light green, except the apices of the tibiae and the tarsi which are black; the antennae are black, except the first segments and the base of the other which are yellow.

The egg is greyish-white, subcylindrical, surface reticulate, 1.2 mm high and 0.7 mm wide.

The neonate larva has black head and thorax; green abdomen with a black patch on each tergite; the body is covered with hair especially at its margins.

Distribution. This insect is distributed throughout the entire palearctic region. Damage is recorded in particular from northern European countries — Finland, Germany, Poland, USSR — and from Italy and Cyprus.

Hosts. Several plants belonging to various families serve as food, including Leguminosae, Graminaceae, Cruciferae, Umbelliferae, Solanaceae, Malvaceae, and others.

The following crops were recorded as being damaged by *D. baccarum:* tobacco, cabbage, beet, cotton, and cereals.

Type of Injury. The bug may feed on the stem but usually prefers the young and soft parts of the plant. Thus it attacks germinating plants, soft seed in pods, and other parts in the stage of development. The damage may therefore be very pronounced. In green wheat, for instance, the kernels are sucked dry.

Life History. Very soon after emerging from their winter quarters the bugs mate, and a few days later egg laying begins. In Poland the oviposition is from mid-June to the end of August, while in Germany (TISCHLER, 1937) it is from end of May to July with the peak occurring in June. The eggs are laid on the underside of the leaf in groups of 20—40. The eggs hatch within 10 days. Thus, larvae in Germany appear in mid-June, while the adults of the new generation appear from the end of July to mid-August. Thus the total life history lasts about 2 months under summer conditions in Germany.

The new adults do not mate nor produce during the first summer, but after a period of feeding retire to their winter quarters for hibernation. In Cyprus these quarters are under stones in the high mountains. MORRIS (1929b) states that they hibernate on a mountain over 2000 metres high and in the spring come down to the valleys. Thus its life history is similar to that of *Eurygaster.*

Limiting Factors. According to MORRIS (1929a) the adults are attacked by the fly *Gymnosoma rotundatum* L.; about 50% of the hibernating bugs are parasitized. The eggs are attacked by species of *Telenomus.* In the USSR (POLIVANOVA, 1957) *T. sokolowi* infests the pest to a great extent — 80–100% of egg batches may be parasitized by this genus.

Cicadellidae (Jassidae)

This family is an extensive group, including many species of diverse forms. They are distinguished from other Homoptera of the Auchenorhynchi group in that their antennae are inserted in the frons between the eyes and the hind tibiae are armed with a row of spines below. The eggs are inserted into the plant tissue.

Some members of the genus *Empoasca* are notorious pests of various crops. A few polyphagous species occur in the Near East; their biology is more or less similar. *Empoasca lybica* has been studied more fundamentally than other species and will be discussed here.

Empoasca lybica Bergevin

[Syn.: *Chlorita signata* HAUPT.]

Description. Length of body 3—3.5 mm; entirely green; tegmina narrow, green with longitudinal veins; hind wings broad, transparent, with fewer veins. As in other members of the family, hind legs are larger and their tibiae are armed with spines. The egg is white, banana shaped, not longer than 1 mm, and is inserted in the plant tissue. The neonate larva is 0.7 mm, but the fifth instar nymph is 2.5—3 mm long. During the first two instars the colour is translucent white; it becomes pale green and finally green.

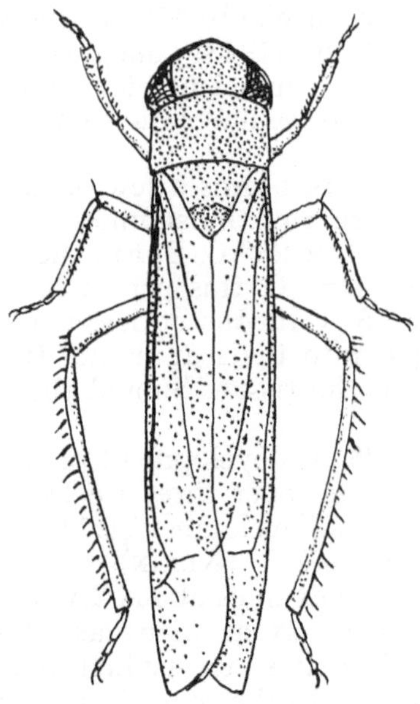

Fig. 17. *Empoasca lybica.*

This insect is conspicuous because of its liveliness and its agile oblique way of walking.

Distribution. The insect is typical of Northern Africa. Its most southern boundaries are in Central Africa while in the north it reaches Spain and Israel. Records of damage done by this pest come from Tripoli, Libya, Egypt, Sudan, Eritrea, Somalia, Tanganyika, and Uganda. Its damage in Israel is less than in the African countries.

Hosts. Many plants are listed as hosts to this insect. Among cultivated plants it is most injurious to cotton, okra, eggplant, tomato, pepper, and potato. It is found on beets, radish, sunflower, alfalfa, and *Ricinus*, and may also attack sorghum and sugarcane.

Type of Injury. The damage consists in sucking the sap from the plant. No doubt along with the sucking enzymes are secreted into the plant tissue. As a result of this sucking and secretion, necrotic spots may appear on the leaf; some writers refer to these spots as "Cicada burns". Not all the plants react in the same way. Burns may not appear in beets and pepper for example but are conspicuous in potatoes and eggplants. The pest may also cause leaf curl and defoliation and may cause blossoms and young fruit to drop.

The transmission of virus diseases to various plants is pointed out by many writers. In Spain, according to CASTRO (1943), grapevines suffer most. In Israel, potatoes and eggplants are most affected, while in the Sudan it is cotton. According to COWLAND (1949) infested cotton plants produce only 45% of the yield of non-infested plants. According to KLEIN (1948) young eggplants, when attacked in midsummer, never give sufficient yield to cover the cost of production. In young tomato seedlings the stem hardens and the entire plant fades.

Life History. Copulation begins a few days after the last moult. The pre-oviposition period in the summer is quite short when the daily temperature averages 25° C. But in the winter when the temperature drops to 14—15° C, this period lasts about 20 days. At a temperature below 15° C the insect does not reproduce.

The egg is inserted into the leaf tissue. Its well-being depends upon the humidity in the tissue. Should the leaf wilt or dry, the egg dies.

In the summer, at 25° C development of the larva takes 10—14 days, while in the winter, over three weeks pass before the larva matures. The threshold of development is 11—12° C and the thermal constant is 280 day degrees C° (KLEIN).

In the summer the adult lives about 2 weeks, in the winter two months. One female may lay as many as 50—60 eggs during her lifetime. Egg-laying decreases at very high temperatures.

Number of Generations. In the coastal plain of Israel the development of a generation, including all stages, lasts three weeks in the summer, four weeks in the spring or autumn, and about 10—11 weeks in the winter. Consequently, about 8 generations may develop during the year in that region. In a warmer climate the number is naturally greater.

Limiting Factors. With the approach of winter and falling temperatures, the development of the pest slows down. As mentioned above at 15° C reproduction may cease altogether — bringing about a gradual reduction in the population of the species. At still lower temperatures — below 10° C — the mortality of the larvae is great, so that the population of the insect during the winter dwindles almost completely. In fact no insect is found on cultivated crops though rare individuals may be found on weeds. In colder climates, such as Spain, a decrease in population takes place as early as October.

High temperatures also reduce the population effectively. At 41° C the mortality of the larvae is very great; at 30° C no oviposition occurs. For this reason the pest is found only in the autumn in the Sudan.

Control. Control measures should be taken before the population of the pest has been built up; it is naturally easier to control a sparse population. This pest is controlled by the substance veratrine (a powder containing at least 5% of ground sabadilla seeds or a wettable powder containing 50% of sabadilla seed meal) at the rate of 150—200 g active ingredient per dunam.

In Spain, according to CASTRO, a powder containing one part pyrethrum and four parts sulphur was effective against this pest on vines.

According to JOYCE (1956), DDT controls the pest on cotton in the Sudan; but it caused an increase of the white fly as its parasites *Eretmoceros* and *Prospaltella* were exterminated. For this reason rogor, which kills both *Empoasca* and the white fly has replaced DDT.

In Israel, plots of cotton treated with arsenate were infested with the pest while adjacent plots treated with endrin were free of it.

Other species of this genus in the Near East are:

Empoasca decipiens PAOLI

Distribution: Egypt, Morocco, Baluchistan, Israel, Cyprus, Italy, France, Spain, Czechoslovakia, Germany, England.

Hosts: Cucurbitaceae, Solanaceae, sweet potato, lettuce, and others.

Empoasca decedens PAOLI

Distribution: Egypt, Israel, Cyprus, Czechoslovakia, France.

Hosts: On cotton (in Egypt), otherwise on various fruit trees.

Empoasca distinguenda PAOLI

Distribution: Transvaal, Congo, Egypt, Israel, Italy.

Hosts: Cotton, bean, maize, *Ricinus*.

Whiteflies — Aleyrodidae

These Homoptera are small or even minute insects, related to Aphidae and Coccidae. In their immature stages they are scale-like, flat, stationary, and closely attached to their food plant; but unlike Coccidae, both sexes of adults are winged and free. They have two pairs of wings like aphids but these are covered with a waxy white dust giving them a white appearance; hence the name. The antennae have seven segments, and the eyes are constricted in their middle or even divided.

This is not a large family but some species are notorious pests, such as the citrus whitefly, the greenhouse whitefly, and the tobacco whitefly which is discussed below.

The Tobacco Whitefly — Bemisia tabaci Gennad

[Syn.: *B. gossypiperda* M. & L.]

Plants infested with whiteflies become "sooty" due to the black fumagine which develops on the sweet excretion; when stirred the insects rise in a cloud and then settle down again. In the Near East the most common whitefly is *Bemisia tabaci* GENNAD.

Description. This is a minute species whose body is not more than 1 mm long; it has two pairs of wings; the base colour is lemon-yellow but due to the fine waxy meal which covers it, the body and wings look white. The third antennal segment is as long as all the following segments together; the end of this segment bears a seta and near it two sensory pores; the last segment is longer than the one preceding it. The male is slightly smaller than the female, and bears at the end of the abdomen a pair of claspers longer than the

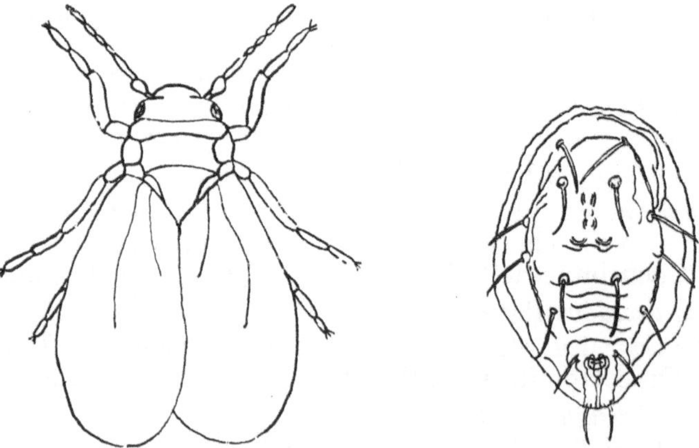

Fig. 18. *Bemisia tabaci* – a. adult. b. nymph. after KLEIN.

abdominal segment preceding it. The egg is elliptical and attached to a short petiole which is inserted into the tissue. The length of the egg without the petiole is 0.26 mm, and its width is 0.12 mm. The neonate larva is a crawler, very much like the larvae of bugs or Coccidae; after it thrusts its proboscis into the plant tissue it remains in this situation throughout its life.

The larva in its fourth instar is yellowish-white; is elliptical; has wide, flat, asymmetrical margins; the dorsum bears two tubercles and 8 pairs of setae; the length of the body is 0.75 mm.

Distribution. From the records in the literature it is evident that *B. tabaci* is distributed in countries around the Mediterranean basin. If we include however the records for *B. "gossypiperda"*, which is a synonym for *B. tabaci*, the range of distribution of the species widens and includes also Sudan, Iraq, Pakistan, and India.

Hosts. The list of plants which serve as food to *B. tabaci* is extremely long. KLEIN (1948) enumerates 49 plants from Palestine alone. This is not the place to name all the hosts, but mention should be made of some of the families of plants of economic importance that are attacked by this pest. These are Malvaceae, Cucurbitaceae, Solanaceae, Leguminosae and Convolvulaceae. In Israel the following crops suffer most: Cucumber, vegetable marrow, cotton, sweet potato, tomato and egg plant. In India, Pakistan, Iraq and Sudan, cotton is the most severely affected crop.

Type of Injury. The insect feeds on the plant by sucking its sap; thousands of such feeders reduce the nourishment of the host. In the process of piercing the plant tissue, the insect injects enzymes which apparently have an ill effect upon the leaves and hasten the plant tissue's drying process. The insect can also transmit virus diseases to tomatoes and cucumbers. Before adequate controls were found against this pest, tomatoes and cucumbers could not be grown during certain periods in the inner valleys of Israel.

Life History. The data presented herewith were taken mostly from AVIDOV.

The egg is laid as a rule on the lower surface of the leaf. Oviposition can only take place at temperatures above 14.5° C. With the aid of the ovipositor the female inserts the petiole of the egg into the tissue and it remains fastened in this way, standing over the surface of the leaf. The incubation period lasts 4—8 days at 26° C, or 8—22 days at 20° C. At a temperature below 12.5° C the egg does not hatch. Soon after the larva begins to feed, a waxy secretion covers its body. During the last nymphal instar it remains quiescent for some time; some writers refer to this stage as the "pupa".

The development of the larval stages lasts during the summer 7—11 days at 26° C, 14 days at 20° C; while in the winter the larval stages take 30—60 days at an average temperature of 16° C. Thus the threshold of development of the larva is 12° C, and the thermal constant 140 day degrees Centrigrade.

Two to four days pass before the adults reproduce; during the winter however the pre-oviposition period may last as long as three weeks. On the average one female lays about 200 eggs. Fertile females lay more – one female attained a record of 300 eggs in the course of 21 days. During the summer the female can live from 15—30 days and the male 7—14 days; in the winter the female lives about eleven weeks and the male eight.

Number of Generations. During the summer in Israel, from May

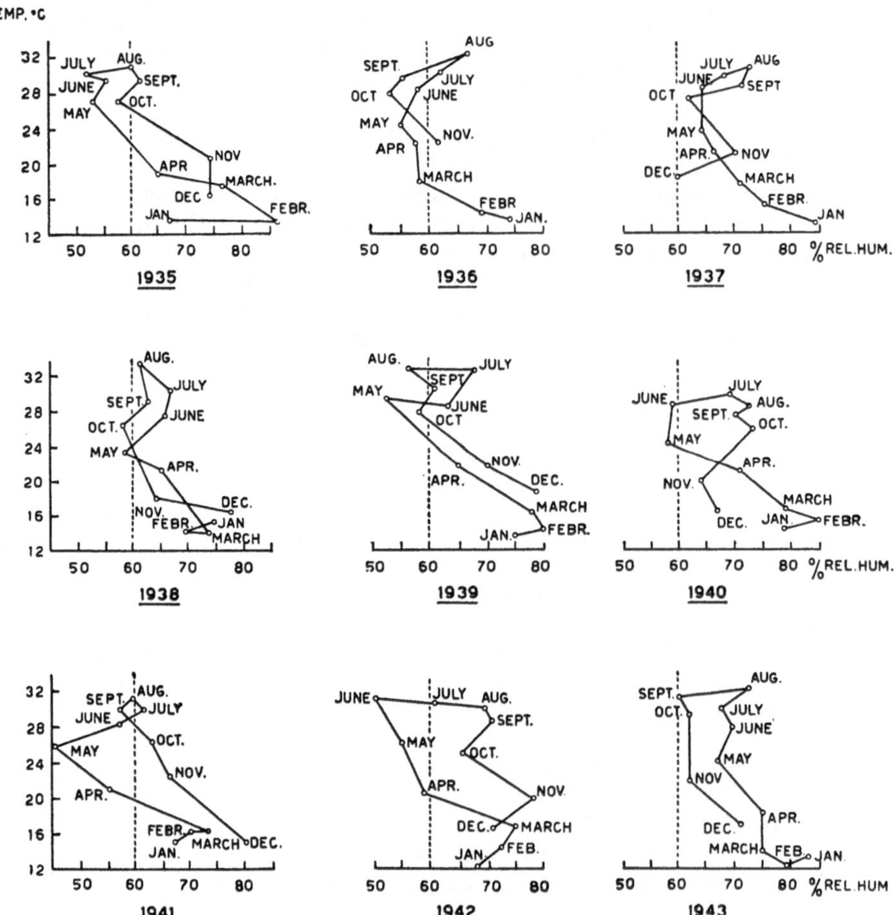

Fig. 19. — Climograms of the Jordan valley – Israel during the years 1935–'43 incl. During the years 37, 38, 39, 40 and 43 *Bemisia* populations were high because humidity during June–Sept. was above 60% R.H. after AVIDOV.

to October inclusive, this pest may raise about nine generations, each lasting 2—3 weeks; three more generations are raised during the cooler six months.

Ecological Notes. The population of this pest is not equally high each year. AVIDOV followed its infestation in the Jordan Valley for nine years. His conclusions are as follows:

a. In that locality the peak of infestation occurs in July-September;

b. When the humidity during these months is above 60% R.H., the population of the insect rises to high levels. This happened in '37, '38, '39, '40 and '43. When the humidity is below 60%, the population is low as was the case in '35 and '36; Fig. 19.

c. Should the months May and June be arid with frequent khamseen* days, the blow to the existing population in those months is so severe that it remains at a low level even if the humidity is above 60% during the peak months (1941 and 1942).

The effects of favourable high humidity explain why the pest remains on the lower side of the leaves, and why low-spreading plants such as the cucumber and sweet potato harbour the pest more than erect plants.

Further observations in the Jordan Valley showed that tomato seedlings are less infested with the pest when the soil around them is covered with sawdust. Temperature measurements showed that sawdust-treated soil radiates more heat than non-treated which absorbs much of the heat. The maximum daily temperature over the sawdust reached 51° C; this was about 6 degrees higher than over the non-treated soils. This temperature apparently proved to be fatal to the eggs and prevented egg-laying. Laboratory experiments showed that the mortality of adults increased suddenly when the temperature reached the 46—47° C level. (AVIDOV).

Control. The often recommended nicotine sulphate never gave satisfactory control of the pest. Also chlorinated hydrocarbons, DDT, dieldrin, and others gave insufficient control. The organophosphorus insecticides proved to be the most efficient. Of the three which were tried, parathion, diazinon and malathion, the first was the most effective. The newer systemic substances rogor and phosphamidon also proved to be quite satisfactory.

LITERATURE

Z. AVIDOV, 1956. *Ktavim* 7(1), *25-41*.
RUIZ A. CASTRO, 1943. *Bol. Pat. veg. Ent. agric.* **12,** *143-189*.
J. W. COWLAND & C. J. EDWARDS, 1949. *Bull. ent. Res.* **40,** *83-86*.

* Khamseen = a hot, dry wind from the desert.

R. J. V. Joyce, 1956. *Bull. ent. Res.* **47,** *399-413.*
M. Kamal, 1937. *Bull. Soc. Roy. ent. d'Egypte* **1937,** *175-207.*
H. Z. Klein, 1948. *Bull. ent. Res.* **38,** *579-584.*
C. S. Misra & K. S. Lamba, 1929. *Bull. Agr. Res. Inst.* No. 196. 7 pp.
W. L. Morgan, 1948. *Agr. Gaz. N.S.W.* **59,** 8, *421-422.*
H. M. Morris, 1929. *Cyprus agric. J.* **XXIV,** 4, *149-150.*
H. M. Morris, 1929b. *Rep. Dept. Agr. Cyprus* 1928, *43-44,* Nicosia.
E. N. Polivanova, 1957. *Dokl. Akad. Nauk SSSR* **112,** 3, *538-541.*
K. Shweig, 1954. Veg. Pests, Hassadeh Co. 56-58.
F. Silvestri, 1939. Compendio di Entom. appl. Portici, *227-234.*
E. Swirski, 1959. Notes on Homoptera in Israel (in Hebrew). Agr. Res. St. Rehovot, Bull. **22,** 33 pp.
W. Tischler, 1937. *Arb. physiol. angew. Ent. Bull.* **4,** 3, *193-231.*
D. O. Wolfenbarger, 1947. *Florida Ent.* **29,** 4, *37-44.*

Aphids – Aphididae

In northern countries aphids are extremely important agricultural pests. In the countries of the Near East, however, they are of lesser importance. Many species are monophagous and they will be discussed in the chapter dealing with the respective hosts. Those that are polyphagous will be dealt with here. Some outstanding features characterize this family as a whole.

Biological Notes

1. The eggs may hatch before they are laid, and in this way nymphs are born. This phenomenon is called viviparity.
2. The bearing of offspring may occur without necessitating fertilization by the male — this is called parthenogenesis.
3. The viviparous females may be winged or wingless — alate or apterate. Wing development depends upon the climate, the season, photoperiodicity, the type of food, and other factors. Winged females may migrate to various secondary hosts.

After several parthenogenetic generations have been produced in the course of a summer, their number depending upon the temperature, sexual forms appear. These are usually winged and they migrate to the primary hosts where sexuparae are born which lay eggs. Very few eggs are laid by each female, often only one. Aphids may have thus a primary host and various secondary hosts.

The optimal values of temperature, humidity, and host availability for aphids in the Near East are formed principally towards the end of the winter and in the spring. The heat and drought typical to this region, especially to its southern part, are not favourable for aphids. Although many species may be found there all year round, the peak in the number of individuals and species occurs in the spring — March to May. As mentioned above, in countries where cold hard winters prevail most aphids produce, with the approach of winter, a gamic generation. The eggs of this generation overwinter and give rise in the spring to parthenogenetic females. In the warm climate of the southern countries of the Near East — Egypt, Transjordan, Israel and others many aphids reproduce parthenogenetically throughout the year, as long as the temperature allows reproduction. The number of generations depends upon the availability of food and on the temperature. It is noteworthy that secondary hosts of many of the species are absent in these countries. Furthermore, in Europe the migration of aphids from one host to another are more regulated than is the case in these warmer countries (BODENHEIMER, 1935).

As stated, during the warm summer many aphids cannot be found. It may be surmised that many seek shelter in more favourable microclimates. Thus, *Rh. maidis* in the summer is found in the rolled corn leaves or in the space between leaves where humidity conditions are more favourable. It may also be surmised that some species migrate to countries with more favourable climates.

Migration. Aphids were found at high altitudes, as high as 3000 ft., and as far as 400 km away from land. It was assumed that since aphids are small and light animals they are carried by the wind as passive objects.

At this point the approach of JOHNSON (1954) to this problem should be noted. If we follow the line of his various papers and their accompanying graphs, his ideas may be summed up as follows:

1. Contrary to the accepted view expressed by several entomologists, aphids are not mere passive objects torn away from their hosts and carried by the wind, as is the case with crawlers of coccids and rust mites.
2. The winged aphids are migratory insects and of their own volition leave the plant in an upward direction. They are aided by currents of air.
3. On higher levels they may lose their ability to fly and then may be carried by the wind.
4. This migration takes place in flushes and their time of occurrence and intensity depend upon the number of aphids which become alate prior to the flush.
5. According to JOHNSON's experiments the peak of maturation of alates occurs in the morning with another peak late in the day.
6. Alates usually migrate about 24 hours after maturation; in exceptional cases they migrate on the same day.

Biological Equilibrium. As stated above, the damage caused by aphids in the Near East does not attain proportions that necessitate its control. In the summer a sharp decrease in population takes place because of climatic conditions. Another factor which should not be overlooked is the biological balance maintained by their enemies. Aphids are preyed upon by many insects — Lady-bird beetles (Coccinellidae), Syrphid flies (Syrphidae), Gallmidges (Cecidomyiidae) and others. They serve also as hosts to many parasites. Species of the following genera of parasites were raised in Israel: *Lysiphlebus, Diaeretus, Asaphes, Alloxysta, Praon* and *Pachyneuron*. The importance of this factor — that is of enemies of aphids — was never properly estimated until the extensive use of synthetic insecticides was adopted. Many of these substances when used against other pests killed off the enemies of aphids and spared their hosts. These hosts increased in such numbers that the crop situation became worse than it was before the use of the insecticide.

Selective insecticides should be adopted and efforts should be made to restore the biological balance.

Chemical Control. The individual aphid may be easily killed, but it is hard to exterminate an aphid population once it has established itself in the field. This is due to the fact that the aphids are well hidden. They are found either on the underside of the leaf, or the leaves may curl over them, or they are hidden in well sheltered places where it is difficult to reach them. Nicotine at a concentration of 0.1% a.i. gives good control. Also, lindane, parathion and diazinon are good aphicides, even when given at far lower concentrations. In cases of severe infestation these substances are applied at the rate of 25—30 g per dunam. It is urged that these substances be employed several days before harvest to prevent ill effects of residues.

Aphids may also be killed by systemic insecticides which act as stomach poisons. Metaisosystox, phosphamidon and rogor are very satisfactory as aphicides. To some extent they kill by contact during the first day or two, and then may also affect beneficial predators and parasites. It must be cautioned that these insecticides as well should be applied some time before the harvest.

Drawbacks of Insecticides. Two other factors should be taken into consideration. Many reports during the past decade point out that aphids gradually become resistant to insecticides. In California *Myzus persicae, Aphis gossypii* and *Therioaphis maculata* have become resistant to parathion, malathion and others, making these species more difficult to control (Stern & Reynolds, 1958).

In addition, these non-selective insecticides have offset the natural balance by killing off the predators and parasites which kept the aphids at low levels.

A study carried out by Bartlett (1958) showed that parathion, malathion and rotenone are the most injurious insecticides against *Hippodamia*, both larvae and adults, while nicotine sulphate, schradan and demeton are the least severe.

The Cotton Aphid — Aphis gossypii Kalt.

This species is called the cotton aphid. Its scientific name implies the same meaning, but in reality it is very polyphagous. In some countries it is known as "the melon aphid".

Description. The agamic apterate: Length of body 1.8 mm; width of abdomen 0.9 mm. Colour greenish-yellow; greenish-grey; or dark-green to greenish-black. Head round, as long as wide, eyes dark brown; the antennae inserted in the vertex and not elevated on tubercles. Antennae shorter than body, do not reach the base of the cornicles. Colour of antennal segments as described in table XIII. Prothorax very short; abdomen soft and smooth, both bearing lateral tubercles; sometimes the abdomen has dorsal brown spots. Length of cornicles 0.14—0.23 of length of body; they are longer than the third antennal segment. Cauda 0.4—0.8 of length of cornicles, colour dark.

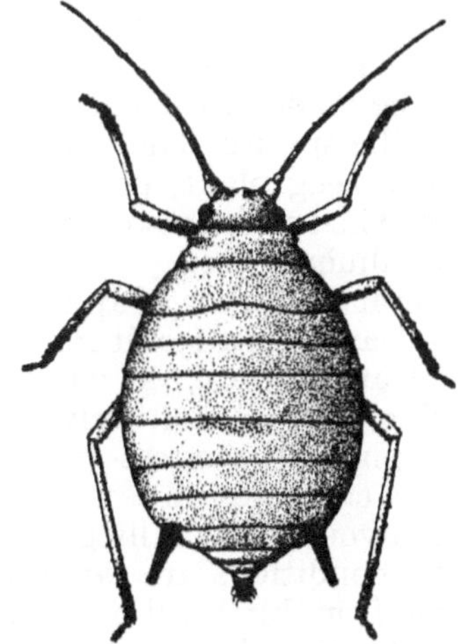

Fig. 20. *Aphis gossypii* – after SILVESTRI.

The parthenogenetic alate is somewhat smaller, 1.35 mm long; abdomen 0.65 mm wide. Wing span 5.1 mm; head broader than long; length of antennal segments as in table XIII.

Distribution. This species is found in most countries having a subtropical or temperate climate. In the Western Hemisphere it is found in the U.S., Mexico, the West Indies, Brazil, Hawaii, and Australia. In the Old World it is known to occur in Japan, China, Formosa, India, Java, and the USSR. In the Near East it occurs

Table XIII.

Description of antennal segments of *A. gossypii* KALT.

No. Segt.	Apterate		Alate	
	Colour	Length (mm)	Colour	Length (mm)
I	Dark	0.08	Dark	0.07
II	Dark	0.06	Dark	0.04
III	Yellow	0.32	Dark, base light	0.22
IV	Yellow	0.25	Yellow, half dark	0.17
V	Yellow, tip dark	0.21	Half dark	0.17
VI	Dark	0.35	Dark	0.37

in Egypt, Israel, Jordan, Lebanon, Iraq, Turkey, and Cyprus; it is found also in Italy, France and Morocco.

Hosts. As stated above, this aphid is very polyphagous, very many plants serving it as food. A list of all the known hosts cannot be given here. Mention will be made of the cultivated crops which are liable to be attacked by the species. They are melon, watermelon, cucumber, cotton, hibiscus, egg plant, potato, sesame, asparagus, garlic, maize, and others. The species also infests fruit trees such as the pomegranate, anona, almond, etc.

Injury. Feeding by several thousands of aphids weakens the plants and causes curling of the leaves. As a result infested plants bear less fruit or do not bear fruit at all; often, as with watermelons under heavy infestation, the plant dwindles entirely.

Biology. The melon aphid reproduces parthenogenetically. In spite of searches in various countries no sexual forms of this species were found. The bearing of young asexually goes on uninterruptedly as long as environmental conditions are favourable. The development period of the female from birth till her bearing young is very short. PADDOCK (1919), who reared several generations, states that at a temperature of 28—30° C, as little as four days were required for its complete development. In the spring at an average temperature of 25—28° C, as long as 10—12 days were needed. These are extreme cases of outdoor breedings. As a rule, 5—9 days are required for the female to mature. According to the same author the female continues to bear young until a short time before she dies. The reproduction period may last from 3—40 days as may be seen from the accompanying table XIV. The number of offspring per female are also given. It is noteworthy that 2/3 of all the females bore from 50—100 larvae, while only 15% bore 100—150 young. (Table XIV).

Table XIV.

Period of development and reproduction of *A. gossypii* KALT. (PADDOCK, 1919)

Number of females	Reproduction period-days	Number of females	Number of offsprings
5	3— 7	10	1— 50
12	8—15	21	51— 75
29	16—25	18	76—100
11	26—35	6	101—125
3	36—41	5	126—154

The average rate of reproduction is 3—12 per day. The biggest number of larvae born by a female was 20 per day, while the average is only 9—12.

The average length of life of one female may be obtained if we add several days to the reproduction period. Some females lived more than 50 days while the average was 30 days.

PADDOCK raised many generations in succession. He chose one female from the first batch of each generation and continued for a whole year. The rearings were outdoors, except during very cold weather when they took place indoors. In this way he raised 60 generations. Overlapping was at times so complicated that a females may have been alive when her great granddaughter of the sixth generation was born.

Phenology in Israel. In Israel this aphid is prevalent throughout the year. SWIRSKY (BODENHEIMER & SWIRSKI, 1957) and other entomologists collected it during the winter months on wild and cultivated plants. However, the population is sparse during these months and the species is almost inactive. Only towards the spring with the rise in temperature does the insect become more active with a resulting rise in population. As one host plant ages winged females develop and these migrate to other hosts.

The peak of the population of this species is during the spring; in the summer it dwindles, and in the autumn it rises again. Many other aphids disappear entirely during the summer months — their peak being the early spring. *Aphis gossypii* is one of the few exceptions and its abundance during the summer depends upon the climate. A mild, humid summer increases its population during these months.

PADDOCK in his paper on *Aphis gossypii* writes in detail about the relationship between ants and aphids. He describes how one ant, having found an egg of a syrphid fly, tore it off the leaf and threw it to the ground. The present author also observed an ant — *Tapinoma* sp. — discovering a *Scymnus* larva in the aphid colony on an egg plant leaf, removing it and throwing it to the ground.

Ants take care of aphids in many other ways. They carry them for food supply, for shelter against cold, and probably use their presence to scare off predators and parasites.

The Peach Aphid — Myzus persicae Sulz.

The *Myzus persicae* is a very polyphagous aphid, migrating from one plant to another throughout the summer. Its economic importance is due not only to direct injury, but also to its being a vector to certain viruses. In particular it is important as a vector to diseases in potatoes and other plants of the same family.

Description. Agamic apterate female: Body greenish-yellow or yellow. Over it one median and two lateral longitudinal brown lines — often obsolete. Tip of antennae, cornicles, and cauda slightly darker than rest of body. Third segment of antennae has no olfactory pores. The cornicles are directed one against the other. Their length is twice that of the cauda which bears three pairs of setae. Length of body 2.40 mm; length of antennae 2 mm; length of cornicles 0.4 mm; cauda 0.2 mm.

Agamic alate female; Head and thorax brown, abdomen greenish; a brown plate is on each of 3—6 abdominal segments; cornicles and tip of abdomen brownish-green; antennae brown. At the base of antennae two tuberculi. Third segment of antennae ⅓ longer than fourth. On third segment 7—15 olfactory pores. Fourth segment slightly longer than fifth. Length of body 2.3 mm; length of antennae 2 mm; cornicle 0.34 mm; cauda 0.20 mm.

Distribution. This is a cosmopolitan insect. Although of Palaearctic origin, it is found in North America, Australia and India. In the Near East it is found in North Africa, Italy, the Balkans, Asia Minor, Lebanon, Syria, Iraq, Jordan and Israel.

Type of Injury. When thousands of aphids are stationed on a plant, sucking its nourishment, they of course weaken the plant. They may also cause the leaves to curl. In addition, this species is a notorious vector of various diseases, especially of the Solanaceous plants. Thus, the leaf wilt virus, leaf roll virus, yellow vein virus and others are known to be transmitted by this aphid.

Food Plants. The first host upon which the species begins its annual life cycle is the peach tree or some other stone fruit tree such as the plum or cherry. From these hosts the aphid migrates afterwards to annual plants. In the literature about 400 plants belonging to some 200 genera are listed as hosts to this species. This is not the place to list all the known food plants; mention is made only of the cultivated crops which are attacked by this species, some of them important, and belonging to the Solanaceae, Malvaceae and Compositae families. As a rule this aphid prefers young, soft leaves to feed upon.

Life History. The biology of the aphid differs with the country and climate of its residence. The most natural course is the one which takes place in Europe — and it is described first. Throughout the summer several generations are produced parthenogenetically. Towards the autumn or beginning of winter sexual stages are produced — the ginoparae and males. The ginoparae which are alate migrate to the peach tree. There they give birth to sexual females, the oviparae, which mate with the males that have also migrated to the peach trees. After mating the females lay a few eggs (4—6) which overwinter. As a rule these eggs are laid in buds, in folds, or between leaves. At the end of the winter or in the spring the eggs hatch; these are parthenogenetic apterates and are considered as fundatrices. Two parthenogenetic generations are produced on the peach tree; they then migrate to annual plants and the cycle is renewed.

The number of generations and the length of each generation depend upon the climate. Thus in France there are fewer generations than in Italy, the sexuals appearing as early as August—September. In India oviposition takes place in December and January.

In some countries in the Near East, Israel for instance, the diapause — in the form of a winter egg — is not prevalent, and the species is capable of continuing its parthenogenetic reproduction uninterruptedly and continuously.

Ecological Notes. Myzus persicae is a native of countries with a cold climate. It is capable of developing even at 6° C. The threshold of development lies between 3—4.5° C. However, at this lower temperature it lives about 43 days while at 24° C—26° C it lives only 7 days. It is apparent that 28° C is detrimental to the species as the mortality of nymphs increases at this point. At 30° C none of the aphids develop.

The parthenogenetic female begins to produce the day after the last moult; however, at 10° C the preproduction period lasts 4—5 days. The number of offspring per female depends upon her food and temperature. Under favourable conditions of food at 25° C one female gives birth to 60 larvae on the average, and the average length of life is 25 days.

Limiting Factors. Several entomologists in Israel note that *M. persicae* is found in that country throughout the year. It should be pointed out, however, that its population is high in the spring only. During the summer it is rare. SWIRSKI collected all the data on its presence in Israel and found that in 110 observations, only 5 were noted during the summer months July-October — i.e. about one in each of the summer months. Only after December does its population begin to increase, so that in April alone 60 reports were noted. In May a sudden drop in the population takes place and it continues at low level till the end of the summer (BODENHEIMER & SWIRSKI, 1957).

The primary factor of this course of events is surely the temperature. Some entomologists mention other factors, such as migration of alates and the presence of parasites. Without slighting these factors, in particular the role of parasites, temperature nevertheless remains the prime factor.

In view of this a few facts should be mentioned. BROADBENT, who has made a fundamental study of this species (BROADBENT & HOLLINGS, 1951), finds that without food and at a temperature of 38.5° C the female will die within an hour. However, at the same temperature, if females are allowed to remain on the host plant, they live for several days but do no reproduce. Females which remained on the host plant at 37.5° C did give birth to a few nymphs. When placed at a temperature of 42—43°, females dropped immediately to the ground.

Further records show that the mortality of females due to high temperature was greater when the females were first placed in a relatively low humidity. At 33—35° C females become restless and drop to the ground.

These records are noted because they help to explain the course of events in Israel. Temperature often reaches high levels which interfere with the welfare of the species, preventing normal reproduction or normal development and even causing starvation. Whichever the case is, the population then is close to zero. Records showing similar behaviour of this species in Australia were also obtained.

The Bean Aphid — Aphis fabae Scop., S. lat.

In various legumes, horse beans and vetch in particular, heavy populations of black aphids may be found. These are colonies of *Aphis fabae* SCOP. In the literature *Aphis fabae* and *Aphis rumicis* are often confused as they resemble each other superficially. The names are confused and often considered synonyms for the same species. In recent years Miss JONES (1942) studied the two species thoroughly and found that they are distinct and independent species. *A. rumicis* is monophagous, feeding on *Rumex* only, while *A. fabae* is a polyphagous species feeding on many plants, in particular upon legumes.

Some taxonomists claim that *A. fabae* includes several species which are much alike and hard to distinguish. In the following discussion the species will be considered in its wider sense.

Description. The apterous parthenogenetic female is globular, 2—2.5 mm long, coloured black with a greenish tinge. The antennae are shorter than 2/3 of the body length; they are black with the exception of the fourth and part of the fifth segments which are light. The cornicles are black and short, 0.25—0.35 mm. The cauda is cone shaped. In its last instar the nymph has white spots over the abdomen, typical to this species.

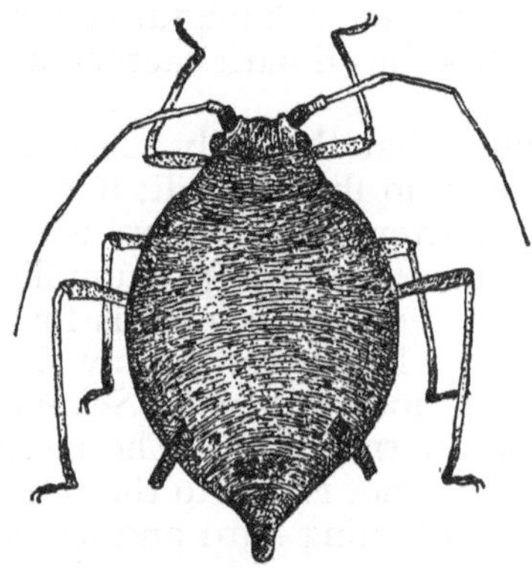

Fig. 21. *Aphis fabae.*

Distribution. This species is very widely distributed, from the arctic to the tropical region. In the Mediterranean area it was recorded in Cyprus, Lebanon, Israel, Jordan, Iraq, Egypt and Turkey.

Food Plants. There are over 200 plants enumerated as hosts to this species. In the Near East damage is done to bean, pea, vetch, horse bean and beet. Other favoured genera are *Carduus, Chrysanthemum* and *Brassica.*

Type of Injury. Naturally, a heavy attack of thousands of aphids weakens the plant. JUDENKO et al. (1952) found that an infested field yielded about 40% less than a non-infested field.

Life History. In North and Central Europe the sexual female lays eggs on *Evonymus* and *Viburnum.* The eggs hatch in the spring and asexual generations develop upon these primary hosts. In May the flight of apterous females to other secondary hosts takes place, and towards the autumn sexual alates return to the primary hosts.

In warmer countries the sexual forms are absent and the species reproduces parthenogenetically all the year, as long as the prevailing temperature allows reproduction.

A great deal of our knowledge regarding the flight of aphids was acquired from studies on this species. Thus MÜLLER & UNGER (1951) found that the temperature threshold of flight of this species is 14 ° C, while the optimal temperature is 21 ° C during December and 23—30 ° C in August. These authors also noticed that the approach of an alate aphid to the plant is not in a straight line but rather in a spiral line; its rings become smaller and smaller as it approaches the plant; and the apex of the cone is the plant which is the source of odour. Furthermore, the alighting of an individual upon a plant does not mean it settles there. It may migrate to another, in search of another host. This tendency of migration is greater as the temperature rises, and the distance of such migration may extend hundreds of metres. On such flights the climatic factors of course take a great toll.

JOHNSON (1954) found that the daily schedule of the flight is as follows: at night there is no flight at all; it begins at sunrise, and increases until midday; after midday there is a decrease, probably due to temperature rising above the optimum flight temperature.

Ecological Notes. In Israel *Aphis fabae* is found in particular in winter. Its climax is from December to May. In summer its population dwindles and it is found then on *Solanum nigrum* growing in the cool shade of the citrus grove. The general decrease of its population during the summer is due to the rise of the temperature and to the host plants becoming hard and less succulent.

Aphis craccivora Koch

[Syn.: *A. laburni* KOCH., *A. medicaginis* KOCH., *A. rotmiae* MACCH., *A. leguminosae* THEOB.]

Aphis craccivora, widely known as *A. laburni,* is found among other species on leguminous plants. It is also found on many other plants.

Description. The agamic apterate female is 15—25 mm long. The body is black, the head and thorax darker than the abdomen; the legs are black except the tarsi which are light. Cornicles are black, their length 0.35—0.40 mm; antennae black except the first and second segments which are light. The antennae are 2/3 the length of the body. The young nymphs are dark green; as they mature they become black.

Distribution. This species is found almost throughout the entire world: in Europe, the United States, Central and East Asia, Formosa, Ceylon, Java, and Sumatra. In the Near East it was recorded in Egypt, Israel, Jordan, Lebanon, Iraq, and Turkey.

Hosts. This species too has a long list of hosts of various families, herbaceous plants and shrubs as well. It may be found on *Gossypium* and even on citrus, but mainly it feeds on leguminous plants. In Egypt and Israel it is found on alfalfa, clover, vetch, chick pea, and bean.

Type of Injury. Like other aphids, it sucks the nourishing fluid from the plants. In leguminous crops its tendency is to feed upon flowers and pods, which increases the injury.

Life History. In the Mediterranean countries no sexual forms were found. The agamic females reproduce all year round asexually, as long as the temperature allows it. According to SWIRSKY, who studied the population of this species on *Spartium junceum*, the peak of the population occurs in April-May (BODENHEIMER & SWIRSKI, 1957).

The Root Louse — Trifidaphis phaseoli Pass.

This species is a member of the Fordini (like *Forda* and *Geoica* spp.) which live part of the year on the roots of plants and the other part on the leaves of *Pistacia* on which they cause gall formation.

Description. The agamic apterate female has an elliptical body 1.80 mm long and 1.50 mm wide. Its colour is yellowish but is covered with a white waxy meal. The antenna is 0.60 mm long and has only 5 segments.

Distribution. This species is distributed in Asia, Europe and America. In Africa it was reported from Rhodesia. In the Near East it was reported from Egypt, Israel, and Turkey.

Hosts. This is a polyphagous species. It was found on the roots of various hosts belonging to different families — Papilionaceae, Solanaceae, Cruciferae, Graminaceae, etc. The primary host is *Pistacia.* In Russia it was found on *P. mutica,* and in Israel on *P. atlantica.*

Life History. Throughout the summer agamic generations develop on the roots of various plants. At the end of the summer alate individuals develop which migrate to the *Pistacia* leaves. These give birth to a sexual generation. The females lay eggs which hatch the following spring. The offspring of these agamic females cause the gall formation and produce in turn a few generations within the fall. Finally, a winged generation develops and leaves the gall (which has in the meantime dried and split) to find roots of other plants and to continue the production of agamic generations.

In countries with a colder climate where no *Pistacia* is found, the species produces asexually all year round.

Scale Insects – Coccidae

The Coccidae is a large family which includes insects of various morphological forms and habits. The females of many species are stationary, legless and wingless, and are protected in one way or another by a layer of waxy secretion. As a rule most species inhabit trees and bushes but some attack grasses. In the Pseudococcinae the females are free-moving, having legs but no wings. The body is covered with a waxy meal — hence the name "mealy-bug". The male has one pair of wings and vestigial mouth parts.

The Citrus Mealybug — Planococcus citri Risso

This insect is more notorious as a pest to fruit trees and vines than to field crops. Often, however, it is found in the roots of some crops causing distinct injury. It may at times penetrate into the storage room and cause potatoes, pumpkins and other fruit to be covered with mealy patches which are dense colonies of the pest.

Description. The body of the female is oblong, about 2.5—5 mm long; it is covered with a white waxy meal which gives the body a white appearance although its colour is actually yellowish or greyish. The margins of the body are decorated with 17 pairs of waxy short filaments, equi-distant from each other and equal in length except for the last which is slightly longer. The antenna is composed of 8 segments. The eggs are laid between waxy threads which are simultaneously secreted with the eggs. The eggs are yellowish, dimensions 0.3 × 0.2 mm.

The larvae — called crawlers — resemble the parent mother except that they are more motile. When hatched they are 0.4 mm long and yellowish in colour, but soon the waxy meal secreted over the body renders its appearance white.

The male, unlike the female, possesses one pair of wings, well developed legs, and long antennae. Its body is one mm long and 0.2—0.3 mm broad. The colour is yellowish-brown; the antennae and legs are lighter. The head bears four large ocelli, compound eyes being absent; the antennae are composed of 10 segments.

Distribution. This is a tropical and subtropical insect, inhabiting warm countries in all continents. In colder countries it exists in greenhouses.

Hosts. This insect has a long list of plants serving as its hosts. It feeds on all parts of the plant — the blossom, the soft fruit, leaves, twigs and stems. It feeds also on the roots of certain plants. Although fruit trees are the prevalent hosts the pest also infests potatoes, tomatoes, pumpkins, peanuts, and other plants.

Type of Injury. The damage is done by sucking of the plant juice. A heavy infestation of tomato seedlings causes them to wilt and subsequently to die. In larger plants such infestation causes a reduction of fruit. Infested potatoes in storage shrink and wilt.

Life History. During the cold winter months the insect is hardly active. It is hidden in cracks and crevices as larva, adult or egg. In the spring the eggs hatch and the larvae and adults become active and crawl to tender parts of the plant to feed.

The female larva moults 3 times before it becomes adult, while the male larva spins a cocoon after the second moult and pupates. It goes through 2 stages of pupa and finally emerges as an adult.

The length of development depends on both food and temperature. Insects which were reared on potato sprouts at a temperature of 24° C completed their development in 30—45 days; at 18° C in 56—65 days; while at 15° C in 60—133 days. These data include the respective pre-oviposition periods.

According to BODENHEIMER (1929) the threshold of development is 8.4° C and the number of day-degrees is 525 for fast-developing females.

The highest number of eggs was laid during June (400 per female) while in January only 60 eggs per female were produced. These figures were obtained from females reared on citrus. When reared on potato sprouts the reproduction is reduced to half the amount.

According to the above figures, fast females may raise about 8 generations in the coastal plain of Israel while lagging females raise but five generations. Since the egg-laying period is very prolonged there is much overlapping of generations.

Ecological Notes. The eggs and larvae are very sensitive to low relative humidity. In Israel this pest is most troublesome when the humidity during spring and summer is high; when dry desert winds are frequent, the pest is less troublesome. For this reason vineyards and citrus groves along the southern coast of Cyprus are heavily infested with *P. citri*. This also explains why, during the hot summer months, the pest is found in the ground on the roots of plants, usually in irrigated soil or in areas near leaking water-pipe lines and the like.

Parasites and Predators. Several predators such as *Scymnus* spp. and *Sympherobius* feed upon the species, and a few parasites, *Anagyrus kivuensis*, *Leptomastidea* and others, destroy the pest and no doubt reduce its numbers. Even so, the population often becomes unbearable and control measures are necessary.

Control. Parathion, diazinon and malathion kill the pest in various degrees. The best of these is parathion. Tomato seedlings with badly infested roots were relieved of the pest when a small amount of a solution containing 0.05% a.i. of parathion was poured around the stem. Potato seeds may be dipped in 0.1% solution of parathion. Care should be taken about toxic residues.

LITERATURE

B. R. BARTLETT, 1958. *J. econ. Ent.* **51** (3), *374-378*.
F. S. BODENHEIMER, 1929. *Z. angew. Ent.* **XIII,** *67-136*.
F. S. BODENHEIMER, 1935. VI. Int. Congr. of Ent., Madrid, *49-58*.
F. S. BODENHEIMER & E. SWIRSKI, 1957. The Aphidoidea of the Middle East. The Weizmann Science Press of Israel, Jerusalem, 378 pp.
L. BROADBENT & M. HOLLINGS, 1951. *Ann. appl. Biol.* **38,** *577-581*.
C. G. JOHNSON, 1954. *Biol. Rev.* **20,** *87-118*.
M. G. JONES, 1942. *Bull. ent. Res.* **33,** *5-20*.
E. JUDENKO, C. G. JOHNSON & L. R. TAYLOR, 1952. *Plant Path.* **1** (2), *60-63*.
H. J. MÜLLER & K. UNGER, 1951. *Züchter* **21,** *1-30, 76-83*.
F. B. PADDOCK, 1919, *Texas Agr. exp. Sta. Bull.* **257,** *7-54*.
V. M. STERN & H. T. REYNOLDS, 1958. *J. econ. Ent.* **51** (3), *312-316*.

The characteristics of these insects are indicated in the Greek names of the order: "Bristle winged" or "bladder footed". The wings are composed of a main petiole bearing bristles arranged feather fashion, and the feet have reduced claws and bladder-like foot tips. There are two main groups — Terebrantia (the females have a saw-like ovipositor), and Tubulifera (females having a tubular tip on the abdomen). The insects are as a rule phytophagous, feeding on the sap of the plant cells. Some species are predators feeding on other small insects and mites. Many insects inhabit flowers. Some of the thrips species are notorious pests.

The Onion Thrips — Thrips tabaci Lindeman

Thripidae

In Israel a large number of species of thrips may be found on the flowers of wild and cultivated plants. They are not easily noticed, since they are extremely small and hide among the various parts of the flower.

Most of the species of this order are not harmful, but there are some which are known as pests of various cultivated plants, feeding as they do not only on the flowering parts but on the fruit and leaves. These species can be very injurious indeed. The best-known and most common, which is polyphagous, is the onion thrips, or, as it is also called, the tobacco thrips.

Description. The female is 0.9 mm long; has a flat and narrow body of a light grey colour, with posterior segments of a darker hue; the head is quadrangular, wider than long. The mandible consists of 3 segments, of which the second is the shortest and the third the longest; the antennae have seven segments; the prothorax is one and a half times wider than it is long. The female has two pairs of straight wings, one margin of which is feathered; the wings are of the same colour as the body but somewhat darker. At the tip of the abdomen, the female has its ovipositor which consists of two small, slightly bent saws some 0.18 mm in length.

The egg is rounded, oblong somewhat curved bean shaped, translucent-white in colour and 0.25 mm long.

The larva resembles the adult. It is light-yellow in colour. The pre-pupa, which is an intermediate stage between the larva and the pupa, resembles the latter but has smaller eyes, antennae are straight instead of curved backwards, and smaller wing pads.

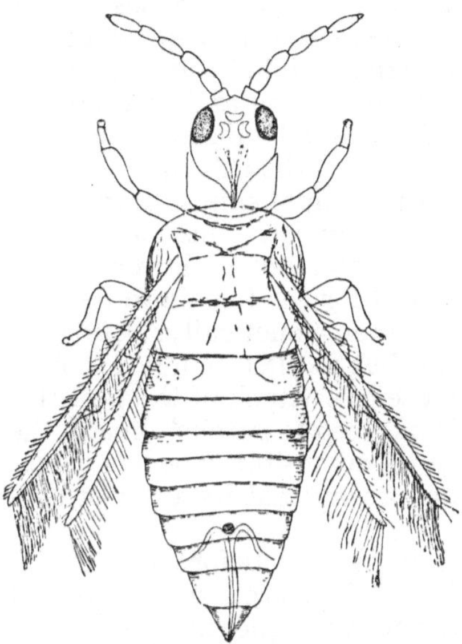

Fig. 22. *Thrips tabaci* (after RIVNAY). Courtesy Hassadeh.

The pupa is 0.7 mm long; has red eyes; antennae curved backwards; wing pads extending to the eighth segment and clavate bristles on the abdominal segments.

Distribution. This pest is cosmopolitan. In Southern and Central Europe it occurs in the open, while in the colder countries of Europe it appears in hothouses. It has been reported from Hungary, Austria, the Alps up to a height of 2000 metres, Canada, the U.S.A., Mexico, Bermuda, and the West Indies. In Asia the pest is known to occur in the following countries: Israel, India, Java, Sumatra, Formosa, Japan and Korea. It has also been reported from Australia and Hawaii and is prevalent in the Mediterranean countries.

Hosts. The onion thrips inhabits many wild flowers; in addition, it feeds on the leaves of a variety of plants that are not botanically related. In the literature, hundreds of host plants are mentioned. Here only those crops will be referred to which are liable to be harmed by the pest to such a degree that control measures become mandatory. They are onions, garlic, cabbage, cauliflower, cucumbers and other cucurbitaceae, tomatoes, potatoes, tobacco, beets, beans, peas, groundnuts, parsley, celery, and cotton.

Type of Injury. The injury to the plant is inflicted by the adults and larvae which suck the sap and chlorophyll of the epidermal layers. The destruction of the chlorophyll reduces the productive potential of the leaves and at times causes them to wither. Infested onion leaves result in decreased crop value. In young leaves, feeding by the pest causes pitting, while in young fruit silver spots appear which render the fruit unsightly and misshapen.

Life History. The following details are taken from the publications of EDDY & CLARKE (1930) in the U.S. and SAKIMURA (1937) in Japan. After emerging from the pupal skin, a number of days pass before the female begins to oviposit. This pre-oviposition period lasts some three days in summer but at lower temperatures it may be longer. There is no need for the female to be fertilized by a male as the eggs develop parthenogenetically. EDDY & CLARKE did find some male specimens, but SAKIMURA failed to find male adults in the Tokyo area during a severe attack of the pest there. When about to oviposit, the female pierces the parenchyma of the plant upon which it feeds by means of its ovipositor. The egg is laid in a pouch under the epidermis in such a way that its anterior end faces the opening of the pouch.

In summer 4—5 days pass before the egg hatches. SAKIMURA maintains that 3.6—4.5 days are required for hatching.

The young larva emerges squeezing its body out of the opening of the egg. It becomes active soon after hatching, and within a short time starts feeding on the tissue sap of its host plant.

The onion thrips passes through two larval instars and two pupal stages — the pre-pupa and the pupa. The larval stage lasts 5—6½ days at a temperature of 23—26 degrees C°. SAKIMURA mentions a range of 6.4—13.5 days without specifying temperature.

Each of the larval stages develops in half this time. The pre-pupal stage extends over some 1.5 days, while the pupal stage lasts 3.5—4.5 days or, according to SAKIMURA, 3—6 days.

The female may lay up to 12 eggs a day, though the average daily number of eggs laid is no more than six. Thus, a female may lay about 80 eggs during her lifetime. SAKIMURA obtained a higher maximum figure of 119 eggs. 20—30 eggs may be reckoned as an average number for every female.

Number of Generations. If we add the pre-oviposition period to the length of time of development of the thrips we see that a single generation requires 18 days to develop, at a temperature of 26° C. At lower temperatures more time is needed for one cycle. EDDY & CLARKE bred 6 generations from June to the end of September; SAKIMURA bred 10 generations in one year in the Tokyo area. Since the female lives longer than the time needed for one generation to develop, the generations overlap.

Ecological Notes. No investigational work has been done on this subject and no experimental data are available with regard to the effect of heat or moisture on this insect. However, observations have shown that the onion thrips is extremely sensitive to dryness and becomes troublesome only in the countries of high humidity and moderate heat.

In America, for example, onion fields in New England suffer badly since humidity in that area is high and summer temperatures are not low enough to check the development of the pest. For the very same reason the insect is considered very harmful in Japan and Hawaii. In the Near East where humidity is lower, the thrips thrives in places and at times when favourable conditions of humidity prevail. The annual course of events is about as follows:

From February onwards, when the temperature begins to rise and vegetation to develop, thrips appear in ever increasing numbers. Practically every flower harbours a number of thrips species, including the onion thrips. During the rainy months of February and March three generations of the insect may develop and consequently its numbers increase enormously. Subsequently dry winds begin to blow from a south-easterly direction and conditions tend to become unfavourable for the thrips. In many instances vegetation dries up within a short period as a result of the early cessation of rains, and thousands of insects suddenly find themselves without hosts and under adverse conditions of temperature and humidity. They then migrate to irrigated fields and attack cotton seedlings, groundnut plants and other crops, as only there do they find suitable hosts and a moist soil where favourable conditions of humidity prevail.

As the season advances temperatures reach a pitch which limits further development of the insect, and the attack of the pest on cotton and groundnut fields comes to a halt. Thrips continue to survive on fields whose vegetation shades the soil surface as this creates a micro-climate of moderate temperature and high humidity such as is found in cucumber fields.

The thrips then hide among the folds of the big leaves or between the spaces of onion leaves. In such spots the thrips continue to exist even during the hot dry summer days.

Control. A variety of contact poisons are lethal to the pest. Nicotine, rotenone and others are effective when applied methodically. Similarly, many of the synthetic insecticides will destroy the thrips, and both larvae and adults may be killed by dusting or spraying with one of these materials. It should be remembered, however, that the efficacy of the control measures depends on the residual effect of the insecticide used, as the eggs inside the plant parenchyma are not affected by the treatment and a new generation can soon develop from these eggs.

The following table XV contains interesting data of experiments conducted in Massachusetts, U.S.A. Results indicate that nicotine and rotenone gave a satisfactory kill as did D.D.T.

But the residues of this latter substance continued to act longer and continued to destroy all larvae hatching from the eggs several days after application — hence a better control.

For the same reason good results were obtained in Israel following treatment with dieldrin. Reports from Australia show that there too dieldrin was the best of all the materials tried. Other insecticides used were, in order of their efficacy, endrin, DDT, parathion, BHC, chlordane, and toxaphen. The addition of a wetting agent increases the effectiveness of the pesticides.

Table XV.

Results of Experiments on Thrips Destruction (after ALEXANDER et al., 1946)

Insecticide	DDT- 0.4% spraying	DDT- 5% dusting	Derris- 2 kg spraying	Nicotine 1.5 kg spraying	Rotenone spraying
% kills immediate count	89	90	90	91	81
% kill count after 4 days	85	98	38	71	24

The Red Thrips — Retithrips syriacus Mayet

[Syn.: *Retithrips aegyptiacus* MARCH.]

Heliothripidae

This thrips is best known as a pest to trees, shrubs and vines, but occasionally it attacks other plants as well; thus *Ricinus* is often severely infested with it.

Description. Like other members of this family the surface of the body is reticulate; it is spindle-shaped, about 1 mm long, and brown. Behind the antennae there is a distinct crest bearing the ocelli.

The larva is red and the last abdominal segment is tubular and bears 18 setae which form a basket into which the faeces are collected.

Distribution. This species is distributed through tropical Africa; its range extends northwards to Israel, Syria, and Libya. So far it has been recorded from Ghana, the Congo, Uganda, Tanganyika, Somalia, Sudan, Egypt, Tripoli, Israel and Syria. It has been recorded also from Brazil.

Hosts. This is a polyphagous species; it most frequently attacks walnut, persimmon, guava, eucalyptus and avocado trees, vines and roses. The *Ricinus*, *Jatropha* and *Acalypha* plants of the Euphorbiaceae are also infested.

Type of Injury. Like the tobacco thrips it sucks the sap from the cells.

Life History. With the aid of the saw-like ovipositor the female inserts the egg into the tissue of the leaf. The incubation period lasts 10—12 days at 27—30° C or 19—24 days at 24° C.

The larval period in its two stages lasts 7 days at 27—30° C or 30 days at 17° C. The two pupal stages complete their development in 2—4 days at 28—30° C, and in 15 days at 15—16° C.

Number of Generations. According to RIVNAY (1939) this thrip raises four generations in Israel during the summer, one each during the spring and autumn, and one during the winter — in all, seven generations.

Limiting Factors. During the colder months of the winter (15—17° C), about 80% of the pupae perish (RIVNAY). The adults of the remainder also die in large numbers; since they do not reproduce at this temperature, a very small percentage of the autumn population survives to begin the new spring generation. By June, after the third generation, a considerable population has been built up. At this point it begins to suffer from the low relative humidity and the population diminishes — to rise again at the end of the summer.

LITERATURE

C. P. ALEXANDER et al., 1946. *Bull. Mass. Agr. exp. Sta.* **436,** *33-43.*

C. O. EDDY & W. H. CLARKE, 1930. *J. ecol. Ent.* **XXIII,** *704-708.*

T. PASSLOW, 1957. *Queensl. Dept. Agr. & Stock Bull.* **106,** *1-20. J. agric. Sci.* **14,** 2, *53-72.*

K. A. RAHMAN & ANSHI LAL BATRA, 1945. *Indian J. agric. Sci.* **14,** 4, *308-310.*

E. RIVNAY, 1939. *Bull. Soc. Fuad 1er d'Ent.* **1939,** *150-181.*

K. SAKIMURA, 1937. *Oyo Dobuts Zasshi* **9,** 1, *1-24.*

The Spider Mites — Tetranychidae

The mites belong to the class Arachnida which includes spiders and scorpions among others. Their common characteristic is a body composed of two sections — the cephalothorax and abdomen and four pairs of legs on the cephalothorax.

In the order Acarina to which the mites belong there is no line of demarcation between the two parts of the body so that it seems as if it is composed of one piece. Acarina include the ticks — which are the larger species and which are parasitic, and the smaller mites.

Many of the spider mites are phytophagous and some of them are quite notorious pests.

Mites reproduce through oviposition. The eggs are spherical, about 0.1 mm in diameter or slightly larger. Parthenogenesis may occur, but non-fertile eggs give rise to males only. The first instar larvae have three pairs of legs only. After the first moult the animal has four pairs of legs. The immature stages are the larva, the protonymph, and the deutonymph. In the male this latter stage is missing. Between each instar there is a resting stage before moulting.

The adults feed by sucking the contents of the plant cells. The beginning of an infestation shows minute pale spots on the green leaf. As infestation advances, these spots increase in size and number; they gradually coalesce into larger patches until the entire leaf becomes pale or rusty brown and withers. In very large leaves such as those of *Ricinus* or even cotton where the infestation is limited to areas between the main veins, only these places are affected. Some mites cause defoliation of the plant.

Lately mites have become quite troublesome to the farmer, more so than in the past. This is due to the offsetting of the biological balance by extensive application of non-selective insecticides which have exterminated their enemies. Their increase has brought about extensive research in the use of acaricides. The mites however show remarkable strength and surprising resources in overcoming these acaricides — and today many species are resistant to several pesticides which formerly were quite effective. Some red mites spin webs like true spiders and these webs serve as protection, making the control application more difficult.

The Red Spider Mite — Tetranychus (telarius) cinnabarinus Pr. and B.

This is a very widely distributed species. Lately, the opinion has been expressed (PRITCHARD & BAKER, 1955) that this is a polytypic species which, among others, includes *T. cinnabarinus*. Acarologists in Israel claim that it is a most common species on field crops. HASSAN et al. (1959) claim that this species is one of the most common in Egypt.

Description. Female: length of body 0.45 mm; width — 0.30 mm; length of legs 0.25 mm. Colour of abdomen dark-red with two lateral brown spots on its tip; cephalothorax yellow; legs yellow with reddish tips. Male: length of body 0.35 mm; width 0.20 mm; colour yellow; legs yellow with reddish tips. The egg is spherical, 0.1 mm in diametre, its colour pale brown. Immature stages yellow with dark spots.

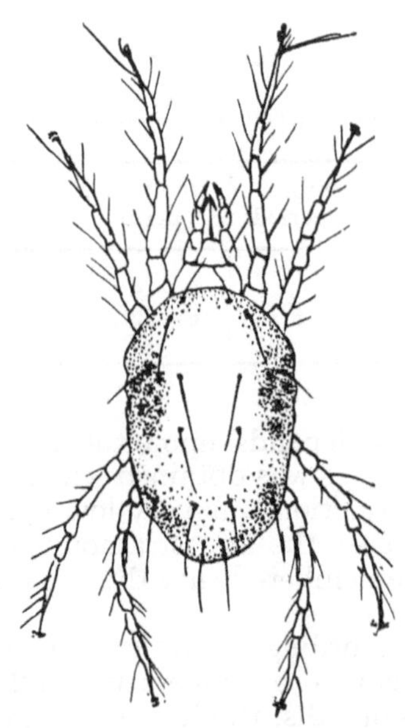

Fig. 23. *Tetranychus cinnabarinus*.

Distribution. This species is distributed throughout the world. It is common in northern countries and extends to the subtropical regions of North and South Africa. In America it extends from Canada to Mexico. It also inhabits Hawaii, Australia, Indonesia, and the Asiatic countries of the Middle East.

Hosts. In the literature many plants are listed as hosts to this species. In Palestine alone KLEIN (1936) mentions 60 host-plants. The cultivated crops which are mostly attacked by this species in this country are cotton, egg plant, tomato, pepper, watermelon, bean, peanut, beet, and *Ricinus*.

Type of Injury. The type of injury has been described in a previous paragraph of this chapter. As to damage, heavy attacks on cotton in Israel caused a reduction of 10% of the crop.

Life History. The eggs are laid singly on the leaf, as a rule on the lower surface between the main veins or in corners. The incubation period is very short, about one or two days, and the development of the larval stages very fast — only 7—8 days in the summer. The data of development at various degrees of temperature (according to KLEIN) are tabulated in table XVI. The threshold of development was calculated from these data and was found to be 8° C with a thermal constant of 170 day-degrees Centrigrade.

Table XVI.

Development in days of *T. (telarius) cinnabarinus* (after KLEIN, 1936)

Season and aver. temp. °C	Preoviposition period	Incubation period	Larval development	Length of life of adult
Summer 24—26	1	2— 4	7—8	15
Winter 12—16	3—5	6—12 (9)	17	30

The number of eggs depends upon the food and temperature. The largest number of eggs was obtained when the females were fed on bean leaves. Reproduction was very low when the females were reared on citrus leaves — the largest amount of eggs being 24 per female. It should be remembered that the egg is about ¼ the length of the female body.

Ecological Notes. According to KLEIN the lowest mortality of eggs, namely 10%, occurs at a temperature of 25—26° C. The more the temperature rises or falls, the higher the mortality rate. Thus, in the cold winter months when the temperature is between 13—19° C, a third of the number of eggs do not hatch. The situation with larvae is somewhat different. The lowest mortality is at 22—23° C, while at 25—26° C a third of the number of larvae do not mature. The slower development and higher mortality of the immature stages at lower temperatures explain why the population of this species is sparse during the winter months.

Table XVII gives further ecological data of this species as compared with another species discussed below. It explains why the range of distribution of *T. cinnabarinus* is wider than that of *E. banksi*.

Table XVII.

Ecological data on *T. cinnabarinus* and *E. banksi.* (after KLEIN, 1938)

Physiological state in °C	*E. banksi*	*T. cinnabarinus*
Freezing Point	6.9	2— 7
Threshold of development	10	8
Normal activity	18—39	16—37
Heat paralysis	43.5	49
Death at high temperature	47.5	52

The Oriental Spider Mite — Eutetranychus banksi McG.

[Syn.: *Tetranychus rusti* McG., *Eutetranychus clarki* McG., *Anychus verganii* BECH., *Anychus orientalis* (ZAHER) KLEIN, *A. ricini* R.S.]

Description. Female: body spherical; 0.45 mm long; 0.3 mm wide; abdomen dark brown; cephalothorax and legs yellowish-red. Male: body tapering posteriorly; 0.35 mm long; 0.25 mm wide; colour of abdomen reddish.

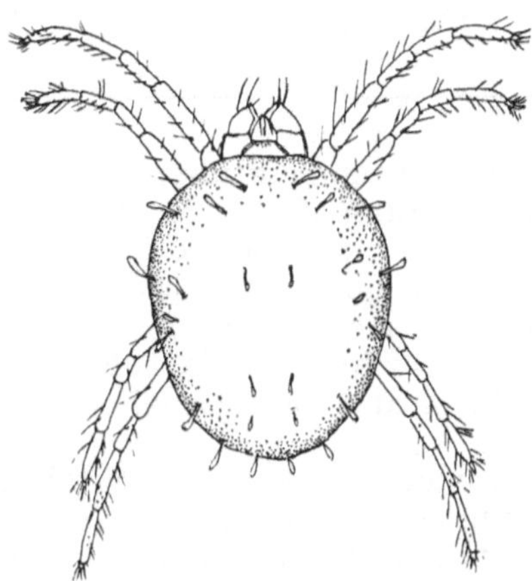

Fig. 24. *Eutetranychus banksi.*

The egg is spherical, 0.1 mm long, translucent and pale brown.

Distribution. This species is distributed in the tropical and subtropical regions of Asia, Africa and America. Damage has been recorded from India, Jordan and Israel, from Egypt, Libya and Mauritius, and from Texas and Argentine.

Hosts. As a rule trees are infested — citrus (the most common host); date palms; *Morus alba; Ficus macrophylla; Plumeria alba;*

Carica papaya; and *Ricinus*. HASSAN et al. (1959) reported that cotton in Egypt is a host.

Type of Injury. Injury is similar to that caused by other mites, and has already been described.

Life History. Like other species this one too develops fast and is capable of raising many generations during the year when conditions are favourable. The life cycle from egg to egg — according to KLEIN (1936) who studied its biology — is 10—12 days at a temperature of 27° C. The development and other biological data at various degrees of temperature are given in table XVIII. The length of development of the various instars is about the same except that the deutonymph lasts a little longer than the others.

Table XVIII.

Development in days of *E. banksi* at various degrees of temperature

Temperature in °C	Preoviposition period	Incubation period	Larval development	Length of life
27	1—2	2—4	8—10	12
17—24	3—4		15	
15—16	6—8	12	20—30	20
13	8	35		

During the winter, at a temperature range of 16—20° C, productive females lay an average of 18—20 eggs; in the summer, at a temperature of 22—27° C, productive females lay about 35 eggs. These records of KLEIN agree with later records of KLAPPERICH (1957) who reared the species in Jordan.

KLEIN, on the basis of his records, calculated the threshold of development to be 10° C and the thermal constant 180 day-degrees Centigrade.

Limiting Factors. There is usually a certain mortality of eggs. During the summer in Israel when the average temperature is about 25° C, only 7% of the eggs fail to hatch; while in the winter, at an average temperature of 17° C, as many as 30% do not hatch.

According to KLEIN, further limiting factors are as follows: A climate whose maximum temperature is above 34° C with an average of 31° C (such as prevails in Jericho during the summer); a relative humidity below 40%; a temperature below 16° C (reproduction is then reduced by half and mortality is about 50%). At 13° C the mortality of the larvae is 75%, and still worse is the 10° C temperature when development ceases altogether.

Phenology. KLEIN followed the phenology of this spider mite in Palestine and records that in Jericho the mite is prevalent during the winter only from September to March. The population peak is in February while in the summer it is not to be found. Around the lake of Tiberias it appears in April and is prevalent all the summer until early winter. In the coastal plain the mite appears in June and the peak of its activity is from August to October. These differences are caused by variations in the climates of these localities.

This may well be illustrated by employing the data of limiting factors on a graph in which the x line is graded with temperature degrees, and the y line with the relative humidity percentage. The favourable and unfavourable zones are drawn according to the following data: Shortest developing period 21—27° C and 60—70% R.H.; zone where insect can exist 18—30 ° C and 55% R.H. Fig. 25 presents the phenology of the two spider mites under discussion. The above detailed phenology is illustrated and compared with that of *cinnabarinus* (KLEIN, 1938).

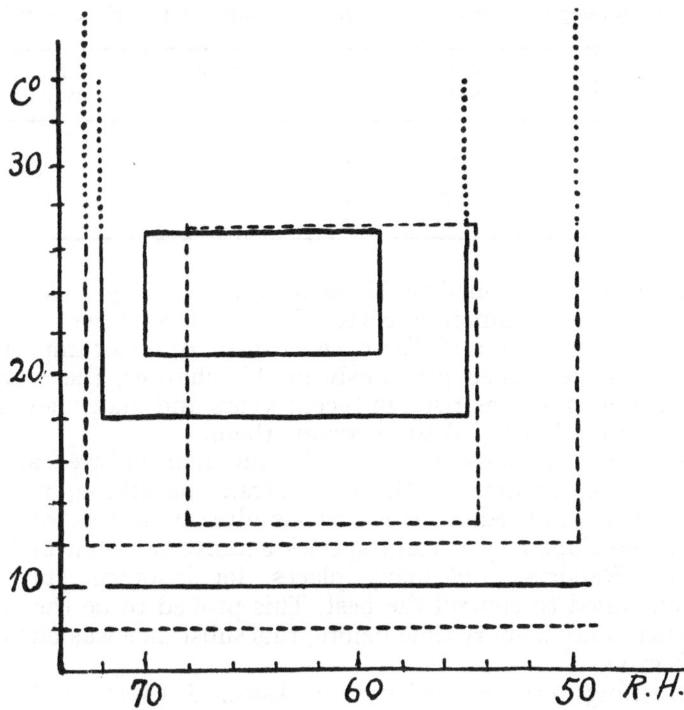

Fig. 25. Temperature – humidity zones of Spider mites. Straight line for *Eutetranychus banksi*: broken line for *Tetranychus cinnabarinus*. The lower lines at 8 and 10° C are the respective thresholds of development. The rectangles demark the zones of the respective most favourable conditions. The outer lines demark the limits of the lower temperature and humidity for their normal existence. After KLEIN.

Other Injurious Species of Acarina

HASSAN (1959) enumerates 5 species of mites on cotton; the two most important are those discussed above. SAYED in 1946 described the new species *Eotetranychus cucurbitacearum*. According to PRITCHARD & BAKER (1955) this is one of the species in the *T. telarius* complex. It may be quite similar or even identical to *cinnabarinus*. Which ever the case, biological data obtained by HASSAN & ZAHER (1956) in Egypt about this species is quoted here:

The number of eggs per female is 72 in the winter, 111 in the summer and 152 in the autumn. These figures are entirely different from those obtained by KLEIN with *cinnabarinus* in Palestine. Of course it should be remembered that KLEIN reared the species on a non-favourable host, namely citrus. Other data are given in table XIX.

Table XIX.

Temperature-development records of *E. cucurbitacearum* (after HASSAN & ZAHER)

Temp. in °C \ Period in days	Preoviposition period	Incubation period	Larval development
26	0.5	4	5
19	1	8	9
13	3.5	14	14.5

Other species mentioned by HASSAN et al. are *Oligonychus terminalis* SAYED, also common on cotton during the summer; *Brevipalpus obovatus* DONNAD; and *Tarsonemus* spec., of lesser importance.

Control. As mentioned previously in this chapter, the economic importance of mites has risen in recent years and many new acaricides have been developed to overcome them.

Some of the famous acaricides in the market are: aramite, chlorobenzilate, eradex, kelthane, ovotran, parathion, phenkapton, sulfanone, and tedion. However, as already stated, resistance has been developed by certain species against common and good acaricides. Reports from many places, for instance, state that parathion failed to control the pest. This proved to be the case in Israel where only a short time before, this substance was one of the most effective.

In screening tests carried out by ASHER & CWILICH (1961) in Israel, over 40 substances were tried. Of them kelthane was the most satisfactory against adults of *T. cinnabarinus* while tedion was effective against its immature stages. Both substances remained effective for about 3 weeks on cotton plants. In Egypt HASSAN et al. found that in 1956 and 1957 of all the substances tried, metasystox and gusation gave the most satisfactory kill and the highest yield in cotton.

In more recent trials in Israel morocide proved to be very effective against mites.

Now, after many instances where predators of mites have been exterminated, entomiologists realize more than ever before the important role these predators play in maintaining the biological balance of mites. In several research institutes acarologists are engaged in importing, rearing and distributing predators of mites in an attempt to reestablish the natural balance.

The predators include lady-bird beetles; *Scymnus* spp. mites of the genera *Typhlodromus; Mediolata* and *Anystis;* lace wings, *Chrysopa;* the anthocorid *Orius;* and some predatory thrips such as the *Scolothrips* sp.

LITERATURE

S. Asher & R. Cwilich, 1961. Progress Report on acaricides (in Hebrew).

A. S. Hassan & M. A. Zaher, 1956. *Bull. Soc. Ent. Egypte*, **40**, *301-320*.

A. S. Hassan, A. K. M. Nahal & E. A. El Badry, 1959. *Bull. Soc. Ent. Egypte*, **43**, *357-365*.

B. J. Klapperich, 1957. *F.A.O. Plant Prot. Bull.* **5**, 4, *51-58*.

H. Z. Klein, 1936. The Jewish Agency for Palestine. *Agr. Res. Sta., Rehovot, Bull.* **21**, *3-61*.

H. Z. Klein, 1938. *Bull. ent. Res.* **29**, *37-40*.

A. E. Pritchard & E. W. Baker, 1955. Pacific Coast ent. Soc., San Francisco, 472 pp.

M. T. Sayed, 1946. *Bull. Soc. Fouad I Ent.*, **30**, *78-97*.

Owlets-Phalaenidae (Agrotidae, Noctuidae)

The moths of this family are richly covered with hair-like scales around the head and thorax. The forewings are usually narrower than the hind wings and are usually of a somber coloration. During the day the moths cling to tree-trunks and the like, their wings covering the entire body and their colour merging with that of their surroundings. They are active at night, though a few are active also during the day.

The larvae are naked, the hair and setae either rudimentary or absent. Most of the larvae feed on plants.

This group of insects ranks among the first in amount of damage caused to field crops and losses inflicted upon the farmer. In some crops in the Near East members of this group are first in importance. Among the reasons for their troublesomeness are:

1. Because they are polyphagous the food supply is available all the year round; 2. They are capable of raising several generations a year; 3. They are migrants and in accordance with the climate and season they move from one country to another. Often they appear unexpectedly in masses, in the darkness of the night, and inflict their blow upon the field before the farmer has a chance to apply any measures even if available.

Details about the habits, ecology, and migration of each of the species under discussion are given in the following paragraphs.

Cutworms — *Agrotis* and allied genera

The cutworms are naked caterpillars which appear greasy and shiny, usually grey or brownish-green. Sometimes the body is marked with spots or maculae; when touched, they curl.

During the latter stages most of the species feed close to or beneath the surface of the ground, cutting the stem or other succulent parts of the plant. A few may feed above the ground and even climb bushes and young trees.

The Greasy Cutworm — Agrotis ypsilon Rott

This is a cosmopolitan pest; in Egypt it is the most troublesome cutworm.

Description. Its body length is 20 mm and its wingspan 40—50 mm. Its colour is brown with dark brown forewings. Along the side margin are black undulating lines. Near them is a black wedge-shaped marking and on the disk there are several irregular black dots. On the basic third of the disk is the typical oval concentric orbicular spot. The hind wings are light with black veins and margins.

Fig. 26. *Agrotis ypsilon* (left) *Agrotis segetum* (right).

The egg is round, slightly flat and its diameter is 0.5 mm, with radial minute ribs carved over the surface. Its colour is bright yellow which then turns reddish-yellow.

The neonate larva is bright green, covered with small tubercles each bearing a seta. The mature larva is smooth, cylindrical and dark green in colour. The ventrum is brighter and over the dorsum, along the median line, two bright lines run longitudinally. On each body segment there are four more conspicuous spots on each side. The prolegs of the abdomen each bear 16—20 hooks arranged in a single arc. On the frons and vertex there are two triangles with apices against each other.

Fig. 27. Typical position of an *Agrotis* larva.

The pupa is 18 mm long, brown, and its cell is not very deep in the ground.

Distribution. The moth is a cosmopolitan insect and occurs on every continent: in Asia, from India and Ceylon to Korea and Manchuria; in Africa from the Cape to the Mediterranean shores; in Europe from the south of England in the west to Caucasus in the east; it occurs also in America and Australia.

Food Plants. Crumb counts about 30 food plants upon which the larvae feed. This list is surely incomplete. In each country one certain crop is more subject to attack than others; thus in Egypt, clover and cotton are most injured; in Israel, clover, winter cereals, and beets; in France, beets; in India, potato and onions.

Type of Injury. The most conspicuous damage is caused by the mature larvae. These hide in the ground near the food plant, and cut the plant at its rootcrown or somewhat above it. Having destroyed one plant the larva moves to another. Often the plant is not consumed entirely by the larvae, but is left to wither and dry. On other occasions, particularly in cases of severe outbreaks, when the caterpillars appear in hordes, as happened in Israel in 1942, the entire plant is consumed, and the field, after they have finished with it, looks as if mowed.

Life History. The eggs are laid singly or in small groups. The incubation period during the summer, at a temperature of 24° C, lasts 3—4 days. The lower the temperature the longer the incubation period (see Table XX) and also the greater the percentage of mortality. This latter depends also upon the humidity. According to RIVNAY, eggs did not develop when placed at a constant humidity below 77% R.H.

The larva, in its early stages, consumes food 3 times its own weight. In later instars the food consumed daily amounts to 30% more than its own weight, while in the last instar it consumes about 3 grams of clover leaves before it is ready to pupate (BISHARA, 1932). According to BISHARA, who reared the insect in Egypt under natural conditions, the larva completed its development in 18 days at an average temperature of 27° C, or in 30 days at an average temperature of 20° C (Table XX). RIVNAY, who reared the larvae at various constant temperatures, obtained somewhat different data as presented in Table XXI.

Table XX.

Development periods (in days) of various instars at different temperatures (after BISHARA, 1932)

Temperature °C	Incubation period	Larval period	Pupal period	Preovip. period	Total
27	3—4	18	10	3—4	24—36
20	6	37	16	6	65
15	13	65	35	14	127

Before pupating the larva builds a cell about 5—12 cm deep in the ground; two or three days later it pupates. The pupal period at various degrees of temperature is given in Tables XX and XXI.

Table XXI.

Development of larvae and pupae at various constant temperature (after RIVNAY)

	Period in days at constant temperatures of °C				Threshold of devel-opment	Thermal constant day-de-grees °C
	30	26	22	18		
Larva	17	22	27.5	39	9° C	357
Pupa	8.5	11	17	22.5	9.5°C	200
(Range)	(7—10)	(10—12)	(15—19)	(19—26)		

Three or four days after emergence the female begins to lay. This pre-oviposition period may last two weeks in the winter at 15° C. According to BISHARA one female may lay as many as 1500 eggs, with an average of 1000 eggs. If larvae and pupae are reared under favourable temperatures many females may lay over 2000 eggs. RIVNAY reared females which laid over 2500 eggs. According to BISHARA the male lives a short while, while the female lives about two weeks. Table XXII shows the length of life of 30 males and 30 females reared at 23—26° C.

Table XXII.

Length of life of adult moths (after RIVNAY)

Number of days / Sexes	1—5	7—10	11—14	15—16	17—19	20—21
Females	4	7	10	6	2	1
Males	7	2	12	3	4	2

The oviposition period lasts several days. As a rule most of the eggs are laid during the earlier part of the period.

Attractive Factors. It often happens that one field may be more infested than a neighbouring field. It was noticed in Israel and elsewhere that certain factors may attract laying moths:

a. Heavy soils are frequently more infested than sandy soils.

b. Irrigated fields are frequented by laying females more than non-irrigated ones. This was noticed by BISHARA (1932) in Egypt and by CRUMB (1929) in Tennessee. In Egypt it was observed that caterpillars may remain 15 hours submerged in water and then return to normal activity.

c. Moths are more attracted to a field with a dense growth of weeds than to sparsely growing vegetation. On many occasions the invasion of a crop was traced to the weeds growing along the margins of the field.

d. According to BISHARA, newly ploughed fields are more liable to attack than non-cultivated.

Phenology. In Egypt the moth occurs all year round except in the summer. In the delta region the moth is absent during July and August (BISHARA (1932); HASSANEIN (1956)) while in more southern localities, from May to September, and in the Upper Egypt, in the Assuan neighbourhood, the moth is absent from April to October. In India FLETCHER (1915, 1925), too, states that the moth is absent from the valleys during July and August. In Israel, according to RIVNAY, the moth is absent from mid-June to mid-August in the coastal plain, and from early May to September in the Jordan Valley.

Limiting Factors. According to BISHARA, high temperature of the soil in Egypt during the summer is directly detrimental to the larvae and pupae and none survive it. FLETCHER (1916) claimed that, while the insect continued to breed in India throughout the summer, adults, which emerged in July and August, laid eggs which did not hatch. Similarly, RIVNAY found that the moths which emerged in Israel during May and June laid non-viable eggs. In experimental rearings the latter author found that when larvae and pupae were

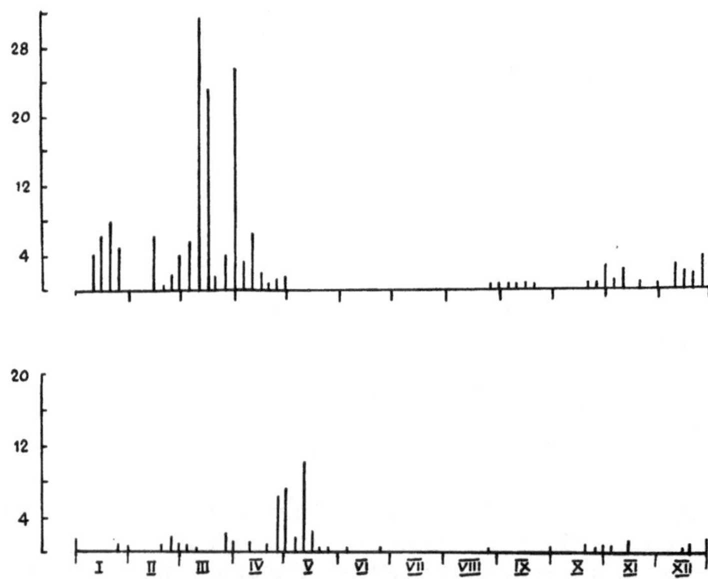

Fig. 28. Occurrence of *Agrotis ypsilon* in Israel during 1959 – after light trapping in the Jordan Valley (above) and at Rehovoth (below). Lines indicate individuals caught during one night (based on 5 day averages).

reared at a temperature close to, or above 30° C, many of the moths were either barren altogether or laid non-viable eggs.

Migration. Early in this century entomologists both in Egypt and India found it difficult to explain how this insect bridges the spring and autumn generation over the summer when no traces of it were found. It was certain that the moth does not enter a period of diapause as do some other moths of this family. The idea was expressed that the moth was a migrant, and evidence to this effect was brought by WILLIAMS (1925) in Egypt and by FLETCHER (1925) in India. FLETCHER expressed the opinion that the moths spend the hot summer in the mountains from which they invade the plains when the weather permits.

In Egypt, apparently, moths from Upper Egypt migrate northward, and the following generation migrates still further north when too warm weather sets in also there. In southern Europe the moth is found in July.

In Israel larvae of this moth are found from September to May — few may be found in June. The occurrence of the adults based on light trappings is depicted in Fig. 28. It is noticed that there is a period during the summer when the moth is not found in Israel at all. This period is longer in the Jordan Valley than it is in the Coastal Plain.

A very dense invasion took place in Israel in 1942. Larvae were found in hordes until late December. They entirely devoured the vegetation in the field and left only the bare ground. When they finished with one section of the field they advanced further each night, mowing another section of the crop.

LITERATURE

IBRAHIM BISHARA, 1932. *Egypt Ministry Agr., Tech. Bull.* **114,** 55 pp.

S. E. CRUMB, 1929. *U.S.D.A. Tech. Bull.* **88,** 177 pp.

T. B. FLETCHER, 1915. Ann. Rep. Bd. Scientific Advice for India 1914-1915 – *Econ. Zool. 1-15.*

T. B. FLETCHER, 1925. *Bull. ent. Res.* **16,** *177-181.*

M. H. HASSANEIN, 1956. *Bull. Soc. Ent. Egypte* **40,** *463-477.*

E. RIVNAY, Unpublished Notes.

C. B. WILLIAMS, 1925. *Trans Ent. Soc. London.* **1924,** *439-456.*

C. B. WILLIAMS, 1958. Insect migrations, I-XIII, 235 pp., The New Naturalist, Collins, London.

C. B. WILLIAMS, M. E. COCKBILL, M. E. GIBBS & DOWNES, 1942. *Trans. R. ent. Soc.* **92,** (1) *101-283.*

The Turnip Moth — Agrotis segetum Schiff

[Syn: *Euxoa segetum* SCHIFF.]

An equally notorious pest in the Near East, and in particular in the more northern territories, is *A. segetum.*

Description. The adult is 15—18 mm long, its wingspan about

35—40 mm; its fore-wings are grey-brown. Across the disc, about a fourth from the base, there is a wavy dark line and about a third from the base, there is the slightly oblong orbicular "eye"-like spot. Near it, more distad, is the reniform spot. Near the side margins wavy lines cross the wing. The hind wings are pearly white; the veins and a line along the margin are light-brown.

The egg is spherical, slightly flattened, 0.5 mm in diameter and its surface is engraved with radial ribs.

The neonate larva is light green, decorated with many black dots, each bearing a seta. It has only 3 pairs of abdominal legs. The mature caterpillar is about 40—50 mm long. Its colour is greyish-green or brownish-green, with darker inconspicuous longitudinal lines. Black dots are scattered all over the body, but eight spots, the setiferous tubercles on each segment are more conspicuous. On the head there are two triangles with their apices against each other. The vertex triangle is slightly narrower than the frons triangle. The abdominal prolegs bear about 10—12 hooks arranged in an arc. The spiracles are black.

Distribution. Although this moth was recorded in many parts of the world, it seems that damage is reported particularly from Europe. It inhabits Europe from the Volga in the east, to the British Isles in the west, from Norway and Finland in the north, to Italy and Spain in the south. In Asia it was recorded in Japan, Korea, Iran, Uzbekistan, Turkmenistan, Iraq, Turkey, Lebanon, Syria and Israel. In Africa it was recorded in Morocco and Cyrenaica in the north, Uganda, Rhodesia, Kenya, Nyassaland and S. Africa. There is one record from Bahia in Brazil.

Hosts. The list of plants which are subject to attack is very large. Of the cultivated crops the following serve as food: Rye, wheat and barley, spinach and beet, potato and tobacco, lettuce and sunflower, cucumbers and melons, onion and garlic, cabbage, cotton and flax. However, from the point of view of the insect not all the food plants are of equal value.

Type of Injury. The neonate larvae are found on the upper parts of the plant and their injury is not as severe as that caused by the mature caterpillars. These hide during the day in the ground, and feed at night. They cut the stem of the plant and thus destroy it without necessarily eating it. But a voracious dense population destroys the entire crop. The literature is rich in incidents where entire crops of over thousands of acres were entirely destroyed. REKACH (1933), for instance, states that in 1929 in Transcaucasus 66% of sesame plants, 50% of cotton and 35—50% of hibiscus plants were destroyed. NEUWIRTH (1932) complains that in 1931 in Czechoslovakia 80—100% of the beet plants were destroyed.

Life History. The eggs are laid either on the plants, close to the ground, or in the ground if it is moist. They are laid singly or in small groups. Their well-being depends upon humidity and temperature. Eggs fail to develop at a relative humidity below 60% and at a temperature below 11° C (KOZANCHIKOV, 1935). The incubation period lasted 6—7 days in the Volga region (SAKHAROV, 1931) and 9—12 days under Israeli summer conditions (SHWEIG, 1954). At constant temperatures the incubation period is given in Table XXIII. which is according to the findings of KOZANCHIKOV (1935).

The neonate larva, because of the absence of two pairs of abdominal legs, looks like a geometrid caterpillar. It feeds during the day on the soft tissue on the underside of the leaf. After the fourth moult it changes its habits; it becomes nocturnal and lives in the ground. Its food habits are also different. It cuts the stems of plants or petioles and may drag the cut plant below the surface to feed upon it. It may feed also on tubers and the like.

The larva has about 6 instars, each lasting a week or more, the last two instars being a little shorter. The larval period lasts about 42 days under Israeli spring conditions (SHWEIG, 1954) or under Volga summer conditions (SAKHAROV). The quiescent period of the larva before pupating is called the prepupal stage. Its length also depends upon the temperature, and at its lower levels may last two or three weeks.

The pupal period lasts, under favourable conditions, about two weeks (Table XXIII).

It takes a few days before the adults begin to reproduce; under spring conditions in Israel it took 4—5 days.

Table XXIII.

Development of *Agrotis segetum* at various degrees of temperature (after KOZANCHIKOV, 1935)

Temp. in C°	Development in days			
	Egg	Larva	Prepupa	Pupa
30	3.8	33	4	10
29				11.5
25	4	36	4	—
23	—	—	—	15
22.5	6	43	5	—
20	12	—	—	—
19	—	—	—	24
15	18.5	76	16	—
12	—	136	—	—

The number of eggs laid by the individual is subject to the food of the adult and the food of the larva as well. Furthermore, conditions of humidity and temperature which prevail during the larval and pupal stages affect the fecundity of the forthcoming females. For this reason, the number of eggs laid by one individual, as quoted by various writers, varies from 300 to over 2000 eggs.

KOZANCHIKOV et al. (1936) found that *Atriplex*-fed insects laid 7—10 times more eggs than moths which, during their larval period, fed on potato leaves. In 1937 the same author (KOZANCHIKOV et al., 1937) found that potato-fed moths produced about 8—10 times more eggs than when maize leaves were the food of the larvae.

On the other hand, SKOBLO (1937) found no difference in the fecundity of females whether they fed as larvae on *Atriplex*, potato, beet, onion, or *Sonchus*. The food of the adults, too, has effects upon egg-laying. SKOBLO (1937, 1939) found that females when fed a 10% solution of saccharose produced four times as many eggs as those fed on water only.

Females which in their pupal stage were kept at 19—20° C and 75% R.H. produced the highest number of eggs and lived longest. Any deviation from these conditions during the pupal stage caused a decrease in the egg production.

The practical importance of these findings may be seen in the field. When there are abundant rains during the pupal period, the moths are more fertile (PRIKHODKINA, 1939). Furthermore, moths from overwintering generations produced 500—2000 eggs, while moths of the second generation, who in the pupal stage passed through a dry and warm period, laid 100—300 eggs. Also, in the North Caucasus it was found that outbreaks occur when a damp September, which is favourable for the larvae, is followed by a damp May, when the pupae develop (KREITER, 1937).

Fecundity depends also upon the temperature at which the adult is kept. Maximum egg-laying occurred when the moths were kept at 12.5° C, while at 32° C no oviposition took place. At 28° C fewer females became fertile than at 7—20° C (KOZANCHIKOV et al., 1937).

Number of Generations. In view of the effects of temperature upon the length of development, the number of generations which the moth can raise differs in the various countries. Thus, in Norway, in Estonia, in Northern Russia and in Siberia, one generation develops during the year, while in Northern France and the Voronezh region two generations may develop. In the Seine Basin, it was pointed out (BALACHOWSKY & MESNIL, 1936) that some of the offspring of the first generation hibernate and emerge as adults only the following spring, while others of the same brood continue to develop and raise a second generation the same summer. It should be pointed out that only larvae of the 5th instar and above it can hibernate (SAKHAROV, 1931). In the Caucasus, Uzbekistan and Hungary three generations develop, while in Israel, according to SHWEIG (1954), four generations may develop. The first is on the wing in

March-April (from the overwintering larvae and pupae), the second is in early May-June, the third late July-August, and the fourth late October. The greatest damage is caused by the larvae in the spring. The larvae of the first generation are of little economic importance.

Factors affecting density of population

Weeds and soil. Many authors, who try to explain the reason for outbreaks of this pest in certain localities, mention weeds as a factor which attracts laying moths. This has been noted with other insects of this family.

Apparently the quality of the soil is also a factor which influences egg-laying. GRIVANOV (1937) points out that fields of light soil with soft loose granules or sandy black soils were far more infested than fields of a hard, compact layer of soil.

Humidity and temperature. This insect is susceptible to changes in the atmospheric and soil humidity. KOZANCHIKOV (1936a) found that 60% of the eggs died when kept at a relative humidity of 60% R.H. and a temperature of 25° C. At a lower humidity the mortality is far greater. Pupae kept at a low relative humidity produced sterile moths. Also when larvae were kept at a too high relative humidity sterile moths were produced. It is therefore understood that the moth thrives in moderate humid conditions. In fact SHCHELKANOVETZ (1926) has shown that in the Voronezh region the pest is more abundant when rains were above normal with at least 50 mm during one of the months or all of May, June and July. Also PYATNITZKII & KLISHEVICH (1936) pointed out that favourable conditions for this insect include a temperature of 15—17° C and rain above 50 mm during May-June. When the temperature rose above 18° C and the rain fell below 50 mm the pest was checked. However, there are certain optimum rainy conditions favouring the moth. Azov (1929), for instance, found that in the Don region 40 mm is the optimum. Any increase in rain above this point decreases the moth population so that by 80 mm of rain its activity is at a minimum. Similarly, a reduction from this optimum causes also a reduction in the activity of the pest so that at 15 mm of rain hardly any infestation takes place. The reason is that low humidity affects the insect directly, while too much rain causes fungi and bacteria to reduce the larval population. Thus POPOV (1931) found that one hour of rain resulted in a high percentage of larvae mortality in the three first instars. A temperature below freezing is injurious to the larvae, and, in fact, in the lower Volga region, when the temperature falls below freezing, the damage to agriculture in the following summer is slight (SAKHAROV, 1928). However, when the drop in temperature to below freezing is sudden, and the caterpillars become cold-hardy, they can survive the cold temperatures which prevail below the snow covering the northern regions of Russia (LOZINA-LOZINSKII, 1935).

The optimal temperature of the pest is between 20—25° C. At a temperature of 10° C the eggs do not survive and larvae do not complete their development. Temperatures above the optimum are harmful as follows: 30% of the eggs die at 30° C and 55% at 32.5° C. 66% of the larvae die at 31° C while 97% of the pupae died at 34° C (KOZANCHIKOV, 1935). Furthermore, as the temperature rises, the fertilization of the moths is impaired; at 28° C only 20% of the females were fertilized, while no oviposition took place at 32° C (KOZANCHIKOV et al., 1937).

LITERATURE

Z. AZOV, 1929. Defense des Plantes V, 1928 no. 5-6, *569-571*, Leningrad.

A. BALACHOWSKY & L. MESNIL, 1936. Les Insectes Nuisibles aux Plantes Cultivées. *1634-1641*, Paris

R. P. GRIVANOV, 1937. *Leningrad Acad. agric. Sci., 16-22.*

I. V. KOZANCHIKOV, 1935. *Z. angew. Ent.* **22,** 3, *452-462.*

I. V. KOZANCHIKOV, 1936a. *Leningrad Acad. agric. Sci.*, p. *54-58.*

I. V. KOZANCHIKOV, 1936b. *Trav. Inst. Zool. Acad. Sci. SSSR* **4** (2), *313-388.*

I. V. KOZANCHIKOV, 1936c. *Bull. Plant. Prot.* (1) no. 19, 36 pp., Leningrad.

I. V. KOZANCHIKOV et al., 1936. *Leningrad Acad. agric. Sci., 51-52.*

I. V. KOZANCHIKOV et al., 1937. *Leningrad Acad. agric. Sci., 28-30.*

E. A. KREITER, 1937. *Leningrad Acad. agric. Sci., 30-33.*

L. K. LOZINA-LOZINSKII. 1935. *Plant Prot.* 1935, 1, *5-22*, Leningrad.

F. NEUWIRTH, 1932. *Z. Zuckerind. Csl. Repub.* 1931-1932, *345-349*; *353-357*, Prague.

P. V. POPOV, 1931. *Plant Prot.* **7,** (1930) 4-6, *227-234*, Leningrad.

T. D. PRIKHODKINA, 1939. *Nauch. Zap. sokh. Prom.* **15** (3-4), *177-183*, Kiev.

G. K. PYATNITZKII & N. V. KLISHEVICH, 1936. *Plant Prot.* 1936, fasc. 9, *83-91*, Leningrad.

V. N. REKACH, 1933. *Trud. Zakavk. nauch. issled. Khlopk. Inst.* **40,** 44 pp., Tiflis.

N. SAKHAROV, 1931. Gosudarst sel'sko Khoz. Izt. R.S.F. 5 R 2nd. 64 pp., Moscow.

N. SAKHAROV, 1928. *Zh. opuitu Agron. Yu Vostoka* **6** (2) – reprint 20 pp. Saratov.

Y. P. SHCHELKANOVETZ, 1926. *Zapiski Voronezhsk S. Kh. Inst.* **V** – reprint 8 pp. Voronezh.

K. SHWEIG, 1954. Vegetable Pests and their Control., Hassadeh – 150 pp.

I. S. SKOBLO, 1937. *Leningrad Acad. agric. Sci., 34-36.*

I. S. SKOBLO, 1939. *Plant Prot.* 18, *142-144.*

Agrotis spinifera Hb.

[Syn.: *Euxoa spinifera* HB.]

Another cutworm which is found in Israel, either alone, or together with *Ag. segetum* and *A. ypsilon* is *A. spinifera.*

Description

Adults: The length of the body is 13—17 mm, the wingspan about 30 mm. The colour of the body is light brown, and the forewings are light brown or brown; close to the base there is a spine-like or wedge-like dark marking with the point directed sidewards.

Another similar marking is situated in the opposite direction with its base at the reniform spot. There is another dark area at the side margin. The hind wings are white; often faint brown lines run along the side margin.

Larva: When mature, the larva may reach 30 mm in length. The upper part of the body is dark grey-green, the lower side greenish-white. Darker bands with a rusty tinge, one median dorsal and two lateral, run along the body. The setigerous tubercles are small, the posterior as large as the spiracles.

Distribution. According to WILTSHIRE (1957), this is a Paleotropical insect and occurs in Iraq throughout the country. It is recorded as a pest from Uganda, Egypt and Israel.

Hosts. Larvae were found near onions, beets, peanuts, and sesame.

Type of Injury: As the two other *Agrotis* spp. mentioned above.

Life History. The moth was reared in Israel, and the following data were obtained by RIVNAY:

The pre-oviposition period lasted from 2—6 days at 26° C. The incubation period during the winter or spring lasted 6—7 days, and 4 days at 26° C. The development of the larvae in the winter, at the average temperature of 14—15.5° C lasted 77—92 days, while at the controlled temperature of 26° C 25—36 days. The pupa completed its development in 11—12 days at 26° C. The number of eggs laid by one female was 330 on the average, and 613 maximum.

Phenology. So far larvae have been found in Israel during the period from December to June. The adults, however, were on the wing all the year round. Their number increased during the spring and early summer. Also in Egypt it was trapped throughout the year and peaks occurring in the spring and autumn.

The Large Yellow Underwing — Triphaena pronuba L.

[Syn.: *Agrotis pronuba* L.]

Description. The length of the moth is 23—27 mm, its wingspan 50—60 mm. This moth may be easily recognized by the conspicuous hind wings which are orange-yellow, and the broad brown band which runs parallel to and near its side margins. The forewings are dark-brown with a rusty tinge. The two typical spots are as follows: the round orbicular is pale and the other reniform one is dark with a white rim around it.

The larva is about 50 mm long; its colour is apple-green, or greenish-brown. The head is red with two dark maculae in front. A dorsal white line runs along the body from head to end. Each abdominal segment, except the last, has a triangular brown patch on each side with a white line bordering it.

Distribution. The species is distributed throughout Euro-Siberia.

It is found from England in the west to Japan in the east; from Finland to North Africa. Not everywhere does it rate, however, as a pest. Damage was reported from England, France, Germany, Danmark, Czechoslovakia and South Russia. In Africa records come from Algeria, Tunisia and Cyrenaica, and in Asia from Turkey and Israel.

Hosts. Several plants may serve as food for the larva. Of cultivated crops the following were damaged: lettuce, asparagus, grapes, tobacco, potato, beet, peas, and cotton.

The *Life History* of the species, its length of development, number of generations and habits are similar to those of *Agrotis segetum*.

The Variegated Cutworm — Peridroma saucia Hb.

[Syn.: *Lycophotia margaritosa; Rhyacia margaritosa* HAW.]

Another cutworm of cosmopolitan distribution found in the Mediterranean countries is *P. saucia* HB.

Description. The adult looks very much like *A. ypsilon*. It is the same size — wingspan 40—50 mm and the forewings are the same colour except that the black wedge-like marking typical to *A. ypsilon* is missing. Also, the reniform and orbicular markings are not distinct. The hind wings, which are dark greyish at the margins and veins, (of individuals caught in Israel) are darker than those of *A. ypsilon*. The general appearance of the larva and its size is like that of *Agrotis ypsilon*. But, upon close examination there are some distinct differences. The body colour above the spiracles is dark-green, mottled with minute yellowish spots. It does not possess the dark setiferous tubercles on the body segment, but dark bands run longitudinally over the spiracles. On the eighth abdominal segment there is a dark, large, W-like marking and behind it a pale-yellowish area. The number of hooks on the abdominal legs is about 20. The pupa measures 15—23 mm in length and is brown.

Distribution. According to CRUMB (1929), this species is distributed in America from "Alaska to Patagonia", and from coast to coast. It occurs in Europe and the Mediterranean countries. It is found in Israel, but WILTSHIRE does not record it in Iraq.

Hosts. The list of food plants is extremely large, and there is no need to enumerate them here. Generally speaking, it may attack almost any crop under cultivation.

Type of Injury. This species does not confine itself to the roots, as other mature cutworms, but has the habit of climbing even on young trees. It may thus defoliate entire plants. But "when occurring in large numbers the larvae may assume the armyworm habit and devour the roots of grasses, potato tubers in the soil, and even the bark of trees" (CRUMB, 1929).

In the Near East this moth does not reach such dense populations as reported from the Northern Pacific States in America nor is its damage as great. In Israel, for instance, the type of damage it causes is like that of *A. ypsilon* but to a far lesser degree. As to the damage done in the Pacific states in 1900 it is worth quoting DOANE & BRODIE — (from CRUMB, 1929), as follows: "By July 15th the outbreak had become so general, and the number of cutworms was so great, that it seemed as if they would carry everything before them, for they had by this time literally become an invading army, marching on from garden to garden, from field to field, from orchard to orchard, eating every green plant that came in their path".

Life History. The life history of this species is very similar to that of *A. ypsilon*. The data on development of the various stages, according to CRUMB, are as follows:

The incubation period of the egg lasted 10—14 days at a temperature of 13—15° C, and 5 days at 20—23° C. The larva completed its development in 33—40 days at 15° C, and 23—28 days at 23—24° C. The pupal period was 18 days at 20° C and 12—14 days at 26—27° C.

In rearings at constant temperatures in Rehovot (RIVNAY), the following data were obtained: The larva developed in 24—29 days at 22° C, and 18—22 days at 26° C; the pupa developed in 12 days at 26° C. The temperature of 30° C seemed to be harmful as the percentage of larvae which survived was very low, and the pupae all died.

Control of Cutworms

Most cutworm moths lay eggs on weeds, and when the larvae become mature, migrate to cultivated crops. Elimination of weeds is therefore essential for the control of cutworm attacks.

Digging ditches around the field is recommended in some localities, to prevent migration of larvae from infested fields.

The young larvae which feed on the foliage, may be exterminated by spraying the plants with insecticides, as discussed below.

Mature cutworms which live in the ground and feed on root-crowns and parts of the plant close to the ground, may be killed with bait. A bait consists of a carrier, a stomach poison and water; an attractant may be added, as well as a substance to prolong its fresh condition. The carrier may be wheat bran, rice, pressed cotton seed, pressed peanut cake, etc. CRUMB finds that wheat bran may be even more attractive than the natural food plant. The bran may be diluted with sawdust, but then its attractiveness is diminished. Bran-sawdust may attract 30—70% less caterpillars than pure bran.

The poison may be arsenicals, fluorides, or synthetic insecticides. Arsenicals and fluorides are given at the rate of 2—4% of the weight of the bran, and the synthetic insecticides at the rate of 0.5—1%.

As an additional attractant it is customary to add molasses. CRUMB finds that corn syrup becomes more attractive when fermented, while ethyl acetate and nitrobenzene are rated by him as distinctly attractive. Contrary to the opinion of some entomologists, lemon juice or orange peel are not attractive — according to CRUMB's experiments.

In order to keep the bait moist for a long period, lubricating oil is added at the rate of 8 litres to 50 kg of bran. Sufficient water is added to make dry mash.

The following are excerpts from the recommendations issued by the Israel Ministry of Agriculture, February 1961:

Before seeding or planting be sure the field is pest-free. Weeds should be eradicated from the field a month in advance. Should there be signs of caterpillar injury on weeds, recently sprouted, in the field or around it, they should be sprayed with endrin 60—80 cc a.i./dunam, or DDT 50% W.P. 250 cc a.i./dunam. For mature cutworms in the ground bait should be applied at the rate of 4—6 kg/dunam. The bait should be applied close to the plants. Apply every 5—6 days while the attack continues.

If plants are young, and it is possible to cover their stems with control substances — dusting with DDT or spraying as described above, may provide satisfactory control.

In small lots caterpillars may be collected by hand-picking. Some entomologists in the U.S. and Israel recommend the application of insecticides to the surface of the ground, followed by overhead irrigation. In Israel, for instance 1.5 kg toxaphene or 65 cc endrin a.i./ dunam gave satisfactory control. This method has some serious drawbacks. Toxic residues may remain in the soil for a long time, and the biological balance of animals living in the ground may be upset.

LITERATURE

S. E. CRUMB, 1929. Tobacco Cutworm. *U.S.D.A. Tech. Bull.* **88**, 177 pp.

H. HARGREAVES, 1922. Annual Report of the Govt. Entomologist 1921 – *Uganda Agr. Ann. Report* 1921, *57-64.*

M. H. HASSANEIN, 1956. *Bull. Soc. Ent. Egypte* **40**, *463-477.*

E. RIVNAY, 1960. Unpublished Notes.

E. P. WILTSHIRE, 1957. The Lepidoptera of Iraq. Ministry of Agriculture, Govt. of Iraq, 162 pp.

The Egyptian Cotton Leaf Worm — Prodenia litura F.

Prodenia litura is one of the most notorious pests of field crops in the Near East, particularly in Egypt and Israel. This polyphagous insect attacks several crops of various types, some of which are of major economic importance. Fodder crops, truck crops, and, in particular, industrial crops such as sugar beet and cotton, are subject to its attack, and the losses inflicted upon the farmer are enormous.

Description. The adult is a brownish-grey moth; its body length is 15—20 mm, and its wing span 30—40 mm. The brownish-grey wings are striped with yellowish-grey stripes in different patterns (see fig. 29). In the male the stripes are broader and more conspicuous. The hind wings are creamy in colour. Dark lines run along the veins, while the margins of the wings are greyish.

Fig. 29. *Prodenia litura,* adult.

The egg is spherical, slightly flat, 0.6 mm in diameter, with minute ribs running from the center to the periphery. Its colour is cream with an iridescent lustre. The eggs are laid in clusters on the surface of the leaf, and are covered with scales from the female body. When the number in the cluster is great, the eggs are arranged in two layers, one over the other.

The neonate larva is bright green. Its head and thoracic plate are shining black and setigerous tubercles are scattered over the body.

Fig. 30. *Prodenia litura,* larva.

Upon moulting the larva changes its colour. In its last stadium the larva is about 40 mm long. Its colour is greenish-brown — but, depending upon environmental conditions, the colour may be also greenish-yellow or black-grey mottled with minute dark dots. Light and dark bands run laterally along the body. On each body segment (excepting the prothorax) two dorsal black spots are situated on each side. Those of the first and eighth abdominal segments are larger and more conspicuous. The dots, depending upon the individual, may be more or less pronounced; often they are absent.

The pupa is 15—20 mm long, 5 mm wide, brown, and on its tip it bears two small spines. Pupation takes place in the ground in oblong cells where the pupa is obliquely situated.

Distribution. The occurrence of *Prodenia litura* is reported from all continents excepting the American. However, not everywhere does it occur or cause damage in equal measures.

In Asia it occurs in the countries and islands along the Pacific and Indian Oceans. In these countries the most damage is recorded from Indonesia, India, Ceylon and the Philippines. Less damage is recorded from Formosa, Japan, China and Korea.

In Australia, Queensland suffers more damage than other states.

In Africa it is spread over the entire continent — from the Red Sea to the Atlantic coasts, from the Mediterranean Sea to South Africa. However, it seems to be more centered in the north eastern part of the continent and, of all the African countries, it is evident that Egypt suffers more damage than any other country. It is also extremely troublesome in Israel and Jordan, less so in Lebanon and Syria, and still less in Iraq or Cyprus.

Hosts. As mentioned above, this species is very polyphagous. It attacks trees, vines, citrus seedlings and apple trees, as well as Casuarinas. However, not all hosts are equally attractive and not in every country is damage recorded to the same host. While in some countries tobacco is the crop which suffers the most damage, in others rice may be the subject of attack. In Egypt, cotton and clover suffer most, while in Israel alfalfa, beet, peanuts and cotton are all subject to much damage from this species.

Hosny & Kotby (1960) of Egypt carried out a field experiment in order to study the preference of the moth for oviposition. Castor showed the most distinct preference, cowpea was next, followed by cotton; maize and sunflower were far less attractive. In Israel sunflower is not attacked by the species, but reports from Tanganyika and South Africa claim damage to this plant by *Prodenia*.

Moussa et al. (1960) also carried out laboratory experiments to determine whether there are any differences in the effects of the plants upon the insects. They found that while berseem (clover) shortened the length of development of the larva, cotton prolonged it. Okra, on the other hand, caused the development of moths with the greatest fecundity. On grape vines and maize larvae failed to complete their development.

Type of Injury. The new infestation of the moth larvae is hardly noticeable. The young larvae eat very little and their presence is noticed only upon close examination. However, as the larvae mature, their damage increases and becomes more conspicuous. When they are numerous, complete defoliation of the crop may be the result of their feeding.

In a field of peanuts, several dozens of larvae were found under each bush. On one beet leaf over 30 medium-sized larvae were counted, and on cotton plants many were found, as indicated in fig. 31. Quite often the damage begins at certain spots in the field and increases as the larvae spread. As a rule the larva feeds on every part of the leaf, but often the most susceptible parts of the plant are attacked. In cotton they choose the buds or young squares and destroy them; in tomatoes they gnaw on the fruit and, although only a small part of the fruit is eaten by the larva, it becomes unsuitable for marketing or consumption.

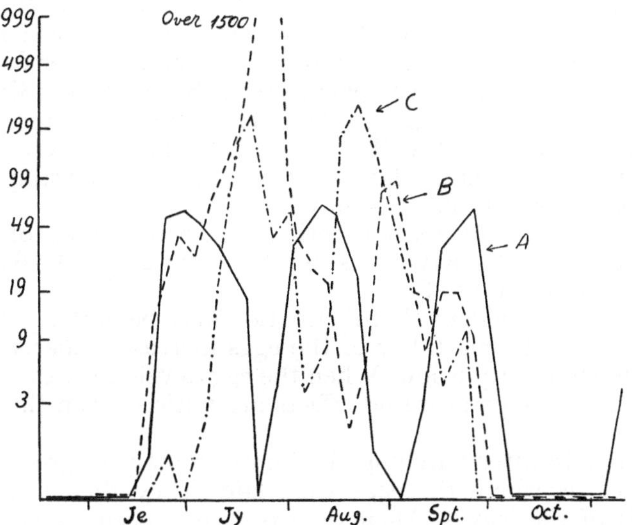

Fig. 31. Infestation of cottonfields by *Prodenia* larvae in Israel during 1959. Lines show number of larvae collected on 60 plants – in each of the three localities Jordan valley (A), coastal plain (B) and Negev (C).

In fodder crops such as alfalfa or clover, the larvae, which remain in the field after the crop has been cut, attack the young buds and thereby prevent the development of the next crop. As a result, weeds thrive and bring about the deterioration of the entire field.

In recent years damage has been caused to grapes. Particularly in irrigated fields, grapes have dropped to the ground around vintage time because their petioles were eaten by the larvae of *Prodenia*. Also brought to our attention were mature peanuts, in which the shell contained a *Prodenia* larva which had eaten the contents of the pod. The soft pods may be entirely consumed.

Life History. Under favourable conditions of temperature the pre-oviposition period is very short. The female may lay even during the night of the mating, or the night afterward. But as the temperature drops this period becomes longer.

HASSAN et al. (1960) who studied the behaviour of the adult found that, while mating may occur at all hours of the night, feeding and oviposition take place early in the morning, one or two hours before sunrise. The eggs are laid in clusters. Each female may lay as many as 3—7 clusters. The number of eggs in each cluster decreases from one oviposition to another, the first cluster containing as a rule the highest number of eggs.

The oviposition period lasts 3—4 days in the summer and 10—12 days towards the end of the autumn.

According to BISHARA (1934) the average number of eggs laid by one female is about 1000, while fertile females may lay as many as 1500. RIVNAY bred several females which each laid over 2000 eggs; among these, two laid over 3000, one 3280, and one 3791 eggs.

Quite often unfertile eggs are laid. This may happen even when the females are placed with males. According to RIVNAY (unpublished notes) when the larvae are bred at unfavourably high or low temperatures, sterile eggs may be laid.

The length of the incubation period at various degrees of temperature as obtained by RIVNAY is: 3—4 days at 26° C, 4—6 at 22° C and 20—26 days at 12° C.

From this it is evident that incubation may be prolonged to over two weeks but the mortality of the eggs increases especially when the relative humidity is low. When the egg development is accelerated by higher temperature no effects are noticed from the relative humidity.

Immediately after hatching the larvae remain in groups in the same place as where the eggs were laid. They then feed on the epidermis of the leaves. Their dispersal may occur a short time later, even the same day in the case of high temperature. Movement from plant to plant is facilitated by swinging on silken threads fastened to the leaf edge.

In their earlier instars the larvae spend the day on the plant, but from the fourth instar and on, they spend the day in the ground and feed at night. According to HASSAN et al. (1960) larvae may feed from before sunset to before sunrise. However, the majority of larvae feed from 8—10 p.m.

Under favourable conditions of temperature, the larva completes its development in about two weeks. According to BISHARA (1934), 12 days was the shortest development period (at 29—35° C). At 24—28° C larvae development lasted 2—3 weeks, while at 18—20° C, 30 days. At controlled temperatures the development period, according to RIVNAY, was as follows: at 22° C — 22.75 days, at 26° C — 15 days, while at 30° C — 14 days.

However, according to this latter author, larvae bred at higher temperatures or at too low temperatures produced weak moths which laid fewer eggs, many of which were sterile.

When ready to pupate the larva chooses a proper site for this purpose. In this respect the water content in the soil is an important factor.

NASR et al. (1960) found that, when four types of soil were offered larvae for pupation, 68% chose soil of 20% water content, 18.9% — soil of 10%, 8.3% — of 30% water content, while only 4% chose dry soil.

The water content of the soil influenced also the time of emergence of the moth. Excessive moisture prolonged the pupal period, while 20% moisture in the soil was the most favourable.

The pupal development lasts 7—8 days at 23—30° C while 14 days are required at 18—20°C.

According to BISHARA 11—12° C is the threshold of development of egg, larva and pupa.

During the summer the moth lives on the average 4—10 days and may lay over one to two thousand eggs. According to RIVNAY, the temperature at which the larva was bred has an influence upon the number of eggs.

Ecological Notes. BISHARA carried out an extensive survey to obtain a clearer picture of the conditions of oviposition. He found that the strongest infestation in a crop takes place 3 days after its irrigation. He states further that in Southern Egypt, where the moth is less widely distributed, they are attracted to irrigated fields only.

For several years BISHARA collected data from moth trappings near Cairo. According to these, *Prodenia* moths are on the wing throughout the year in Egypt, but the densest population occurs during June. About 7 generations develop in Egypt and, while the winter, spring, and early summer generations are quite small, the June and July generations are very dense, and by August — September there is a sudden decrease in the number of moths which were caught.

Three reasons for this picture may be obtained from BISHARA'S paper, as follows:

a. The Climate. During June and July the temperature is quite high — and no doubt this influences the mortality of the moths.

Even surviving moths lay few eggs or they do not lay at all, and the eggs may be sterile.

b. Measures of Control. In Egypt there is a law in force whereby each farmer must collect the egg clusters from his cotton fields. This removal of millions of clusters, each containing dozens of eggs, may reduce the population to some extent. Other measures of control also contribute to this end.

c. Biotic Factors. Under this heading are included virus, bacteria, and fungal diseases, parasites and predators. A polyhedral virus disease is quite frequent in both Egypt and Israel, but it occurs only under certain climatic conditions which are not always prevalent in these countries.

BISHARA mentions a parasite *Tachina larvarum* which attacks *Prodenia* larvae — but it is not sufficiently active. Only 1—5% of parasitism was recorded early and midsummer and 40—50% in the autumn. Other parasites are still less active.

A more extensive study of the natural balance and biological control of this pest was carried out in Egypt by KAMAL (1951). In a survey of the beneficial insects in cotton fields in Egypt he found that about 3/4 of a million are found in one acre during the spring. With the advance of the summer their number decreases. *Prodenia* is attacked in particular during the egg stage and early larval instar. According to the estimate of KAMAL, 19—25% of the eggs are preyed upon, and about 33.5% of the neonate larvae disappear, probably due to enemies. The survey shows further that Coccinellidae are the most important predators, next are the bug *Orius* and spiders. Of the parasites, *Tachina larvarum* L. is the most important. In the fall it may parasitize as much as 50—60% of the *Prodenia* larvae, but it is far less active during the summer when it is most needed. Two *Zele* species parasitize about 6—10% — but in summer they are inactive. *Barylypa humeralis* BRAUNS was found in 3—5% of the population during the spring, less in the summer and about 40% in November.

In addition to the survey, KAMAL made an effort to establish in Egypt parasites from foreign origin. He introduced *Actia nigratula* MALL. (Tachinid.) from Brisbane, Australia, *Microplitis demolitor* WEK. (Braconi.) from Australia, *Telenomus nawaii* ASHM, (Scelionid) from the Fiji Islands and *Telenomus spodopterae* DODD. from Java. All these were liberated in great numbers in various localities but at the time of publication of that paper no definite conclusions could be drawn about the control by these parasites in Egypt.

Periodic Outbreaks. In Israel during the last few years the *Prodenia* has increased to more destructive proportions than in the past. The question is often raised as to what are the causes, and will this situation remain so. There are three factors involved:

1. Climatic. This species is a tropical and subtropical insect.

Low temperature is one of its limitations. In the winter, at 15° C larvae hardly survived. For this reason larvae are not found in more northern countries of the Near East. During the cold winters of Israel the existing larvae and the pupal population decrease enormously. But when the winter is mild, as during the past few years, more individuals survive to start an early and stronger summer population.

2. Food Supply. Before the State of Israel was established, alfalfa was the only food crop which was available on a large scale. Other host crops were grown in limited, small areas. Since the establishment of the State of Israel, three new crops were introduced which are grown over wide areas under irrigation. These hosts are sugar beets, peanuts and cotton. With the increase of this food supply, greater populations of moths could develop in successive generations than when these crops were not available in such large

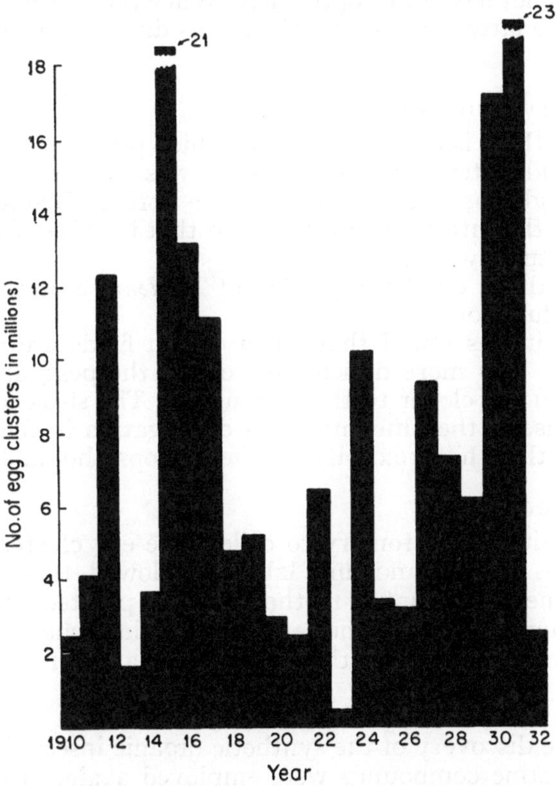

Fig. 32. Comparative abundance of *Prodenia* in Egypt during the years 1910–1932 based on number of egg batches collected. (after BISHARA).

numbers. What is more, these three crops complement each other to form a continuous food supply throughout the year. Peanuts and cotton are available before beets are removed, and vice versa.

The third reason is migration — which is discussed below.

Some entomologists, such as WILLIAMS, PRIESNER and BYTINSKI-SALZ verbally expressed their opinion that *Prodenia* is a migrant, but they had no evidence to support this assumption. Also WILTSHIRE (1957) expressed his opinion that the moth is probably a migrant. RIVNAY, in a paper on the phenology of *Prodenia* in Israel, supports this idea and brings evidence as follows: 1. In Egypt as well as in Israel trappings of the moth show sudden and abrupt increases of population. Such increases, he points out, could not be attributed to the offsprings of a previous generation but rather to invading moths from other places. 2. *Prodenia* moths were trapped in a semi-desert locality — namely at Eilat (Akkaba). These were caught during September, October and November and surely could not have developed there. While the first case is indirect evidence, the latter may be considered as direct evidence.

Control

Preventive measures

In Egypt there is a law since 1913, which prohibits the irrigation of alfalfa fields after May 10th. This law was made in order to deny breeding fields for *Prodenia* during the spring. BISHARA already pointed out the futility of this law, and that in spite of it, chemical control measures were imperative.

The hypothesis on the migration of *Prodenia* adults may throw light upon this problem.

In Israel it was found that when cotton fields are adjacent to clover fields, it is more difficult to control the pest, as the larvae migrate from the clover to the cotton field. The situation is aggravated because of the different times of irrigation in the two fields. In view of this the proximity of these crops should be avoided.

Hand collection

In Egypt it was customary to collect the egg clusters by hand. Where wages are low, and child labour is allowed, this may be considered as one of the control methods. It was pointed out, however, that it is more expensive and less efficient than chemical applications, but it does not upset the biological balance of insects.

Chemical control

Before the discovery of the synthetic organic insecticides, arsenicals and fluorine compounds were employed against the larvae of *Prodenia*. These applications caused an increase in aphid populations.

In the literature there are several references about the efficiency of DDT against *Prodenia*. In Rehovot, laboratory tests were made in 1957 to compare several insecticides. DDT proved to be the most efficient of the five chlorinated hydrocarbons compared. This and endrin were efficient as contact and stomach poisons. dieldrin killed the larvae through the stomach, and was efficient against small larvae only. BHC was efficient, but was of short residual duration. Methoxychlor was not effective at all. Four organo-phosphorus insecticides, which were tried, rated in their killing efficiency in the following order: parathion, diazinon, dipterex, and malathion.

Field experiments carried out in alfalfa fields in 1957 illustrate the behaviour of some synthetic insecticides in various dosages and combinations. The results of these tests are presented in Fig. 33. Further results of 1961 tests are given in Fig. 34.

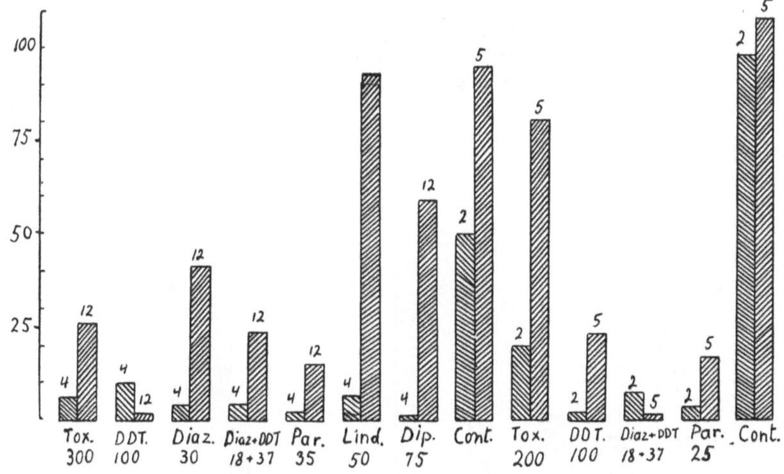

Fig. 33. Number of *Prodenia* larvae survived in an alfalfa field after treatment with insecticide in 1959. The figures over each column indicate the number of days after treatment. The figure below the names of the insecticides indicate grams of active ingredient applied per dunam. The abbreviations are in order as follows: toxaphene, DDT, diazinon, diazinon and DDT, parathion, lindane, dipterex, Control etc.

Later experiments in 1958 and 1959 showed that Sevin is also efficient against *Prodenia*. However, since the above mentioned tests were made, changes have taken place, and the pest has become less responsive to some of the above-mentioned insecticides. In 1960, during the heavy infestation of the cotton fields, the only satisfactory application proved to be that of parathion combined with endrin at the rate of 150 g of the former and 80 g of the latter active ingredients per dunam.

The following are excerpts from recommendations issued by the Department of Agriculture in Israel, February 1961:

1. In order to prevent migration of caterpillars from one field to another, a five meter safety belt may be dusted with gammacid 5 kg per dunam (gammacid contains 2.6% BHC).

2. Should repeated applications be imperative, it is advisable to employ substances of a different group each time. Choice may be made of one of the following groups:

a. Chlorinated hydrocarbons:
DDT 50% W.P. 200 g a.i./d.
DDT 25% E.C. 125 g a.i./d.
DDT E.C. + toxaphene E.C. 100 + 150 a.i./d.
endrin 19.5 E.C. 80 g a.i./d.
DDT 10% dust 5 kg/d. (advisable for low-lying plants such as peanuts).

b. Carbamates:
sevin 50% W.P. 200 g a.i./d.
sevin 10% dust 4 kg/d. (for low-lying plants).

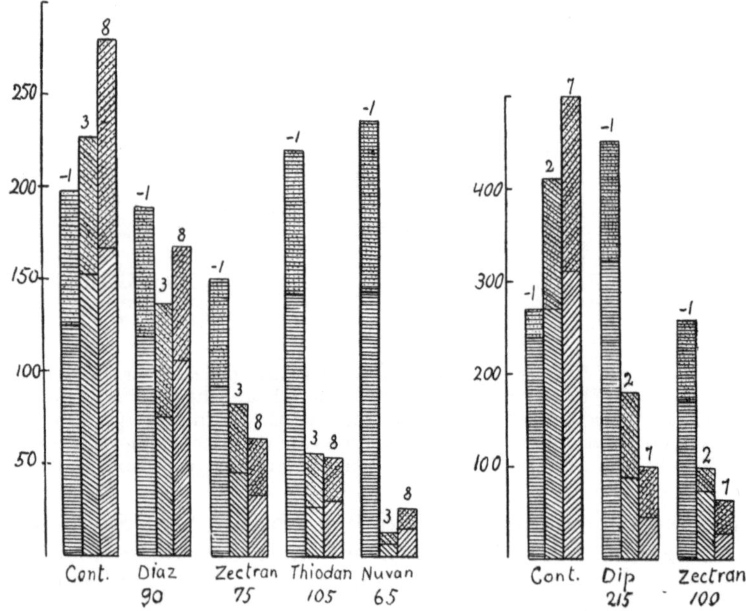

Fig. 34. Number of *Prodenia* larve survived in a Peanut field (left) and Potato field (right) in 1961. The first column for each test indicates number of larvae in the field on the day before application. Lower part of column shows number of small larvae (below 10 cm), upper part larger caterpillars. Other signs as in Fig. 32. After ZIV.

c. Organo-phosphorus insecticides:

parathion has been excluded because of its mammalian toxicity.

diazinon 25% E.C. } 75 g a.i./d.
diazinon 25% W.P. }
dipterex 80% E.C. 75 g a.i./d.
diazinon 5% dust } 3—4 kg/d.
dipterex 5% dust }

d. Whenever possible, baits should be employed (see measures against cutworms). The bran used should be fresh — not mouldy, and should be distributed in the evening.

Further tests during the summer of 1961 have shown that nuvan and zectran are quite efficient against *Prodenia* larvae (ZIV), see Fig. 34.

LITERATURE

I. BISHARA, 1934. *Bull. Soc. Ent. Egypte* **18,** *223-404.*

A. S. HASSAN, M. A. MOUSSA & A. NASR EL SAYED, 1960. *Bull. Soc. Ent. Egypte* **44,** *337-343.*

M. M. HOSNY & F. A. KOTBY, 1960. *Bull. Soc. Ent. Egypte* **44,** *223-234.*

MOHAMED KAMAL, 1951. *Bull. Soc. Fouad 1er Ent.* **35,** *221-270.*

A. M. MOUSSA, M. A. ZAHER & F. KOTBY, 1960. *Bull. Soc. Ent. Egypte* **44,** *241-251.*

A. NASR EL SAYED, M. A. MOUSSA & A. S. HASSAN, 1960. *Bull. Soc. Ent. Egypte* **44,** *377-382.*

E. RIVNAY, 1961. Unpublished Ms.

E. P. WILTSHIRE, 1957. The Lepidoptera of Iraq. Ministry of Agriculture, Govt. of Iraq, 162 pp.

M. ZIV, 1962. Agr. Res. Stat. Rehovoth Progress Report. 371.

The Lesser Army Worm — Laphygma exigua Hbn.

[Syn.: *Caradrina exigua* HBN.]

This insect is one of the important pests of agriculture in the Mediterranean area. In some years, its damage is not so great; but frequently there may be outbreaks in its population when it becomes a very destructive pest.

Description. The wingspread of the adult is 25—28 mm. Its body is grey-brown in colour and so are its fore-wings, which are marked with darker spots. The disc near the lateral border is darker and a double zigzag line runs parallel to it. Near this line there are two light spots, the larger of which is kidney shaped and the smaller of which is round (see Fig. 35). The hind wings are light, except for the anterior border and veins which are grey brown.

The egg has the form of a mandarine orange. Its diameter is less than 0.5 mm and there are small grooves from the center to its circumference. Upon oviposition the colour of the egg is greenish grey and afterwards it turns brown. The larva reaches 30 mm in length, its body is cylindrical and its general colour is green in the first stages. On the ventral side of its body, the green is striped with thin brown lines. On both sides of the body there is a light green

Fig. 35. *Laphygma exigua.*

line underneath the row of spiracles and above it is a wider dark brown one. The dorsal side is light brown and it is also striped with delicate brown bands. In the middle of its body is a dark longitudinal stripe. The colour is typical for larvae of the "migratory" phase. In larvae of the solitary phase (see further on), dark or light green takes the place of the dark brown colour. The abdominal feet have 18—21 hooks which are arranged in a wide arc.

The pupa is reddish brown, the thorax is darker and its abdomen has a greenish tinge. It is 15 mm in length.

Distribution. This insect is, to a great extent, widespread all over the world and causes damage mainly in tropical and subtropical countries. Reports on its damage come from most of the countries of Asia, Africa and Australia. It occurs in New Zealand and Hawaii; in California, Arizona and Florida in the United States. It is found in Italy, Greece, Cyprus, Lebanon, Syria and Israel in the Mediterranean region.

Hosts. Unlike *Laphygma exempta* which feeds only on monocotyledonous plants, *L. exigua* is very polyphagous and feeds on several plants which are not botanically related to one another. In the literature we find that it attacks the following crops: cotton, tomato, tobacco, beet, asparagus, jute, corn and various legumes. In the countries of the Mediterranean area it attacks many additional plants; however, the principal damage is caused to legumes and corn.

Type of Injury. Immediately after hatching, the young larva feeds on the tender parts of the leaves, leaving a fine network of the veins. Later, the more mature larva feeds on the entire leaf, often leaving only the midrib. In this manner the larvae, in case of a serious attack, may destroy all the vegetation in a field. At this stage they migrate in masses and continue their destruction in adjacent fields. Hence, its common name, the "armyworm".

Life History. As a rule the adult appears first in the spring or summer. The date of its appearance depends on climatic and geographic factors. The development of the pupae and the emergence of the moths is accelerated by higher temperatures at the end of the winter. In the southern part of Israel, the adults appeared in 1957 in the middle of April. As for geographical factors, the adult migrates from one country to the next. Therefore, the more distant a country from the center of its distribution, the later it appears there.

The adult females are able to oviposit one day after emerging. In the fields they lay their eggs in clusters of about 100 per cluster, (according to TAYLOR, 1931) and up to 200 eggs (according to CHERIAN & KYLASAM, 1939). The eggs are laid on the plants and are generally covered with hairs and scales from the body of the female. Each female is able to oviposit more than two dozen clusters of eggs. The number of eggs laid by one female may reach several hundred. This number is different in the various generations of the pest and in the different countries in which it is found. Thus, NIKOLSKIĬ (1930) (of Turkmenistan), for example, reports that the maximum number of eggs laid by a female of the second generation was 1700 eggs and that of the third generation, only 500 eggs. The following table includes data on the differences in egg-laying in various countries:

Table XXIV.

Data on the development of *Laphygma* in various countries (Note: The numbers in parentheses are the maximum amount. The other numbers are the average amount)

The Research Worker	The country	Length of development in days			Length of life of adult	Length of egg laying period	Number of eggs	Temperature °C
		egg	larva	pupa				
WILSON (1934)	Florida		11—13	6— 8		4— 5	(1321) 600	26—27
WILSON (1934)	Florida		37					16
FAURE (1943)	South Africa	2	10—11	6— 8	8—10			29
TAYLOR (1931)	Transvaal	3	18	9—48	15			
COWLAND (1930)	Sudan	2	14	6— 7		4— 5		
ZELENSKY (1938)	Morocco	3	17—20	8—11	6—12			
CHERIAN & KYLASAM (1939)	India	2	15	6— 7	15		(1300) 461	
PATEL (1944)	India	2	13—16	5— 7	8—11	2— 4	(1278) 500	
NIKOLSKIĬ (1930)	Turkmenistan				14—17	10—15	1700 (900)	
RIVNAY (1960)	Israel	2—3		10—12		3— 8	(900) 250	24

In Israel most of the adult moths of the first generation lived 11—12 days; a few lived 17—19 days.

In rearings conducted in Israel in the spring of 1957, 72 females of the second generation were raised and their rate of oviposition was examined. The results are given in Table XXV. The average rate of egg-laying was 250 eggs per female. From this table it can be seen that more than one third of the females oviposited 300—500 eggs and only 5% of them laid more than 500 eggs. The length of life of the adult of this generation is also given in this table.

Table XXV.

The fertility and length of life of the adult *Laphygma* moth.

Fertility		Length of life in days		
Number of females	Number of eggs	Number of males	Number of females	Days
12	0	34	31	1— 4
19	1—100	13	13	5— 6
16	101—300	11	8	7— 8
22	301—500	3	8	9—10
3	501—909	4	4	11

The larva hatches 2—3 days after oviposition. In the beginning it feeds on the shell of the egg. Afterwards the cluster of larvae disperses to all parts of the plants. The young larva hangs on a thread which it weaves and the wind casts it from plant to plant.

During its first stages the larva is active during the day, but later it confines its activity to the night time. Mature larvae, wnich did not hide but rather remained on the plants during the daytime, have also been found. The larval period lasts 10—12 days at a temperature of 27—29° C. At a temperature of 16° C it lasts longer than five weeks.

The pupa is usually found in the soil. The larva excavates an elongated cell, binds the soil particles which make up the walls of the cell with threads and pupates in it. In cases where it cannot penetrate the soil, the larva pupates on top of it, underneath dry leaves, etc. The pupal period lasts 6—8 days at a temperature of 27—29° C (Wilson 1934). In Israel it was found to last 10—12 days at a temperature of 24° C.

Number of Generations. Under favourable conditions of temperature, i.e. 27—29° C (according to data in Table XXIV) the development of the insect, including all its stages, takes 23—25 days. In this manner, the moth is able to raise a number of generations during the same summer. Nikolskiĭ reports the occurrence of at least five generations during the summer in Turkmenistan. WILSON established that in Florida six generations developed between the first of May and the twentieth of September, averaging approximately 24 days per generation.

The calculation of the number of generations in a year is not practical because the insect does not overwinter in every place.

Migration. According to existing data, it can be surmised that the *Laphygma,* which is a resident in Israel *Laphygma exigua,* is ismilar in its life history to the species of the same genus *Laphygma frugiperda* in America which was thoroughly studied.

Laphygma frugiperda does not overwinter in the northeastern states of the United States, but rather appears there during the summer, oviposits and raises one or more generations. It is not found there during the winter. There are years in which the moths appear in dense swarms causing serious damage. According to studies conducted in the United States (LUGINBILL, 1928), it was found that the moth overwinters in southern Florida, southern Texas or Mexico and from there invades the north and northwest, and migrates to the other states of the United States. On its way it raises generations of descendants which continue the migration northward.

The life history of the South African armyworm, *Laphygma exempta,* is similar. In South Africa the outbreaks of this pest begin in the northern regions and gradually move southward. Certain entomologists think that here is a chain of migration southward which is accompanied by oviposition in the resting places and is continued by the descendants.

HATTINGH, (1941), who studied the biology and ecology of this pest in order to understand the causes of the above-mentioned outbreaks, states the following combination of factors of food and temperature:

1. In Africa the armyworm feeds on the tender leaves of *Cynodon* and does not eat hard leaves;
2. The development of the larva is possible only under the conditions of temperature prevailing during the months of January-March.
3. The absence of tender vegetation during the above-mentioned months prevents the development of a dense population.
4. As a result of the combination of dry conditions during the months of October-November, which delays the growth, and heavy rainfall in January-February, which stimulates the growth of vegetation during those months, an abundance of food is found in a season favourable for the development of the *Laphygma,* thus causing outbreaks of the pest. This author collected climatic data during the years in which the outbreaks of *Laphygma* occurred which supported this theory. His conclusions were that these out-

breaks are not the result of the migration of moths from other places, but on the contrary, they are due to the increase of the local population as a result of the above combination of favourable conditions. He states further that the pest is found throughout the year in the places in which the outbreaks occur. Furthermore, he states, direct proof of the theory of migration of this species was not found.

These conclusions stimulated MATTHEE to study further this problem. In his paper in 1952 he states that he did not find the larvae and pupae of the armyworm during the months of November and December in the places in which dense flocks of adults appeared during the following January-February. This fact shows that the moths indeed migrated.

However, even without the aid of the data of MATTHEE it is possible to oppose the conclusion on HATTINGH from a simple, logical point of view. A resident population may be reinforced by the invasion of flocks from other localities. The ecological data which HATTINGH collected prevailed probably also in places where the pest originated and caused an increased population, which in turn stimulated migration.

In the meantime new evidences about the migratory habits of *L. exempta* were given by LAIRD (1960) at the XI International Ent. Congress at Vienna.

In the countries of the Mediterranean area, *Laphygma exigua* is the representative of the genus; and, as mentioned above, its biology is probably similar to that of *Laphygma frugiperda* in America and *Laphygma exempta* in South Africa. According to WILLIAMS (1958), this moth often appears in England; however, it is not a serious problem and does not overwinter there. In Israel, the pest is a resident and is found throughout the year, but quite often there are outbreaks when its population increases suddenly and reaches large numbers.

FAURE (1943), who raised this insect under various crowded conditions, found that the colour of the larvae growing in crowded conditions is completely different from those growing alone under solitary conditions. The dark-coloured larvae, which were raised in a crowd, were called the migratory phase and those raised separately, whose colour was green or light green, were called the solitary phase. In Israel, green or dark green larvae are usually found throughout the year. However, in the spring of 1957 a serious general attack occurred which involved various crops throughout the country and in which the colour of the many larvae was brown or dark brown. There is no doubt that this attack commenced with the invasion from more southern regions. It must be added that the winter of 1957 was rainy and the spring was unusually rainy during the month of April.

It is possible, therefore, to surmize that *Laphygma exigua* overwinters in the warm regions of the Mediterranean area and, in years of great increase, it invades cooler northern regions, similar to the habits of its relatives in America and South Africa.

Control. *Laphygma* may be controlled with stomach poisons. All the arsenic compounds and fluorine insecticides, either in the form of a dust or suspension, controlled the pest. With the development of the synthetic organic poisons control of the armyworm was effected by dusting with 2.6% BHC, 2% dieldrin, 10% DDT, 10% toxaphene, and 1% parathion.

As a rule the treatments given against *Prodenia* are effective against this pest too.

LITERATURE

M. C. Cherian & M. S. Kylasam, 1939. *J. Bombay Nat. Hist. Soc.* **41,** *253-260.*
J. W. Cowland, 1930. *Bull. Wellcome Trop. Res. Lab. Sudan Govt., 71-75.*
J. C. Faure, 1943. *Farming in So. Africa,* **18,** (203), *69-78.*
C. C. Hattingh, 1941. *Union S. Afr. Dept. Agr. & For. Sci. Bull.* **217,** *1-50.*
A. Laird, 1960. A Flight of Insects in the Gulf of Aden. XI Int. Ent. Congr.
Ph. Luginbill, 1928. *U.S.D.A. Tech. Bull.* **23,** *1-91.*
J. J. Matthee, 1952. *J. ent. Soc. S. Afr.* **15,** (2), *122-128.*
V. V. Nikolskiĭ, 1930. *Khlopkovoe Delo* **IX** (7-8), *879-889.*
J. S. Patel, 1944, Indian Central Jute Comm. Ann. Rpt. of the Agr. Res. Scheme for the year 1942-1943, 1943-1944, 51 & 38 pp.
E. Rivnay, 1960. Unpublished Notes.
J. S. Taylor, 1931. *Bull. ent. Res.* **22,** *209-210.*
C. B. Williams, 1958. Insect Migration. *The New Naturalist* **36,** Collins, London.
J. S. Wilson, 1934. *Bull. Fla. Agr. Expt. Sta.* **271,** 26 pp.
V. Zelensky, 1938. *Rev. Zool. Agric.* **37,** *137-143.*

The African Cotton Bollworm — Heliothis armigera (Hb.).

[Syn.: *H. Zea* (Boddie), *Chloridea obsoleta* F.]

The genus *Heliothis* includes a few species which may be regarded in some places as pests of field crops of the first rank.

In the literature of many countries one finds dozens of references about their damage, their habits and methods of control. As a rule the larvae of these species choose to feed upon the soft succulent parts of the plant.

Until a few years ago, it was believed that the *Heliothis* species which is distributed in the northern part of the Western Hemisphere is the same as the species which is abundant in Asia and Africa — namely *Heliothis armigera*. Closer examination, however, has revealed that several species are involved — species which are similar superficially but which differ from each other in the structure of their genitalia and in their habits. As a result of the studies of Todd (1955) and Common (1953), it became apparent that the species *H. punctigera* Wllgr. is limited to Australia and Tasmania, and *H. zea* BODDIE is found in the Americas from Canada to Uruguay; while the species *H. armigera* (HB.), is abundant in Southern Europe, Africa, Asia and the eastern part of Australia.

This brought about some confusion, especially when reference to the literature was necessary. Some investigators began to refer to the species prevalent in the Near East as *H. zea*. It is advisable for the sake of clarity to accept PEARSON's suggestion (1958), as follows: Whenever a paper deals with a pest in the Americas from Canada to Uruguay the reference is to *H. zea* (BODDIE), regardless of what name the species was called (whether *obsoleta* or *armigera*). In Europe, Asia and Africa the species referred to is *H. armigera;* in Australia the reference is to *H. punctigera*, except on cotton where *H. armigera* may also be involved. *Chloridea obsoleta* F. is synonymous with *H. zea*.

Description

Adult. Length of body is 14-16 mm; wing span 30—35 mm; colour yellowish-brown. A brown undulating band runs parallel to and a little distant from the side margins. On the forewing (on the disc, near the front margin) there is a brown spot characteristic to the species. On the side of the hind wings there is a brown band broader in its middle and narrower at its edges. Within this brown area there is sometimes a small pale narrow oblong mark.

Fig. 36. *Heliothis armigera.*

The egg is yellowish and shaped like a tangerine; its surface is ribbed radially from the center to the peripherey.

The larva may reach the length of 35—40 mm; its colour is pale green or dark green. Along its body a pale lateral band runs on each side from the thorax to the end; a dark line runs along the middle of this pale band. Another dark line runs dorsally along the length of the body. The body of the neonate larvae bears conspicuous small black setigerous tubercles; these become less noticeable as the larva matures.

The pupa is 15—17 mm long; 5 mm wide; its colour is light-brown. The margins of the spiracles are dark brown, and give the appearance of 6 pairs of dark lateral dots. The end of the abdomen bears two spines, each about 1 mm long.

Distribution. Heliothis armigera is found in the South of Europe and on the entire continent of Africa. In Asia it exists in Cyprus, Israel, Syria, and no doubt in their neighbouring countries, in the Caucasus, Turkestan, India, Malaya, Indo-China, Japan, the Phillipine Islands and Indonesia, Guam, Australia, New Zealand and the Fiji Islands.

Hosts. Many plants are listed as hosts to this species. Mention will be made of the cultivated plants only. These include peanut, pea, bean, alfalfa, chick pea, corn, sorghum, sunflower, tobacco, tomato, cucumber, and cotton plants. In Israel the first plant to be attacked is the safflower (*H. peltigera* is involved too), corn, peanuts and cotton follow.

H. zea is known in the United States by several common names: The "cotton bollworm", the "corn ear worm", the "false tobacco bud worm" and the "tomato fruit worm". Similar names can be applied to this species in the Near East as well.

Type of Injury. In clover, safflower and sunflower the soft leaves are consumed. As soon as buds or flowers appear and soft fruit is present, these are mostly attacked. In tomatoes the fruit is penetrated and rendered unsuitable for the market. In cotton, small squares and soft bolls are gnawed. Older balls may also be penetrated and destroyed. In corn the pest attacks the male flowers and the ears to a lesser extent, while in the U.S. *H. zea* attacks the ears in particular. In that country 75% of the ears may be infested although only 10% of the kernels are destroyed.

Life History. The first appearance of adults depends upon the climate and the geographic position of the country. In a country with a mild climate wehre the pupa can survive the winter, the appearance of the moth depends upon the time of their emergence from the ground. In countries where the pupa does not overwinter, it depends upon the time of the invasion of moths from other countries. In South Africa pupae enter diapause in March-May. These become adults and emerge in August-September. Also, larvae which are not in diapause (because of their slow development) mature during August-October, which are the spring months. In the Sudan, the first attack of larvae is felt during January-February on corn plants (PEARSON, 1958).

In Israel the moth is on the wing in April. In mild winters rare adults may appear even as early as February (as mentioned, the early attack on safflower include larvae of *H. peltigera*).

In this respect it is of interest to make a comparison in the time of appearance of *H. zea* at different latitudes in the U.S. In Georgia they appear in March, in Tennessee and Virginia in May, in Pennsylvania and New York in June and in Canada in July-August. In Canada they migrate every year.

Females may begin to lay two days after their emergence from the pupal cell, but at a temperature of 22° C this pre-oviposition may last over a week.

Eggs are not laid in groups but singly and are scattered over the plants. The oviposition period lasts from two to three weeks.

According to TAYLOR (1936) a female may lay an average of 730 eggs. He recorded the highest number of eggs by one female as 1600.

The eggs are laid on the plant or near it. In an experimental plot of tomatoes 70% of the eggs of *H. zea* were laid near the plants and not upon them (MARCOVITCH & STANLEY, 1941). As stated above, in corn the eggs are laid in the inflorescence or on the tassel. On cotton the eggs are laid on the soft buds, squares, and soft bolls.

The incubation period of the egg lasts 3 days at a temperature of 22° C, and 9 days at 17° C.

According to reports from Russia (quoted from PEARSON, 1958) the threshold of development is at 14° C; at 38° C the eggs do not survive.

The neonate larva consumes the egg shell and then begins to feed upon the plant tissue on which it is located. On the following day it begins to wander in search of succulent parts of the host plant. In tomatoes larvae begin to bore after five days of feeding on the leaves. In peas and chick peas they feed on the leaves until they are about 20 mm long; they then penetrate the pod. In cotton the soft squares and young bolls are attacked immediately.

Under favourable conditions of temperature the larva completes its development period in two to three weeks. The development period of the larvae at various degrees of temperature is given in Table XXVI. It is noted that at the lower temperatures, such as prevail during the winter in the Mediterranean region, almost two months pass before the larva completes its development.

Table XXVI.

The development period of larvae of *Heliothis* spp. at various degrees of temperature

Heliothis armigera		*Heliothis zea*	
Temperature (Centigrade)	Days	Temperature (Centigrade)	Days
17.5	51	15—18	36
22.5	18	19—21	28
		22—24	24

Before pupation the larva penetrates the ground 5—10 cm deep, builds an oblong cell, and pupates within 3—4 days. The pupa is situated obliquely in the cell — head upwards. A non diapausing pupa completes its development under favourable conditions in two weeks. Table XXVII gives the development periods for various places at various degrees of temperature.

Table XXVII.

The development period of pupae of *Heliothis* spp. at various degrees of temperature

H. armigera		*H. zea*	
Temperature (Centigrade) and locality	Days	Temperature (Centigrade)	Days
17 (Russia)	37	16—18	25
23 (Russia)	13	19—21	20
Sudan	14—40		
Tanganyika	12—23		
Uganda	14—37		

Diapause. With the approach of winter many pupae do not complete their development but enter a stage of diapause. Details about the diapause of *H. armigera* are available from studies in Africa (PARSON, 1939). According to these studies the pupae of larvae from April to May do not proceed with their development to the adult stage. Instead 90% of them enter a stage of diapause. These become adults in August—October. The peak (about 50%) occurs in September. The length of this diapause period is not uniform but depends upon ecological circumstances. After diapause the pupa must still go through several stages of development which it cannot do if the temperature is below the threshold of development. It needs several days of favourable temperature in order to mature. During this particular period the temperature and development-time relationship is the same as is the case with non-diapause pupae. As a result of this factor, pupae of different ages may mature simultaneously. In South Rhodesia, for example, pupae from March, April and May may all mature in August-September, which means that pupae 6 months old and 4 months old mature simultaneously, and since non-diapause larvae develop slowly during the winter (about two months), they pupate and mature at about the same time as those in diapause. Thus it is difficult to distinguish between them (PEARSON, 1958).

In the U.S. conditions with regard to *H. zea* are similar. According to WILCOX et al. (1956) the percentage of diapausing pupae changes with the season. Thus in California pupae in May, June and July, all except for one percent, continue their development with no diapause. The percentage of diapausing pupae increases with the advance of summer as follows: 5% in August, 59% in

September, 90% in October, and 99% in November. In such cases, the diapause period of one individual may be 20 times longer than that of another.

In both the U.S. and Africa the cold weather factor is stressed. It is not the low temperature that is the primary factor here but the photoperiodicity. In countries around the Equator, such as Uganda, no diapause occurs in these pupae. Direct evidence of this is furnished by GORYSHIN (1958) who has induced diapause experimentally. He has found that the threshold of diapause is 14 hours day light at 21° C. With the decrease of the number of hours of light the number of diapausing larvae increases.

Migration. It is an established fact which needs no further evidence that *Heliothis armigera* and *Heliothis peltigera* are migrants like their related *H. zea* in America. This alone explains how crops sown in virgin land in the semi-desert of the Negev in Israel were heavily attacked by a pest that was never known there before. Although this moth is a resident of countries around the eastern shores of the Mediterranean, new invasions occur from neighbouring countries. In addition to these long distance movements from one country to another, one must not overlook the migration of the pest within a country, from one crop to another, often quite distant from it.

Thus, in spite of control measures in a particular field, new laying females may come from a field some distance away. But not only the adult moths migrate. The larvae too may leave one field for another, after first consuming what was there — or if they are annoyed by parasites and predators, as pointed out by SLOAN (1941) in Australia.

Control. Pupae in the ground can be exterminated by ploughing in the autumn. BARBER (1937) found that in a ploughed field only 5—10% of adult *H. zea* emerged as compared with the number of moths emerged in a noncultivated field.

In order to avoid migration of larvae from neighbouring areas, weeds must be exterminated. SLOAN suggests a girdle of bait around a field against marching caterpillars.

Bait composed of 10% cryocide in bran gave satisfactory control (MARCOVITCH & STANLEY). WILCOX found that dusting with cryocide gave equally good control. In both cases a reduction of 65—75% of the damage was achieved. Furthermore, there was no difference in results with baits of 5, 10 or 20% of the insecticide.

Dusting with DDT (10% dust) gave more satisfactory results than did dusting with cryocide. In many cases 93—97% reduction was attained. DDT was also superior to several other synthetic insecticides. Toxaphene caused only 85—95% reduction of the damage, while aldrin and dieldrin even less.

In tests at Rehovot it was found that by contact DDT was the most effective of the chlorinated hydrocarbons used, while as stomach poison endrin compared favourably with DDT, and BHC was ineffective.

Recommendations in Israel for 1961 include also the following: On vegetables to be used until 10 days before harvest:

dipterex	80% solution	—	100 g a.i. per dunam;
diazinon	25% emulsion	—	60 g a.i. per dunam;

Until 5 days before harvest:

toxaphene	80% emulsion	—	200 g a.i. per dunam;
sevin	50% W.P.	—	400 g a.i. per dunam.

LITERATURE

G. W. BARBER, 1937. *U.S.D.A. Tech. Bull.* **561.**

I. F. B. COMMON, 1953. *Aust. J. Zool.* **I,** *319-344.*

N. I. GORYSHIN, 1958. The Ecology of Insects, Ed. Danilevskii A.S. Uchen. Zap. Leningrad gosn. Univ. No. **240,** *3-20.*

I. V. KOZANCHIKOV, 1938. *Inst. Plt. Prot. Leningrad* **16,** *27-34.*

S. MARCOVITCH & W. W. STANLEY, 1941. *Univ. of Tenn. Agr. Exp. St. Bull.* **174,** *1-17.*

F. S. PARSON, 1939. *Bull. ent. Res.* **30,** *321-328.*

E. O. PEARSON, 1958. The Insect Pests of Cotton in Tropical Africa. Empire Cotton Growing Corp. and Commonwealth Inst. Entomology, London.

W. J. S. SLOAN, 1941. *Dept. Agr. Stock, Queensland, Pamphlet* **79,** *1-18.*

J. S. TAYLOR, 1936. *Sci. Bull. Dept. Agr. S. Afr.,* **113,** 18 pp.

E. L. TODD, 1955. *J. econ. Ent.* **48,** *600-603.*

J. WILCOX, A. F. HOWLAND, & E. CAMPBELL, 1956. *U.S.D.A. Tech. Bull.* **1147,** *1-47.*

The Plusia group

The genus *Plusia* and allied genera are characterized by the tufts of long hair which cover the body, and by a typical marking on the front wing which is usually bright in colour, often resembling a particular letter or print mark.

The larvae have only three abdominal legs and the body tapers slightly anteriorly. About 10 species are found in the Near East. The most important from an economic point of view will be discussed here.

The 'Silver Y' — Autographa gamma L.

[Syn.: *Plusia gamma* L., *Phytometra gamma* L.]

This is the most common species of this genus.

Description. The moth is of medium size, 13—16 mm long; wing span 30—40 mm. It is easily recognized by the silver marking in the form of the Greek letter gamma or Latin y, hence the Greek name and the English name "silver y". The forewings are dark brown; two light brown patches extend towards the letter y on the disk, one from the front margin and one from the side margin. The wings are decorated with a fringe of hair, grey brown in colour with dark spots; the hind wings are light brown at the base while the third apical is dark brown; the fringe is white with dark spots.

The egg is globular, slightly flat, 0.5 mm in diameter, radial furrows are engraved over the surface, colour yellowish-white.

The larva is 30—40 mm long; width of head 2 mm; metathorax 3 mm; 6th abdominal segment 5 mm; 8th abdominal segment 4 mm wide. The colour of the body is apple-green. Three dark green longitudinal bands run over the dorsum; adjacent to these are two white green bands, one on each side. The spiracles are situated within these bands. Each spiracle is white, encircled by a black border.

The pupa is 18—20 mm long, entirely black, found in a sparsely woven cocoon.

Distribution. If we follow the list of countries where damage has been recorded, we find that its distribution is mainly in Europe and North Africa. It is recorded from England in the west to the Caucasus in the east, and from Finland and Sweden in the north to Malta, Algeria and Egypt in the south. It is found also in Western Asia.

Hosts. The moths are not very selective in sites for egg-laying. In Europe larvae were found on beet, flax, potato, lettuce, cabbage, pea and bean. In Israel the larvae inhabit fields of alfalfa and clover in particular.

Type of Injury. The larvae feed upon the host plant and defoliate it. During severe attacks the crop may be exterminated. Often, after finishing one crop, they wander to another field, much like the army worm. Such records come especially from Europe. In the Near East they are less severe.

Life History. The eggs are laid on the lower surface of the leaf, singly or in small groups. According to CUNIN (1933) the incubation period lasts 10—12 days (temperature is not mentioned). During the Israeli winter at a temperature of 15—16° C, 10 days are required for the egg to hatch. At the lower temperature of 13° C the incubation period is 12—16 days, while at the higher temperatures of 20—30° C the egg hatches within 2—3 days.

In the winter 20—28 days are required for the larva to complete the development while in the spring the larva matures in 15 days.

When ready to pupate, the larva spins a cocoon between the leaves and 3—4 days later pupates. The pupal stage sometimes lasts as long as the larval period. In breedings at room temperature in Rehovot the following records for pupal periods were obtained: 17 days during November; 24 in December; 17 in February; 14 in March; and 6 in June.

Migration. Many papers dealing with this insect refer to it, either directly or indirectly, as a migrant and there is no need to quote again the extensive evidence which has accumulated on this subject. More information and references may be obtained from the books by WILLIAMS.

According to WILLIAMS (1958) the moth resides throughout the year in the Mediterranean region, in particular in North Africa. In the spring and summer it migrates to Central and Northern Europe. On its journeys northward it reaches Finland and may even reach Iceland. Contrary to other moths it travels both by day and night. In England the moth appears all year except during January and February, but the peak of its activity is during May-June. A second peak occurs in August-September. WILLIAMS believes that the spring peak is a result of moths coming from the south, while the second peak consists of moths which developed in England or came from countries still further north on their way southward. Evidence of this assumption is given in several diagrams based on many observations showing that in England *A. gamma* flights are in a northern or westward direction from May to July. During the first 20 days of August few moths fly northwards but most fly southwards and westwards; from the end of August they fly southwards only. Some writers in Europe claim that spring migrants are sexually mature but that autumn fliers are not ready to reproduce.

A. gamma is capable of flying long distances without a stop. Moths were caught at sea 50 km from the nearest coast. (There exists no relationship between flight and wind directions.) The source of energy is the fat stored in the body which in this moth amounts to about 17 mg. As a rule the species appears in small numbers which only collectors and specialists can notice. On certain occasions, however, migrating flocks are dense, and egg-laying is so prolific that the farmer soon becomes aware of their existence. Such outbreaks occurred in 1936 in England, and in 1946 in Central Europe. In Israel the moth was scarce in the winter of 1936, but continued till late in June of that summer.

Phenology in Israel. Israel, like North Africa, is one of the countries where the moth spends the winter.

Light trappings indicate that the adult is found in Israel throughout the winter and is not found there in the summer. The first adults appear in September-October, and their peak is reached in April-May. Larvae are found throughout the winter, the numbers reaching a peak in December-February, and diminishing with the approach of spring. The ages of the caterpillars at the beginning of the season indicate that egg-laying is continuous for some time, probably by successive invasions of flocks.

During a mild winter in Israel the moth may raise about 4—5 generations, but if the winter is cold only 3—4 generations develop.

The Cabbage Looper — Trichoplusia ni Hb.

[Syn.: *Plusia ni* Hb.]

This moth is smaller than *P. gamma*, length of body 10—15 mm; wing span 23—32 mm. The colour is grey-brown, mottled with light and dark markings. On the disk of the wing there is a bright marking in the form of the Greek letter ν or 8; the wing margins and the area around this letter are darker; and a light arc connects the letter with another light spot on the wing. The hind wings are brown, the base lighter than the apical half.

The egg is spherical, yellow with an irridescent lustre, and is engraved with minute radial furrows. Its diameter is 0.5 mm.

The larva is very much like the larva of *P. gamma*. When mature it is 30—40 mm long; the head is straw-yellow, its body green with bright longitudinal lines running along the side of the body (these are bordered with yellow lines); two other lines run parallel to the dorsal median line.

The pupa is light brown, 20 mm long, located in a loose cocoon composed of sparse entangled threads through which the pupa may be seen.

Distribution. The moth is distributed in Southern Europe, North Africa, Central Africa, Western Asia and in America from Canada to Mexico.

Food Plants. Several plants serve as food to this species. In addition to cabbage, it feeds on spinach, beet, lettuce, parsley, celery, potato, tomato, and pea. In Israel it has been found on citrus seedlings and on melon also.

Type of Injury. The larvae feed on the leaves and may thus defoliate the plant. In addition they may attack young fruit as described below. In a watermelon field, over 50% of the first fruit of grapefruit size and larger, had the chlorophyll layer of the peel gnawed away, leaving yellow and white patches (see fig. 37). Such fruit did not mature. Under each fruit several larvae of *Prodenia litura* and *Plusia ni* were found, which were, no doubt, responsible for this damage.

Life History. Our knowledge of the biology of this species, at least in the Near East, is incomplete. Casual breedings at Rehovot show that it has a life history similar to that of *A. gamma*. According to Shweig (1954) the egg develops during the spring within 5—7 days, and the larvae within 14—18 days. During the summer the larvae develop in 10—12 days. The pupal period lasts 7 days in September, 12 in November, and 44 days during January and February. The adults mate 2—3 days after emergence, and egg-laying begins 2 days later. The eggs are laid singly on the underside of the leaf, and one female may lay as many as 400—500.

Migration. Both Williams (1958) and Wiltshire (1957) believe

Fig. 37. A young watermelon-fruit whose chlorophyll has been eaten by *T.ni* larvae.

this species to be a migrant. There is less evidence of this, however, than in the case of *P. gamma*. Its phenology in Israel may offer some indirect evidence.

Phenology in Israel. Observations of many years have established the fact that larvae of *P. ni* are found during May, June, and early July. Following this, there is a period of about 10 weeks when no larvae are found. They begin to appear again at the end of September, and are particularly prevalent during November and December — rarely in January. The adults, on the other hand, are found throughout the summer. They appear in May, their peak of flight is in August; by December only rare individuals are found. It is clear that the autumn larvae are not direct offspring of the larvae of early summer. During the 10 week interval between them, two more generations could have developed if they existed then. Obviously, the moths which emerged in the summer did not lay then in Israel. During the period January-April none of the stages of this species are found in Israel. In the spring the moths appear before the larvae. The renewed spring population is produced therefore by migratory moths coming from other places.

Biological Balance. The larvae of this species are subject to attack by several parasites. In laboratory rearings Tachinids or *Apanteles* Spp. are very often obtained. The larvae collected from the melon field mentioned above were 90% infested with these latter Braconids.

Plusia chalcites Esp.

[Syn.: *Phytometra chalcites* Esp.]

Description. The length of body is 13—27 mm; wing span 28—37 mm. This species may be easily recognized by the golden metallic lustre on the forewings. The print mark is bright and interrupted, making a colon or a semicolon. The hind wings are brown, base lighter than the apical half.

The larva is similar to that of *A. gamma* except that it is slightly broader, and the colour lighter. The spiracles are yellow and lack the black border around them.

Distribution. Records of injury caused by this species come from tropical and subtropical countries. The species is known in Eastern Australia, New Zealand, Malaya, Indonesia, the Fiji Islands and Hawaii. In Africa it was recorded from Sierra Leone and North Africa. It also appears in Southern France, Cyprus and Israel.

Food Plants. Damage was recorded on tobacco, tomato, potato, cabbage, bean, lettuce, and cacao. In Israel it was found on cucumbers, alfalfa and clover.

Life History. Information about the life history of *P. chalcites* is sparse. Casual breedings of the species at Rehovot indicate that it is very much similar to that of the other two species. Thus, the larva develops within 10—12 days during the summer and the pupa within 7 days. In winter the pupal stage lasts one month.

Phenology in Israel. The moths occur in the summer only. The earliest ones appear in May-June; the peak is reached in August; and they become sparse towards the autumn. In the early winter rare specimens may be trapped here and there.

The larvae, however, are found only towards the end of the summer. In other words, the moths of early summer do not lay, as the larvae appear from the end of August to December.

Control measures recommended against *Prodenia* may be applied also against these insects. In the Near East, however, chemical control is seldom necessary.

LITERATURE

G. Cunin, 1933. *Bull. Soc. Hist. Nat. Afr.* N. **24**, (2), *34-42.* Algiers.
V. Kanervo, 1947. *Ann. Ent. fenn.* **13,** (3), *89-104.*
K. Shweig, 1954. Vegetable Pests and Their Control, 149 pp. Hassadeh (in Hebrew).
E. Sylven, 1947. *Medd. Växtskyddsanst.* **48,** 42 pp.

C. B. WILLIAMS, 1958. Insect migration, 235 pp., *The New Naturalist* 36 – Collins London.

E. P. WILTSHIRE, 1957. Lepidoptera of Iraq. Ministry of Agriculture, Govt. Iraq, 162 pp.

The Corn Seed Maggot — Hylemyia cilicrura Rond.

[Syn. *Chortophila cilicrura* ROND.]
(Anthomyiidae, Dip.)

The species of *Hylemyia* and allied genera of agricultural importance, whether polyphagous or restricted to one host, attack the parts of plants close to the ground, or a little below its surface. The eggs of *H. cilicrura* are laid in the ground, and the larvae seek as a rule subterranean parts of plants to feed upon.

Description. The fly is 4—6 mm long. The male is dark grey in colour; on the thorax are a few darker longitudinal lines and on the abdomen one median line. The female is light grey and the lines on the thorax are inconspicuous or missing altogether. The head is brownish grey. The eyes of the male are close to each other; in the female they are separated by a silvery strip one third the width of the head. Over the lunula there is a brown area, the arista bears sparse fine hair, the thorax is convex and is narrower than the head.

The egg is about one mm long and about one third mm wide, cylindrical, slightly bent, posterior end rounded, and anterior end tapering. The colour is white and a fine reticulation over the surface may be seen with the aid of a binocular.

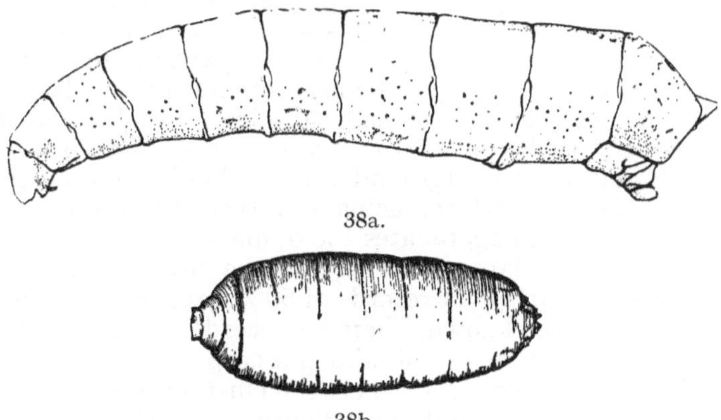

38a.

38b.

Fig. 38. *H. cilicrura* maggot and pupa after REID.

The maggot is white, cylindrical, tapering towards the head, and truncated at the posterior end. When mature it is 5—7 mm long. The last abdominal segment projects in the middle; there are five pairs of papillae around its margin and three on the disc. The posterior spiracles are on slightly elevated buttons and distant from each other three times their own diameter.

Distribution. Hylemyia cilicrura is a cosmopolitan pest. It is found in America from Alaska to South America, in Eurasia from England to Japan, and from Sweden to South Africa.

When discussing the distribution of this insect the chronology of its economic importance is worth noting. In America the species was recorded as early as 1856, but is first mentioned as a pest of primary importance in 1916; in England it was known prior to 1931, but its economic importance was felt only in the fourth decade; in Japan it is mentioned as a pest in the third decade of this century. In Israel it was recorded towards the end of the third decade as a predator on eggs of locusts, but became known as a pest of importance only in the fifth decade.

Food. The larva feeds on any organic matter, whether of vegetable or animal source and either dead or alive. Seeds and underground parts of living plants are attacked. There are certain plants which are subject to attack more than others. The extent of infestation and the choice of food depend on the concurrence of two factors —the activity of the fly, and the ecological circumstances. Thus in France the maggots feed on roots of various cereals which in Israel they do not touch. In the United States corn seed is attacked to a far greater extent than is the case in Israel. The following plants are usually infested: bean, pea, peanut, cucumber, vegetable marrow, melon, watermelon, cotton, corn, wheat, rice, oats, potato, tobacco, cabbage, cauliflower, turnip, beet, and onion.

Type of Injury. The maggot attacks the seed. It may destroy it before germination or it may injure it partially resulting in a weak plant. If the seed escapes the maggot and sprouts, larvae may attack the underground stem and cause the so-called "fainting" and subsequent withering of the plant. Larvae may also attack young roots and even overground leaves, should they be close to the soil as is the case with spinach and lettuce. The injury to fleshy roots causes rot which aggravates the damage.

The extent of the damage varies with the circumstances. In the literature there are frequent records of 50% damage. In Israel even 80% damage has been known. There are records of farmers reseeding their land three times before obtaining a crop.

Life History. Several days pass between emergence and egg-laying. During this period the adults visit flowers or feed on other sweet substances. They can then be seen over newly ploughed fields where they oviposit. The eggs are laid in groups in crevices in the soil or under small particles of earth.

From rearings at Rebovot it is evident that the incubation period is very short. Only one day was required at a temperature of 27° C for the maggot to hatch. Similar results were obtained by HARUKAWA et al. (1933—1934) in Japan, namely one day at 30° C and 8 days at 10° C; and by REID (1940) in Carolina — one day at 27—28° C, 3 days at 16—19° C, and 8 days at 5—7° C.

Immediately after hatching, the larva crawls into the ground and feeds on organic matter. It is believed that micro-organisms serve as food too. After a few days the larva penetrates the seed or other parts of the plant. The maggot goes through three instars; at 15—18° C the first and second instar last but a day or two each while the third lasts longer. The total larval development is 30 days at 10° C (HARUKAWA). REID obtained similar records: 20—28 days at 11—12 °C; and 8—11 days at 24—25° C.

When ready, the maggot leaves the seed and enters the soil to pupate. As a rule this occurs at a depth of 5—6 cm. The pupal period lasts 10 days at 20° C, and 18 days at 17° C.

Records of the pre-oviposition period as found in the literature are given in Table XXVIII.

Table XXVIII.

Pre-oviposition period in days of *H. cilicrura* in various countries

Country	Winter	Summer	Source
U.S.	35—40	17—18	REID (1940)
U.S.S.R.	32	13.5	REKACH (1932)
Japan	60	5—17	HARUKAWA (1933—34)
Israel	19(15° C) 13(22° C)	10(27° C)	YATHOM (1961a)

Oviposition. Fertile females which in total laid a great number of eggs did not produce more than 10—12 eggs per day. In general, the rate of oviposition was far less (YATHOM). According to REID one female lays about 97 eggs. YATHOM obtained different records — of 22 females 10 laid less than 50 eggs each; 5 laid from 50—100; five from 100—200; one laid 213; and another laid 325 eggs in her lifetime.

Length of life of adult. According to REID males lived an average of 29 days and the females 34 days. The records obtained by YATHOM in Israel show that 50% of the flies lived about 20 days while 10% lived over 40 days.

Length of a generation. In Japan a generation, from egg to emergence of adult, lasted 35 days at 10° C; 24—25 days at 20° C; and 16—17 days at 25° C (HARUKAWA). If we add the pre-oviposition period, the length of one generation is 100 days during the winter and 20 days during the summer in Japan. In Carolina the life of one generation was 44—225 days at a range of temperature 7—37° C. In Israel the development of the insect (excluding pre-oviposition period) lasted 46 days at 16° C, 27 at 20° C and 18 days at 26° C. (According to YATHOM). The threshold of development was 8° C and the thermal constant 330 day degrees centrigrade.

Number of Generations. Unlike the onion fly the seed maggot does not enter an obligatory diapause. As the temperature drops the insect develops at a slower pace or remains in a quiescent stage for several weeks until favourable temperature sets in again. High temperatures above the optimum bring the adult to a lethargic state of activity. Thus in England the flies are active from March to October during which time four generations may develop. In Carolina the fly is active in the autumn and spring. Although the fly is prevalent throughout the winter the peaks of its activity are in October-November, and February-March. Three generations are raised during this period (REID). In Japan the fly is found all year round except during the very hot summer months (HARUKAWA). Similar records are given by REKACH from Azerbaidjan. In Israel YATHOM reared eleven generations in the course of one year in the laboratory. In December—January a generation lasted 60—70 days while in the summer only 25 days. In the field, however, this was not the case. As from May the population dwindles, and renewed activity is noticed again only in November, continuing until April. During this period four generations may develop — one in November, one in December-January, and two in February-April. How the fly "bridges" the summer remains obscure in Israel as well as in Carolina and other places of similar climates.

Encouraging Factors.

Soil Conditions. Most authors who discuss this fly mention the fact that it is attracted to wet soils. In Israel which is quite arid, it was noticed that newly ploughed soil attracts the fly more than non-cultivated soil. Also, recently irrigated fields are more subject to attack than those irrigated some time ago.

Reports from the U.S., U.S.S.R., and Japan state that horse manure, fish meal and green manure encourage the increase of the fly population in a field. In Israel too it was observed that fields with rich decaying organic matter had more maggots than neighbouring fields with no such decaying matter. In short, it can be surmized in the words of MILES (1948): "The higher the standard of cultivation the more favourable the conditions are for the development of the flies".

Range of Temperature. The seed maggot is found in countries of extremely diverse climates. This is possible because it can exist within a wide temperature range. REID reared the fly at temperatures ranging from 4° C to 38° C. The adult may even be active at a temperature close to the freezing point. This does not mean, however, that these extreme temperatures are favourable for its development. It can be stated that the fly adapts itself to the most favourable season of the country of its residence. In colder countries it is active in the summer, and in warmer countries in the winter; and if the winter is too cool, it is active in the autumn and spring only.

Limiting Factors

Humidity. Both in Japan and in Israel it was pointed out that in dry non-irrigated soil the fly cannot exist in all its stages. YATHOM reared the eggs of this fly under controlled humidity conditions. The results are indicated in Fig. 39. From this figure it is evident that at 76% R.H., conditions begin to be unfavourable, and below 55% R.H. conditions are detrimental to the eggs of *H. cilicrura.*

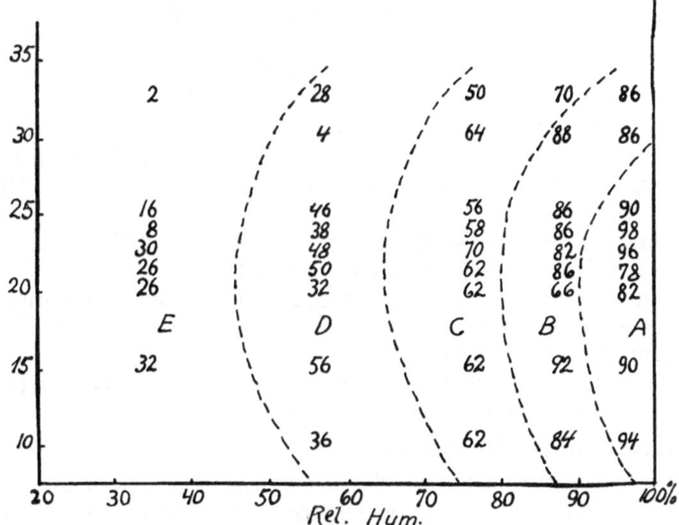

Fig. 39. Temperature – Humidity Zones for eggs of *H. cilicrura.* Figures indicate percentage of eggs hatched in each of the demarked zones. After YATHOM.

Temperature. REID finds that the fly is active within the range of 10—32° C; in other words, the fly is not on the wing at a temperature below 10° C or above 32° C. MILES in England also states that the fly is inactive below 10° C. As mentioned earlier, the fly is not on the wing during the summer in Japan, Azerbaidjan, and Israel. In Israel the fly was reared in the summer in the laboratory where the temperature was maintained at a level below 28° C, but outdoors, where the temperature often rises above 32° C it was impossible. The mortality of the eggs increases as the temperature rises — at 35° C 100% of the eggs died immediately (HARUKAWA).

Control. In the literature we find several preventive measures recommended against this fly. These are early sowing in order to escape the pest, avoiding seeding in a field rich in organic matter, denying the soil green manure, and reducing irrigation — in short denying the soil all the good factors for a good crop.

REKACH (1932) recommended the attraction of flies into a field by means of a bait and destroying them there. Regarding cut seed potatoes it was suggested to dry them to some extent before sowing. All these methods did not yield satisfactory results. Also the various insecticides with which the soil was treated did not yield good control. Lately, a slurry composed of 80% dieldrin as a seed dressing is being used in California (ELMORE, 1953). In Rehovot experiments were carried out trying to find the reasons for only partial control or even failure of these methods (RIVNAY & YATHOM). The results of this study may be summarized as follows:

1. Many of the insecticides which were used in soil treatments against the pest in various countries did not kill the maggots by contact.

2. Seeds treatment with an effective stomach insecticide does not render complete control because the unprotected endosperm becomes infested upon being exposed after the splitting of the seed coat.

3. Treatment of the seed with a non-phytotoxic penetrating or systemic insecticide yielded better control.

4. In order to avoid damage to the plants after sprouting, treatment of soil with an efficient contact insecticide is essential. Parathion proved to be effective.

As a result of this study and certain observations, instructions to farmers were issued in Israel for 1961 as follows:

A. Cultural Control Measures.

a. Fast sprouting of the seed reduces the chances of attack. It is advisable to soak the seed in water for several hours before sowing.

b. Deep seeding is not advisable.

c. It is advisable to sow in wet soil. First irrigate well, then seed. This method attracts far fewer flies than when irrigation comes after seeding.

d. In the spring avoid sowing of land in which green manure was recently turned over.

e. Sow about 30% more seed to allow for damage.

B. Chemical Control Measures.

a. Before sowing, soak the hard shelled seed in a parathion emulsion or VC 13 containing 0.05% a.i. for 6 hours. Caution: use gloves, avoid contact of skin with emulsion, avoid breathing in of fumes. This method should not be applied to shelled peanuts.

b. Treat soil in the furrow before covering it with a 0.05% parathion emulsion at the rate of 30 cc a.i. per dunam.

LITERATURE

J. C. ELMORE, 1953. *J. econ. Ent.* **46,** (6), *1054-1059.*

C. HARUKAWA, R. TAKATO & KUMASHIRO, 1933-34. *Ber Ohara Inst. Landw. Forsch.* **V,** *457-478*; **VI,** *83-111*; *219-253.*

M. MILES, 1948. *Bull. ent. Res.* **38,** *559-573.*

W. J. REID, 1940. *U.S.D.A. Tech. Bull.* **723,** *1-43.*

V. N. REKACH, 1932. *Trud. Zakavk nauchno issled Khlopk. Inst.,* **XVI,** 26 pp., Tiflis.

E. RIVNAY & S. YATHOM, Studies in the control of *H. cilicrura.* Ms.

S. YATHOM 1961a, *Israel J. agric. Res.* (*Ktavim*) **11,** *51-56.*

S. YATHOM, 1961b. Unpublished Notes.

Darkling Beetles- False Wireworms, Tenebrionidae

Several species of this family are pests of agriculture. The adults feed on the stems of annual plants near their crown, on tender seedlings sprouting from the soil or on germinating seeds. As a rule the larvae of these beetles, the so-called false wireworms, feed on wilted plants; sometimes, however, they attack healthy plants. Species of the following genera may be involved:

Pimelia. These are comparatively broad beetles, with the sides of the elytra rounded. Each elytron has four longitudinal ridges which often bear spines or short bristles; in some species the ridges are hardly noticeable. The species of this genus range in size from 15—30 mm.

Adesmia. These are elongate spindle-shaped beetles, with long legs and sluggish movements. The elytra bear rough longitudinal ridges, with rough punctation or reticulation between them. The species range in size between 10—22 mm.

Erodius. Superficially they resemble *Adesmia*, but are less elongate, legs are comparatively shorter, ridges on the elytra are smooth and without reticulation, and the space between them smooth or with dot-like tubercles. The species range in size from 7—16 mm.

Opatrum. They are flat beetles, with sides of the elytra running parallel over two thirds of their length. Pronotum distinctly broader than the elytra. The colour is dull black.

These beetles may attack seedlings of beets, cucumbers, tobacco, and others.

They may be controlled by poisoned bait as described for *Agrotis*.

Opatroides. More notorious is the genus *Opatroides*, the adults of which resemble *Opatrum* except that they are far smaller and have shining black colour; and prothorax is not wider than elytra.

The adults *O. punctulatus* Bruille are found at the beginning of the summer in the upper soil levels, especially in former hay fields. They are capable of eating the sown seeds before they germinate and the newly sprouting plants as well. Complete areas of watermelons may be thus destroyed.

The larvae of the beetle may feed on the roots of corn causing the gradual destruction of the crop. Bundles of pressed alfalfa hay may be turned into powder by the multitudes of larvae which may be harboured within. Complete rows of newly planted pepper seedlings were destroyed by these larvae in Israel.

Hionthis saulcyi RTT. This insect is often found in corn fields. The beetle is smooth and black in colour. The length of its body is 9—11 mm; the width near the thorax is 4 mm and in the posterior third 4.5 mm. Its head is wide, and on both sides of it there is a ridge in the shape of the latin letter V, whose apex is at the base of the mandibles. The eyes are small, flat and inconspicuous, and extend ventrad; they are wider at the upper part and narrower at the lower one. The labrum is hidden under the clypeus whose anterior margin is tri-dentate. The antennae are filiform, with segments equal in width. The anterior end of the prothorax is wider than the posterior one which is convex.

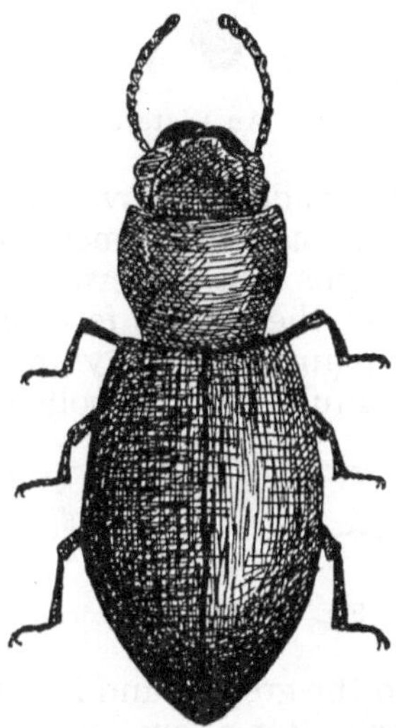

Fig. 40. *Hionthis saulcyi* (after RIVNAY).

This pest is sometimes found on plants; it is, however, usually found in the soil. This beetle is often found together with the Syrian millipede. It seems that conditions of high moisture in the soil suit both arthropods. Soil treatment with 2% aldrin along the seeded rows destroys all the beetles.

Click Beetles, Wireworms, Elateridae

In the Eastern Mediterranean Region wireworms are not as troublesome as in Central Europe or North America. In view of this, and also because the knowledge of the species found in this region is limited, the group will be treated as a whole, and in short.

These beetles are brown or black, or black mottled with grey spots. As a rule, the body is elongated and flat, and the sides of the anterior part of the elytra are parallel. The tip of the body tapers posteriorly. The humeral angles of the prothorax are prolonged posteriorly and fit over the angles of the elytra; when placed on their back they jump with a click into a normal position.

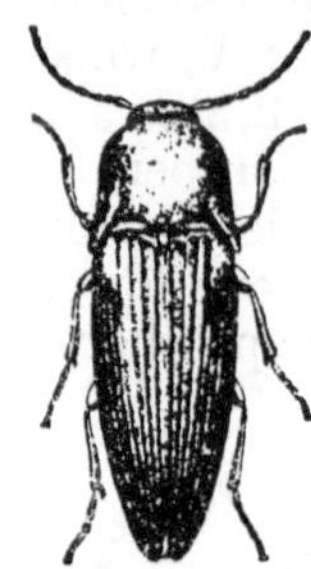

Fig. 41. An Elaterid beetle.

The larvae are various shades of brown, elongate and hard (hence the name "wireworms"). The last segment has a plate bearing engravings or dentate margins in a pattern specific to each species.

As a rule, the larvae of the beetles feed on roots or germinating seeds of graminaceous plants, but they may bore into potatoes, carrots, beets, etc. The adults feed on pollen.

Fig. 42. A wire worm.

The female bores into the ground and lays her eggs near the roots of plants. The eggs hatch after a few days, but the development of the larva lasts for a long period. Some species do not pupate until two or three years after hatching, while other do not complete their development before the end of six years.

During this long period extreme changes take place in the conditions of temperature and moisture in the soil. The larvae of Elateridae wander about vertically or horizontally, and find the most favourable conditions.

Pupation takes place in a cell in the ground. The pupa develops within a few weeks, but the beetle may sometimes remain in the cell, and emerge only after it has hardened and weather is favourable.

The larvae of Elateridae may be killed by various insecticides such as DDT, BHC, chlordane, aldrin, dieldrin, endrin, and parathion: also ethylene dibromide 83% at the rate of 3 litres per dunam was effective.

With the aid of these insecticides the wireworm problem in California and other states has been greatly reduced. However, according to STONE, the reduction of wireworms brought an increase of the seed maggot, since the beetle larvae are also cannibalistic and feed upon these maggots. Soil treatment should therefore be given only where absolutely necessary. In the Near East, especially, treatment may often be superfluous.

LITERATURE

M. W. STONE, 1953. *J. econ. Ent.* **46,** *1100.*

III. FIELD CROP PESTS

WINTER CEREALS

A Key to the Pests of Winter Cereals

A. Denuded areas are seen in the green field.
   1. The roots are bitten and the plant dried; the responsible grubs are in the margins of the denuded areas . . . .: *Phyllopertha* page 157
   2. The entire plant is eaten away
      a. Agent of destruction- metallic dark-blue beetles: *Marseulia* page 154
      b. Agent of destruction- nude Caterpillars in the ground . . . .: *Agrotis* page 88

B. Plants have become yellow
   1. Roots are infested with aphids . . . .: Aphididae page 60

C. Leaves become partly yellow and wither.
   1. In hidden parts of the plant maggots are found . .: Frit fly page 175
   2. In hidden parts *Thrips* are found . . . .: *Thrips* page 74

D. The stem or central leaf becomes yellow and withers; side plants develop from the root.
   1. The lateral leaves are bluish-dark green; at the root crown flat whitish maggots or flax-seed like pupae: Hessian fly page 171
   2. In the root crown cylindrical white maggots or wheat seed-like pupae . . . .: *H. flavibasis* page 178

E. On the leaf surface straw-yellow line or patches.
   1. The responsible pest is on the surface a whitish larvae covered with a viscid fluid . . . .: *Lema* page 151
   2. The responsible pest is within the leaf between the epidermis layers . . . .:
      a. A mined cavity between the epidermis layers, large- often occupying a great part of the leaf; the pest- a brown Lepidopterous caterpillar . .: *Syringopais* page 165
      b. The mined cavity, small — within it a white maggot . . . .: Agromyzidae page 238

F. The stem is easily broken at the time of ear formation: *Cephus* page 162

G. The kernels of the ear are affected:
   1. by white maggots . . . .: Frit fly page 175
   2. by red *Thrips* . . . .: *Thrips* page 74
   3. by green or red bugs . . . .: "Suni" bug, *Aelia* page 142, 145
   4. by green flies . . . .: Aphididae page 60

The "Suni" bug — Eurygaster integriceps Put.
(Pentatomidae, Hemip.)

This is a large bug of a dark grey-brown colour which feeds upon the foliage or soft grains of graminaceous plants. In Israel it appears in limited numbers so the damage is consequently negligible. In the northern and north eastern countries of the Middle East, however, it is present in large numbers and is considered to be one of the worst cereal pests.

Description. The adult bug attains a length of 12—14 mm; its colour is greyish-brown interspersed with black spots. The head is triangular with its base measurement exceeding the length. Two grooves pass along the epicranium. The antennae have five segments, the first being larger and darker than the rest. The egg is barrel-shaped, and opens by means of a "lid", typical of bug eggs. The newly-hatched larva is light-green but turns black several hours later.

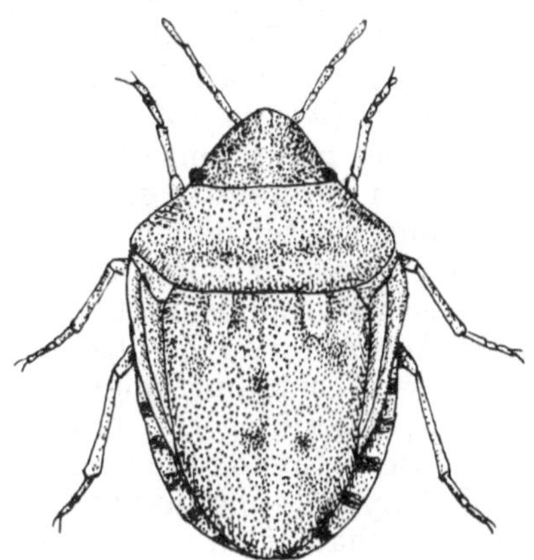

Fig. 43. *Eurygaster integriceps.*

Distribution. Eurygaster integriceps is a severe pest in Turkey, Syria, the Lebanon, Iraq and Iran. It is also found in South Russia, Greece and Israel, but in these countries it is of no great economic importance.

Hosts. The commonest host of *Eurygaster integriceps* is wheat, while barley and rye are attacked to a lesser degree. Occasionally it is found on sorghum and maize without causing any damage.

Type of Injury. The bug sucks the sap, thus weakening the plant. It feeds mainly upon tender leaves. As the ears develop, it attacks the soft young grain and sucks its milky juice — as a result the grain becomes shrunken, light in weight and lacking in starch content.

Life History. In the wheat-fields of Southern Turkey the adults begin to appear at the end of March (ALKAN, 1952) and in more northerly regions at the beginning of April. Field invasion continues until the end of April. The adult bugs feed on plant foliage, mate and begin laying. The female lays her eggs on the leaves in batches of

12—15. A single female is capable of laying 150—180 eggs during her life (ZWÖLFER, 1931—1932). The incubation period lasts 10—20 days, depending upon the temperature. The larval instars last 20—30 days at 20°—29° C. Oviposition begins in mid-April and terminates by the end of May. The larvae emerge throughout May and June. Larvae of all ages are to be found in the field until the end of June, when adults of the new generation appear. The generation which invaded the fields in spring gradually disappears with the termination of laying. At the end of the winter-cereal harvest, adults of the new generation may be found on maize and sorghum plants; however, most of the bugs migrate to the mountains, where they enter into a state of diapause which lasts from the end of summer to the end of winter. They awake only in March of the following year. The mountains to which they migrate have an altitude of at least 1500 meters (ALKAN); ZWÖLFER contends that bugs may also be found on lower mountains of 500 meters altitude. The maximal altitude at which dormant bugs have been found is 2,200 meters (ALKAN). The bugs shelter under bushes such as *Astragalus* (ALKAN) or, as in Iran, under dry oak leaves (SCOTT, 1929). The distance from the breeding-grounds to the places of hibernation may be considerable. According to ZWÖLFER, the bugs migrate a distance of some 120 km, each bug migrating twice during the course of its life. SCOTT, describing this pest in Iran, reports that great numbers of the insects invade the field, their swarms resembling a cloud composed of myriads of individuals. The noise they make while in flight resembles that of an aeroplane. They swoop down upon places of habitation, and settle on houses and roofs. Only afterwards do they congregate in wheat-fields and begin their work of destruction.

Ecology. Detailed data on the ecological factors causing mass multiplication are to be found in the paper by ZWÖLFER (1931—32). According to this writer the optimal temperature for the development of the egg and larva is 20—22° C. In addition, 10—20 mm of rain in May is necessary for normal development, this being the month of maximal oviposition and larval development.

Optimal conditions such as these in two successive years may bring about mass breeding of the pest. The adults, like the larvae, do not tolerate high temperatures, and during the noon hours of hot days they descend into the ground and hide in groups in nooks and crannies. Thus some 200 insects were once found near the roots of a single vine.

The optimal conditions of temperature and humidity necessary for *Eurygaster integriceps* are very different from those prevalent in May in Israel, when the hot "Khamseen" winds cause an increase in temperature and a decrease in humidity. Consequently no mass development of the pest is possible in this country.

Control

Parasites. ZWÖLFER reports that a small parasitic wasp *Telenomus semistriatus* preys upon the eggs of the bug. This wasp lays its eggs on those of the bug. Flies of the Tachinidae family are parasitic on the adult bugs, while various Asilidae feed on them. However, when climatic conditions conducive to bug breeding prevail, biotic control methods are not adequate.

Hand Picking. This is the simplest method, suitable only to certain regions, and recommended as a last resort. The picking must, of course, be done before oviposition, thoroughly and methodically. However, because the adults hide near the ground on the plants during the heat of the day and also because the invasion lasts several weeks, complete eradication by this method cannot be achieved.

Burning of hibernation sites. In Iran, as in Turkey (ALKAN), a method of burning the dry leaves and bush under which dormant adults hide has been suggested. This may reduce the number of bugs, but does not solve the problem completely, since it is almost impossible to find all the hiding-places.

Early crop varieties. Cereal varieties which harden and ripen before oviposition and larvae emergence may escape undamaged. Early varieties therefore are to be used for this purpose.

Chemical control. The bug is a sucking insect and control by stomach poisons is not to be expected, thus, arsenical poison sprays were ineffective. Benzene-hexachloride (TALHOUK, 1951) and other contact insecticides were likewise not efficacious. Amongst the many preparations tested for the purpose parathion has proved the most successful. A dust containing 0.8% active ingedient at a rate of 3.5 kg/dunam* caused 90% mortality. The larvae are especially sensitive to this dust. Dusting should therefore commence at the period when there are most larvae — e.g. before the second moulting — thus the damage done by the larvae, which are the main pest, is avoided, and the development of the next generation prevented. However, the application of insecticides over large areas is not practical because of their high cost, apart from other reasons. Biological means of control are being sought to overcome this serious pest.

Aelia rostrata Boh.

(Pentatomidae, Hemip.)

Description. The red *Aelia rostrata* differs from its relative, *Eurygaster*, in colour and shape. This bug attains a length of 12 mm. The head is longer than it is wide, and together with the prothorax

---

* 4 dunam = approximately 1 acre.

comprises a triangle; its colour is reddish brown, with lighter longitudinal stripes along the body.

Distribution. Aelia rostrata is to be found throughout the Mediterranean countries; its depredations are especially severe in Turkey, Italy and Portugal. It is to be assumed that it also causes damage in other countries with a similar climate, even if this fact has not been recorded. In countries having a hot and dry climate like Israel, this pest presents no problem.

Type of Injury. This bug sucks the sap of the plant foliage and the milky juice of the tender grains.

Life History. The life history of *Aelia* resembles that of *Eurygaster.* In April it, too, emerges from its hibernation quarters, feeds upon the standing crop and lays eggs on the leaves.

The incubation period is of only 4—5 days duration, and the larval instar two weeks. Similarly, after harvest it returns to the mountains, where it hibernates under dry leaves until the arrival of spring. Some workers are of the opinion that not all adults migrate to the wheat fields but that some remain in the mountains, where a local generation is established. The degree of infestation varies every year. About every fifth year the adults appear in larger numbers.

Control. Measures which are efficacious against *Eurygaster* may be expected to control this pest as well.

LITERATURE

BEKIR ALKAN, 1952. *Trans. 9th int. Ent. Cong. Amsterdam* **I,** *623-626.*

A. BALACHOWSKY & L. MESNIL, 1936. Les Insectes Nuisibles aux Plantes Cultivées, Paris *1091-1092.*

SCOTT, 1929. *Ent. Month. Mag.* **65,** *69.*

A. S. TALHOUK, 1951. *Bull. ent. Res.* **42,** *375-377.*

W. ZWÖLFER, 1931 & 1932. *Z. angew. Ent.* **XIX,** (2) *161-187.*

The Greenbug — Schizaphis graminum Rond.

[Syn.: *Toxoptera graminum* ROND.]
(Aphididae, Homop.)

Description. The wingless female is pale or yellowish green in colour. The antennae are dark, except for the three basal segments which are lighter. The cauda and the cornicles are pale, but the latter have dark tips. Its length is 2—2.3 mm. The antennae measure slightly more than half the body length. The cornicles are well-developed and have no reticular region. Their length is about that of the fourth segment of the antennae (0.3 mm) and 1—1.5 times that of the cauda.

The winged parthenogenetic female is similar in size and shape to the wingless type, but with a dark head and thorax, shorter cornicles and antennae about three-quarters the length of the body.

Distribution. This aphid is of European origin, but has spread to many other countries and today is found throughout the American Continent, Siberia, South Africa and the Mediterranean countries.

Hosts. The aphid attacks graminaceous plants only. During its development it passes from one plant to the other according to the physiological condition of the host. It is to be found on species of oats *(Avena)*, wheat *(Triticum)*, barley *(Hordeum)*, *Sorghum*, Bermuda Grass *(Cynodon)*, *Phalaris*, *Irisetum* and others.

Type of Injury. The numerous aphids settling on plants feed upon their sap, weaken them and arrest their development. In a severely infested field, as may be the case in America, places where the pest has totally destroyed the plants can be found; often the infestation begins in patches which gradually increase in size, climatic conditions being favorable.

According to an American estimate, losses due to this pest amount to 3% of the total world wheat crop — 25% of the country's yield in America itself and in South Africa even more.

As regards the Mediterranean area, damage done by this species is not heavy and is hardly economically.

Life History. The aphid reproduces parthenogenetically throughout the summer in Europe and North America, and in the Mediterranean area in winter. The period between "birth" and maturity lasts 7 days at a temperature of 24° C and two to three weeks at lower temperatures. During her lifetime one female may produce 50—60 individuals. In such a manner several generations develop in one season.

With the approach of autumn, in northern countries, the gamic fertilizing males and the laying females appear. The eggs pass through the winter and hatch only in the next spring, when parthenogenetic fundatrices emerge and commence on the summer cycle of asexual reproduction.

In Israel these eggs have not been found, despite a systematic search for them (HARPAZ, 1953). In the summer, however, it is possible to find parthenogenetic females, which apparently bridge the gap between the two winter seasons, eking out a meagre existence on summer cereals in spots sheltered from the hard summer climate.

The English Green Aphid — Sitobium avenae F.

[Syn.: *Macrosiphum granarium* KBY., *Macr. avenae* F.]
(Aphididae, Homop.)

Description. The wingless parthenogenetic female is pale or sometimes dark green; its antennae and cornicles are black. The length of the body is 2.7 mm and the antennae 2.3 mm; the third segment of the latter is longer than the fourth by one third. The length of the cornicles is 0.45 mm longer than the cauda which measures 0.4 mm. The winged female resembles the wingless type in size and colour, but has a darker abdomen.

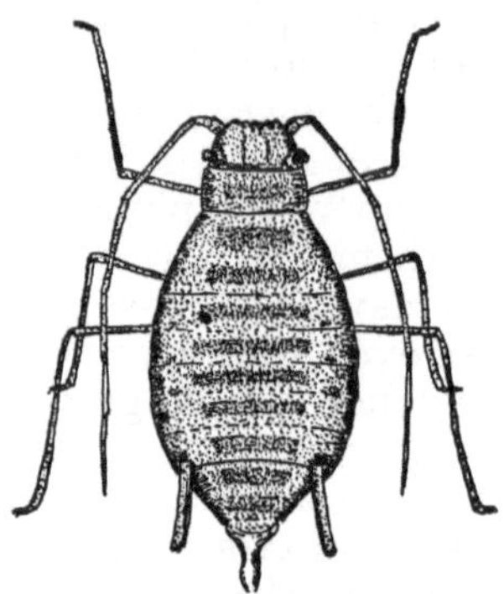

Fig. 44. *Sitobium avenae.*

Distribution. This aphid is found in all Paleo- and Neo-Arctic countries, in Central Africa and in the Mediterranean countries.

Hosts. Various graminaceous winter species serve as hosts to these aphids. They are to be found on both cultivated and wild plants — on oats (*Avena* spp.), barley (*Hordeum* spp.), *Bromus* and *Phalaris*. They do not usually infest summer grains, but they may occasionally be found on *Sorghum* species.

Type of Injury. When the plant is green and succulent, the aphids feed upon its leaves. At the time of ear formation, they migrate to the spikes and feed upon the sap of the grain. This prevents grain maturation and causes shriveling.

Life History. Throughout the summer the females reproduce parthenogenetically, passing from host to host. In Italy, a reproduction rate of 20 individuals per female during a 15-day period has been noted. Sexual females appear at the end of summer and lay a limited number of eggs (about 8 in number) on cultivated cereals, wild plants or onions. These eggs pass through the winter, and parthenogenetic females emerge in the spring. According to SYLVESTRI (1939) the females also lay on *Rubus* species, but it seems likely that *Macrosiphum fragariae,* which will be mentioned below, was mistaken for this species.

In America, it was found that the parthenogenetic nymphs may also pass the winter in a homodynamic arrested activity. Similar observations have been made in Holland and in Italy. In Israel the aphids are to be found throughout the winter from January (when cultivated winter cereals begin to develop) until May(when they mature). Thereafter they disappear (HARPAZ, 1953).

In Israel it is not very clear how the population re-establishes itself with the advent of winter. One possibility is that the nymphs spend the summer in concealment on various graminaceous plants as does *Schizaphis graminum*, or there may be a migration from abroad, as in the case of *Macrosiphum fragariae*.

Macrosiphum fragariae Walk.

[Syn.: *Aphis fragariae* WALK., *Amphorophora rubiella* THEOB.]
(Aphididae, Homop.)

Description. This aphid is very similar to *Sitobium avenae* and it is possible to differentiate between them only by the following characteristics: the body colour of *Macrosiphum fragariae* is green or pinkish brown, the antennae and cornicles are black, the cauda yellowish and the legs light brown. Body length is 2 mm, the antennae being slightly longer than the body. The length of the cornicles is 0.5 mm and that of the cauda 0.26 mm. In *Macrosiphum fragariae* the reticulated region at the tips of the cornicles occupies 1/10—1/15 of its length while in *Sitobium avenae* this region covers only 1/25—1/33.

Distribution. This aphid has been mistaken for *Sitobium avenae*, and in many cases the latter has been recorded in error; thus its distribution is difficult to determine with exactitude. It is, however, certain that it occurs in Western and Central Europe and in the Mediterranean area.

Host and Damage. To the list of hosts mentioned for *Sitobium avenae*, wheat (*Triticum* spp.), *Iresetum* and other species may be added. The damage done by this species resembles that caused by *Sitobium avenae*.

Life History. The life history, too, of this insect resembles that of the *Sitobium avenae*. The one fundamental difference between them is that in the autumn the gamic (sexual) females lay their eggs on *Rubus* and strawberries and, on rare occasions, on rose species. This fact has been noted in Europe, while in warmer countries no *Macrosiphum fragariae* have been found on dicotyledonous hosts. In Israel thorough searches have also been made (SWIRSKI and HARPAZ) but to no effect.

According to HARPAZ (1953), this aphid is most widely distributed in Israel both as regards the number of infested plants and the number of individuals. However, its sojourn here is brief lasting from December to May. During this period it breeds parthenogenetically. After the ripening of the cereals the pest vanishes completely. As for population renewal the following winter, HARPAZ reports that dense swarms of winged aphids may be observed in November descending upon vegetation like a light rain. It is possible that these are the migratory insects returning to Israel from neighbouring or even distant countries where they breed during the summer.

Subterranean aphids

Underground leaf aphids living on graminaceous roots are also to be found in this region. In these cases the graminaceous species is a secondary host, upon which the aphid reproduces parthenogenetically, the primary host being another plant upon which sexual oviposition takes place.

Several such species are often found on cultivated cereals in the Mediterranean region. We shall here mention the most important:

Forda formicaria HEYDEN. This aphid is to be found throughout Europe, Siberia and the Mediterranean region. The gamic female is 0.75 mm long, the length of antennae is 0.20 mm while their colour is yellowish green. The parthenogenetic female attains a length of 2 mm, antennae length being 1 mm, and colour yellowish white.

During the critical winter months, the aphid reproduces parthenogenetically upon barley, oat and wheat roots. In February, winged females appear and alight upon Terebinth trees *(Pistacia terebinthus)*. Here they viviparously produce a gamic generation — mating males and females which lay their eggs on the leaves. The females which hatch from the eggs suck the leaf tissue and cause gall formation. Within these galls they live and reproduce parthenogenetically. The next generation continues in a like manner and this process is repeated all through the summer until the gall dries up. When it splits, it liberates winged females which fly to grain fields and feed upon the various crops.

The second aphid, *Geoica urticularia* PASS., resembles the aforementioned in its life history, but the gall it causes is situated at the base of the leaf and is spherical.

LITERATURE

F. S. BODENHEIMER & E. SWIRSKI, 1957. The Aphidoidea of the Middle East; The Weizmann Science Press of Israel, *1-3B*.

I. HARPAZ, 1953. Ecology, Phenology and Taxonomy of the aphids living on graminaceous plants in Israel, Thesis, Hebrew University.

F. SILVESTRI, 1939. *Compendio di Entomologia Applicata*, **I**, *412-618*.

The Cereal Thrips — Limothrips cerealium Halid.

(Terebrantia, Thys.)

The insect hides among leaves of wheat, and feeds on tender plant tissues. In Europe, excluding the Mediterranean area, it is apt to cause heavy damage.

The female is 1—1.5 mm in length, dark brown in colour, the edges of its legs and antennae being lighter. The head is longer than it is broad, while the prothorax is broader than it is long. The male is 0.9 mm long and is wingless. The larvae are yellowish red in colour with black legs. The development of this insect takes from 30—40 days. It is widespread over Europe, Russia and Western Asia. Specimens of this species have been found on the flowers of citrus and other plants in Israel.

The Ear Thrips — Haplothrips tritici Kurdj.

(Tubulifera, Thys.)

In the spring, small purple insects may often be found among the green grains of the ear; these are the larvae of the ear thrips.

The adults are dark brown in colour, except for the tibia, the tarsus and part of their antennae, which are lighter. The head is longer than it is wide. The last abdominal segment has a tubular sheath which contains the ovipositor. The adults are dormant during part of the summer and winter. They appear in the spring, oviposit among the grains of the ear and die.

The egg, which is 0.4 mm in length, develops in 2—3 days, and from it hatches a small reddish larva. The larval period lasts 30—40 days. In Europe these insects produce one generation a year. Details of the life history of this pest in the Mediterranean area are not known.

Lema melanopa L.

(Chrysomelidae, Col.)

This beetle causes damage mainly in the more northerly countries. In the Mediterranean area it is scarce and the harm it does is negligible. This pest will therefore be discussed but briefly.

Description. The adult beetle is 4—5 mm in length. Its head is blue-black in colour with protruding eyes, separated by a groove from the rest of the head capsule. The antennae are black and are composed of 11 segments, of which the first four are relatively shorter. The prothorax is spherical, orange in colour, with a black anterior border. On the pronotum are three rows of dots. The elytra are parallel, shining and blue-black in colour, with approximately 10 rows of punctures on each. The legs are yellow with black tarsi.

The egg is about 1 mm in length, oblong with rounded ends, and of yellowish colour.

The larva has a thick, oval body, with a dark-brown, round head. On the sides of the head is a yellowish white line and the labrum is red. On the dorsal side of the prothorax are two large grey spots; smaller ones are found on both sides of the mesa- and

Fig. 45. *Lema melanopa* adult – after BALACHOWSKY & MESNIL.

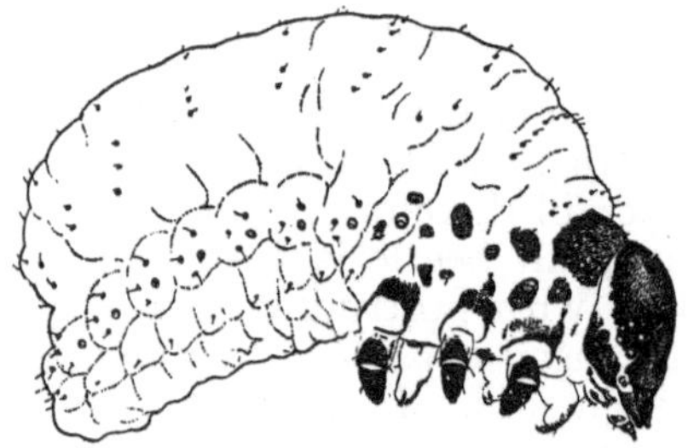

Fig. 46. *Lema melanopa* larva – after BALACHOWSKY & MESNIL.

metathorax. The legs are small, consisting of three segments and terminating in a curved claw. The abdomen is composed of 9 segments and the anus is directed dorsad.

Distribution. This beetle is widespread over Europe, in Siberia and Mediterranean countries. The optimum conditions for its development are found in Central Europe and in Siberia.

Hosts. This insect is a pest of all the winter cereals. It attacks mainly oats and barley, and, in the summer, corn. *Phleum, Dactylis* and *Lolium* may serve as hosts, as well as other grasses. In Israel the adults are often found in fields of clover and vetch.

Type of Injury. The larva chews the leaf tissue in parallel strips running along the veins. When one plant is infested with many larvae, which in case of a large-scale attack may reach 30 in number, the plant turns yellow and withers. In years of heavy infestation, the pest can destroy approximately one-third of the spring-sown grain. Damage of such proportions has been noted only in Hungary and Rumania, but not in Mediterranean countries.

Life History. In Europe the adult beetles appear in the spring and begin to oviposit. The egg-laying season varies in different countries, depending on the climate. In England it takes place at the end of May, in France at the beginning of that month, and in Hungary at the beginning of April. In the Mediterranean area it occurs even earlier, as in Israel, where the beetles oviposit as early as February. The eggs are stuck to the leaves, usually one or two to each leaf. However, in case of a heavy attack, ten or more eggs are found attached along the leaf. The eggs hatch after a week or two, depending on the prevailing temperatures. In Europe, the larvae appear therefore in the summer, and in the Mediterranean area at the end of winter. The newly hatched larva is approximately 2 mm long. It clings by its claws to the leaf on which it feeds, with the head buried in the plant tissue, while its body is perpendicular to the surface of the leaf. The larva is covered with a dark viscous liquid which is its excrement and protects if from drying out. The development of the larva is very quick, — in the laboratory in Rehovot it lasted 10 days at a temperature of 18° C. Since its body is large in comparison with its legs, it is aided in walking on the surface of leaves by the abdominal segments, whose posterior margins press against the leaf while the larva is moving (as does the snake). From time to time the cover of excrement which has become viscous is shed, and fresh excrement covers the body.

When the time comes for it to pupate, the larva falls to the ground and digs to a depth of a few centimetres in its search for a suitable place for pupating. After finding one, it secretes a white spittle-like foam from its mouth which covers its entire body and in hardening, forms a spongy cocoon which protects the insect during the pupal stage. After three of four weeks, the adults emerge; in Europe they appear in July-August and in Israel in April. They do not oviposit in the same year, but enter diapause until the following spring (in Europe) or winter (in Israel).

Ecological Factors. As stated above, in Israel this beetle is very rare, but during warm winters with scanty rains, as in 1959 and 1960, it may become more numerous. Reports from other countries state that the pest is more prevalent in years with less rainfall.

Control. In case of a severe attack, this beetle can be controlled by dusting with synthetic organic insecticides. In this region, there is no need for special measures.

LITERATURE

A. Balachowsky & L. Mesnil, 1936. Les Insectes Nuisibles aux Plantes Cultivées, *788-795.*
E. Rivnay, Unpublished Notes.

Marseulia dilativentris Reiche

(Chrysomelidae, Col.)

This beetle is found in Israel in the hill regions. As a serious pest it is known mainly in the Hills of Ephraim, where it regularly causes much damage. In years when these beetles appear in large numbers, they are liable to destroy whole fields of grains in a day or two. Often the farmer finds denuded patches all over his field which only the day before was verdant and blooming. This is the result of the small blue beetles which chew and eat all the green plants.

Description. The adult beetle is 2—3 mm in length and dark blue in colour. It is oval, with a large head, filiform antennae and a small prothorax. Unlike the other members of its family it has small elytra which do not cover its wide, soft abdomen. No segments are

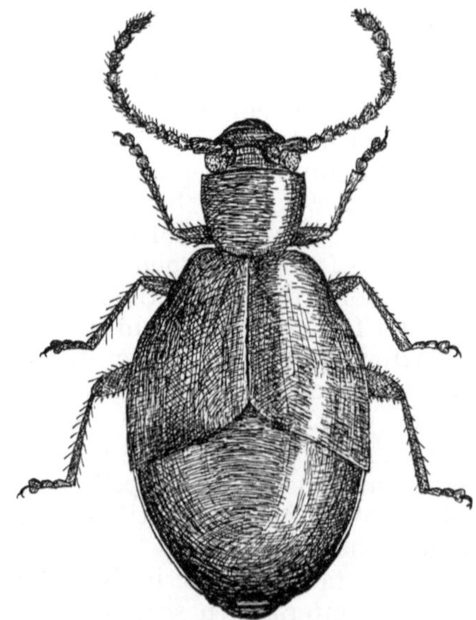

Fig. 47. *Marseulia dilativentris* adult after Rivnay. Courtesy Hassadeh.

visible in the abdomen, which seems to be all of one piece. The egg is small, yolk-yellow and in shape like a slightly elongated ball. The young larva is thread-like, less than 1 mm in length and has a chitinous plate at the posterior end of its body. Its movements are quick and supple. The other larval and pupal stages are at present not known.

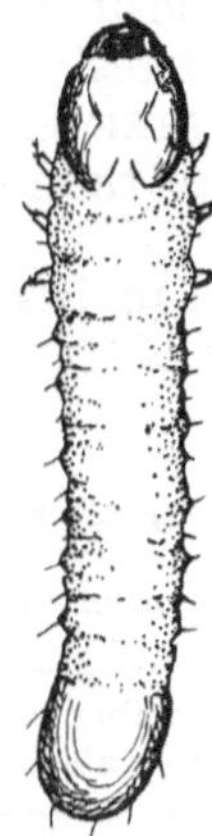

Fig. 48. *Marseulia dilativentris* larva after RIVNAY. Courtesy Hassadeh.

Distribution. This beetle is found only in Lebanon, Syria, Israel and Jordan. Although the pest occurs mainly in mountain regions it is often found also in valleys along the bottom of hills. Thus *Marseulia* has also caused damage in Geva, Sarid, Hephzi-Ba and Tirat Zvi which are in the inner valleys, but near hills. According to BODENHEIMER it is also known in the Jordan Valley (BODENHEIMER & KLEIN, 1928). This pest is usually found in the calcareous soils characteristic of the Israeli mountains in general and of the hills of Ephraim in particular.

Hosts. The beetles devour plants of various families. They are frequently found on the plants of Compositae, Cruciferae, Leguminosae and Gramineae. Most of the damage is caused to the plants of the last two families, among which the pests prefer the soft-leaved ones. They like wheat and barley best, but will also attack oat seedlings. Among the legumes they especially like vetch, but also are found on horse beans and on clover.

Type of Injury. Most of the damage is caused by the adults; the extent depends on the age of the plants and on the time of attack. If the plants are attacked immediately after sprouting, the field is completely destroyed; if the seedlings are able to grow and strike roots first, they may recover. The beetles appear unexpectedly in vast numbers, with very serious results. Because of this pest, the farmers of the Ephraim Hills frequently have to sow their fields a second or even a third time.

Life History. The adult beetles appear in December when the wheat and barley plants reach a height of 5—10 cm, and immediately begin to chew the leaves. It is often possible to find more than a dozen beetles on one small plant. A few days after their appearance, they begin to copulate and oviposit. Beetles raised in the laboratory laid eggs freely on the sides of test tubes as long as the humidity was high. According to BODENHEIMER & KLEIN (1928) the average number of eggs per female is less than twenty and about 14% of the latter do not oviposit.

In an experiment conducted in 1946, thirty females were reared in the laboratory. The number of eggs laid is given in the following table XXIX (RIVNAY).

Table XXIX.

The Rate of Oviposition of the Female *Marseulia*

Number of Eggs per Female	Number of Females	Percent of all the Females
0	9	30.00
— 50	7	23.33
51— 75	7	23.33
76—100	5	16.70
107	1	3.33
113	1	3.33
Total	30	100.00

These figures show that 47% of the beetles laid more than 50 eggs each, while 23% laid more than 70 eggs. In 1947 the experiment was repeated and the egg laying was somewhat lower than that of 1946.

A month after the appearance of the beetles, their population begins to decrease and by the end of January very few may be found in the fields. Of thirty beetles gathered on the eleventh of December 1945 and kept in the laboratory, eight died before the tenth of January, eleven during the second third of that month, six in the last third, and five during the first week of February.

The females oviposit in the soil near the plant, and sometimes on the plant itself. The oviposition period in the laboratory lasts from 14 to 30 days, depending on the conditions of temperature and moisture. The egg must always be kept moist, otherwise it will not hatch.

Details of the life history of the larva and the time of pupation are still not known, nor how the insect spends the summer; but it is definitely known that it lives in the soil. Observations made in infested fields during February and March gave no results; laboratory cultures also failed. On one occasion in the laboratory a young larva was seen boring into the root of a wheat plant. BODENHEIMER supposes that the young larvae fall into summer dormancy and only at the beginning of the following year do they awake and feed on the roots of young wheat plants. Afterwards they pupate and the adults emerge in December. According to this author, the development of the insect must be very rapid. This development should begin in November and end at the beginning of December. If this conjecture is correct, it is still to be determined how and on what the insects feed during their development, because their host plants begin to germinate only at the end of November.

Preventive Measures. The pest spends the summer in the soil either as a larva or pupa. Thorough ploughing during this season turns up and aerates the soil, and exposes the pest to the rays of the sun which destroy many of them. In the settlements of the Ephraim Hills, wherever land is under cultivation, this pest is being eradicated by thorough ploughing. However, in this region untilled slopes and roadside ditches serve as the main stronghold of the insects, from which they invade cultivated fields. Chemical barriers should be practiced here.

Chemical Control. In case of severe attack, the beetles can be controlled by dusting with one of the synthetic insecticides. DDT 5%, BHC 5% or dieldrin 2% control the beetles efficiently, their residual effects are short lived, however. Control is especially difficult when continuous invasion from non-cultivated areas goes on for successive weeks, as previously mentioned above. In such cases the strips of land between the untilled and the cultivated fields should be carefully and frequently dusted.

LITERATURE

F. S. BODENHEIMER & H. Z. KLEIN, 1928. *Z. angew. Ent.* **XIV**, *343-355.*
E. RIVNAY, 1948. Field Crop Pests in Israel. Hassadeh, *1-72* (in Hebrew).

Phyllopertha nazarena Mars.

(Scarabaeidae, Col.)

In the month of January, one can sometimes observe bare patches in fields of wheat and barley on the slopes of hills. An examination of such places shows that the plants have withered because their roots have been eaten by Scarabaeid grubs — the larvae of *Phyllopertha nazarena.*

This beetle was completely unknown as an agricultural pest until about 1936, when it first aroused the anxiety of the Esdraelon Valley farmers in Israel, and a study of it began. But it may have caused damage previously, without the farmers being aware of it.

Description. The body of the adult male is 8—12 mm in length. It has a brown or greenish brown thorax and its abdomen and legs are also brown. The elytra are light brown and along them are dark brown patches of varying sizes; in some cases they cover a large part of the elytra, while in others there are only traces or none at all. The adult female is larger than the male. Its body is 12—13 mm long. Its elytra are wider than those of the male and do not cover the abdomen completely. It resembles the male in colour. The thorax however is always brown. The antennae of the female are smaller than those of the male.

Fig. 49. *Phyllopertha nazarena* adult – after AVIDOV.

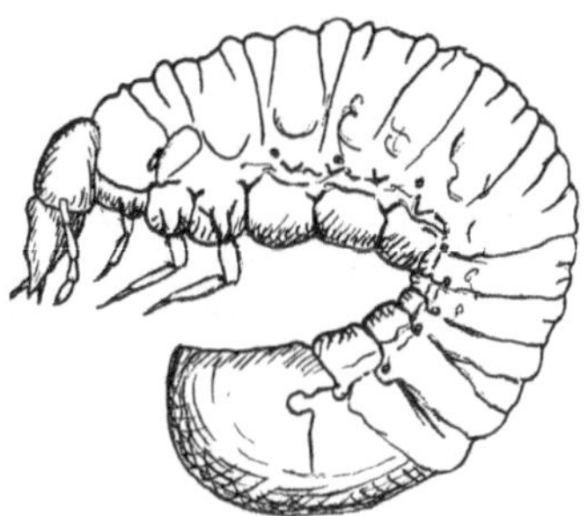

Fig. 50. *Phyllopertha nazarena* larva – after RIVNAY.

The egg is 1.5 mm long and 1 mm wide. It is oval in shape and white in colour. The larva is a white grub with a thick and usually crescent-shaped body (see fig. 50). At the beginning of the larval period its head is 1 mm wide and its body 5 mm long. At the end of this stage, the head of the grub is 4.5 mm wide and its body 30 mm in length. The pupa occupies a cell, 40—70 cm deep in the soil. It is white in colour but eventually turns brown.

Distribution. Phyllopertha occurs in Syria and Jordan and it may be present in Cyprus, too. In Israel it is found in the coastal region, the Northern Negev and in the inland valleys. Only in the latter region is it known to be a serious pest of grain fields.

Hosts. As a rule wheat and other winter cereals serve as food, but should a dicotyledonous crop be sown in an infested field, the roots will be attacked by the grubs. Beets were thus severely damaged in a field in which wheat was grown the year before.

Type of Injury. Severe damage caused by this insect was first discovered at Merhavia in 1936. Yellow patches began appearing in fields of young wheat; the wheat plants, which had reached 10—15 cm in height, were completely withered and their roots gnawed. Each patch was from 250 m² to 1 acre in size sometimes even adjoining one another, so widespread was the attack. Out of a field of 2,500 dunams, 300 were destroyed. This destruction of the crop resulted in further damage; weeds grew on the denuded patches, and harvesting was rendered more difficult and expensive since the infested areas were spread all over the field. In 1938, the normal yield in the region was about 120 kg per dunam, while the average yields of infested wheat fields was only 30 kg per dunam.

Life History. The adult female oviposits in cracks in the soil at a depth of 40—60 cm, where the temperature and low humidity of summer cannot affect the eggs. They remain in a summer dormancy until the rains begin, when the absorption of water causes them to hatch.

The young larvae move to the upper levels of the soil, where cereal roots which serve as their food, are found. They are active from January to March, feeding on roots in the surrounding area. At the end of March or the beginning of April, the larvae descend once more into the soil, where they prepare cells for themselves. Here they remain in a summer dormancy until the approach of winter. With the first rain, the larva awakens, moults and crawls to the upper soil levels. Again they feed on wheat roots. The grub goes through the same cycle during the second and third winters as the first. After the third winter, the larva, whose length has meanwhile increased to 35 mm, descends again and pupates. The pupal stage lasts a week or ten days. The young soft-bodied beetle does not emerge from the pupal cell until the hardening of its body at the end of May.

The life history of the European June bug is very similar to this — only our pest aestivates while the European hibernates. The adult in Israel is active in the early summer — the European in mid- and late summer.

The adult beetles, whose emergence takes place approximately three years after hatching, appear in large numbers on the surface of the fields from eight to ten in the morning and from four to six in the afternoon. It seems that temperatures above 27° C are not suitable for them and they seek shelter in the soil during the warm periods. Most of the beetles on the surface are males, the females are hidden in the soil and seldom emerge; while hundreds of males were collected in the fields, females were never found, and the only females known were laboratory bred. This, according to REITTER, seems to be a feature of *Amphimallus ater* F. in Europe. The males fly over the fields, and at times they alight on a dry stalk with their

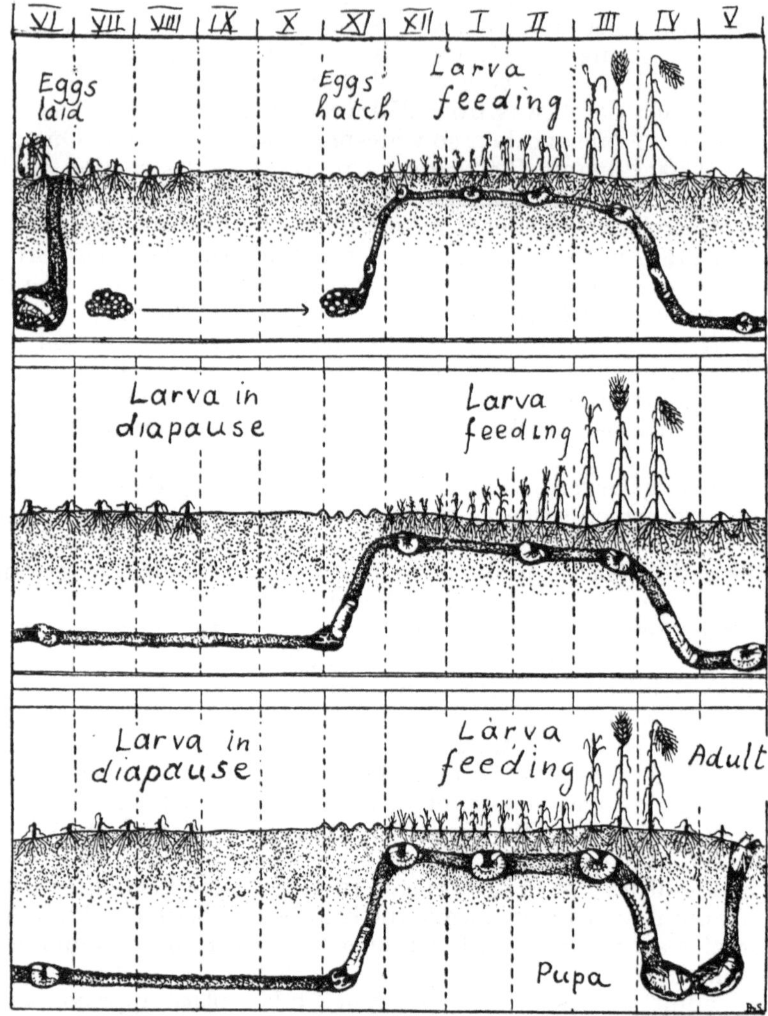

Fig. 51. Life cycle of *Phyllopertha nazarena* in Israel after RIVNAY. Courtesy Hassadeh.

antennae stretched in search of females, and may even enter theri hiding places in order to copulate. The life-span of the adult *Phyllopertha* is very short, lasting only 10—20 days. As early as the end of June it is difficult to find it in the field.

Special pains were taken to find out what the adults feed on, with a possible view of controlling this pest through its food. In laboratory rearings the beetles did not eat any of the leaves of various grasses and trees served as food; neither were any beetles observed eating in the fields. Fifty male beetles were dissected in the laboratory and their stomachs examined. No remains of food were found, but stomach juices only. In the light of these facts, it can be supposed that the adult *Phyllopertha* does not eat, but exists on fats stored in its body during the larval period.

If an infested field is dug-up at the beginning of the winter many larvae will be found — more than one hundred to the square metre. As previously mentioned, the grubs in the soil are of different ages, ranging from young ones only just hatched, to older ones 3 cm in length. They rest during the day and are active at night. The distance covered by a larva is determined by the need to search food and the depth at which it is found, and also by the amount of moisture present in the soil. After rain, when the soil is saturated, the larva tends to ascend to the upper levels and a covering of one or two centimetres deep is sufficient. When the soil dries up, the larvae descend once more to a deeper and moister level. Thus it is always found in an environment of optimal moisture. With the approach of the dry spring in Israel, and the hardening of the wheat-roots, the larva descends to a depth of 40—50 cm.

Ecological Factors. Soil moisture is one of the most important factors governing the occurrence and increase of *Phyllopertha* larvae. There is no doubt that moisture is an essential factor in the hatching of the eggs and the interruption of the larval diapause. It may be assumed that through insufficient moisture the dormancy of the eggs may continue for a second year. It seems also that the larvae awaken from their diapause when the soil-moisture reaches a certain point. There is therefore no large population development in years with low rainfall when the water does not penetrate to the level of the eggs and larvae. In years with high rainfall at the beginning of winter, the number of grubs in the field increases.

On the other hand, excess water in the soil is harmful to the active grubs. In soil which has standing water even for a short time, the grubs die because they are unable to breathe. This phenomenon explains the severe infestation by *Phyllopertha* of fields on the flanks of hills, while in the neighbouring plains the pest is not found at all. The fields of Merhavia, Genigar and Sarid, which are situated on the slopes of hills, are infested, while the fields of Balfouria, Mizra, Tel-Adashim and Kfar Gideon, which are in the plains of the same region, are free from attack.

Control. a. Rotation of Crops. In the fight against pests of cereals, a proper crop rotation may sometimes be helpful. If a certain crop is attacked by a monophagous pest, it can be controlled to a great extent by starvation, i.e. the removal of the plant which serves as its food from the rotation of crops for a certain period. However, there are pests which are able to exist in the soil for a long time without food, and to remain in a dormant state until their food is again available. However, as mentioned above, the grubs of *Phyllopertha* feed on plants other than Graminaceae. Therefore the rotation of crops which is practiced in a number of settlements in the Esdrealon Valley, by which wheat is grown every third or fourth year, cannot serve as a method of controlling this pest. An experiment in this connection was conducted in Merhavia. An infested field was sown with wheat in 1938 and again in 1941. The intensity of attack by the pest was not diminished even though wheat was absent for three years from the field.

b. Baits. This method was tried in America to control two kinds of beetles which had invaded the country — the Japanese and the Oriental, both Scarabaeids, relatives of *Phyllopertha*. There, hundreds of thousands of beetles were caught in traps containing materials emitting an attractive odour. Nevertheless, their population did not decrease in the fields. Our own experiments in controlling *Phyllopertha* showed that it is not attracted by the odour of food, because, as mentioned above, it does not feed at all during the adult stage.

c. Chemical and Biological Control. Experiments with synthetic insecticides which were conducted in 1952 showed that DDT, chlordane, dieldrin, endrin and BHC can kill the larvae. BHC was the most efficient of these materials, and dieldrin came next. In an experiment conducted in 1956, the introduction of aldrin into the soil caused the death of all the two months old larvae. The spreading of 4% aldrin at a rate of 8 kg per dunam gave good results.

However, the use of insecticides should be limited as much as possible, and other, biological means should be studied. Organisms like *Bacillus popiliae* which causes "milky disease" in the Japanese Beetle in North America should be cultivated and tried, should the pest become too troublesome.

LITERATURE

E. Rivnay, 1944. *Bull. Soc. Ent. Fouad., 101-108.*

E. Rivnay, 1948. Field crop Pests in Israel. Hassadeh, *1-72* (in Hebrew).

E. Reitter, 1909. Fauna Germanica. Die Käfer des Deutschen Reiches. II, *332.*

The European Wheat-Stem Sawfly — Cephus pygmaeus L.

(Cephidae, Hym.)

This sawfly, as implied by its name, feeds upon graminaceous plants. It is seldom noticed by the farmer and then only when damage — which is usually slight in the Mediterranean region — has already been caused.

Description. The length of the adult is 10—12 mm. The body is shining black, except for the edge of the femur, the tibia and the tarsi which are yellowish. At the base of the abdomen there is a yellow triangle, and the edges of the third and fifth abdominal segments bear two yellow rings. This description is taken from the relevant literature.

Fig. 52. *Cephus pygmaeus* adult – after BALACHOWSKY & MESNIL.

In Israel, we find sawflies of this species differing slightly from the above. They have yellow front and middle legs, but the external side of the femur is dark brown; the hind legs are light brown except for the coxae, which are yellow, and the external side of the femur, which is dark brown. The ventral side of the pro- and meta-thorax is also yellow.

Fig. 53. *Cephus pygmaeus* larva.

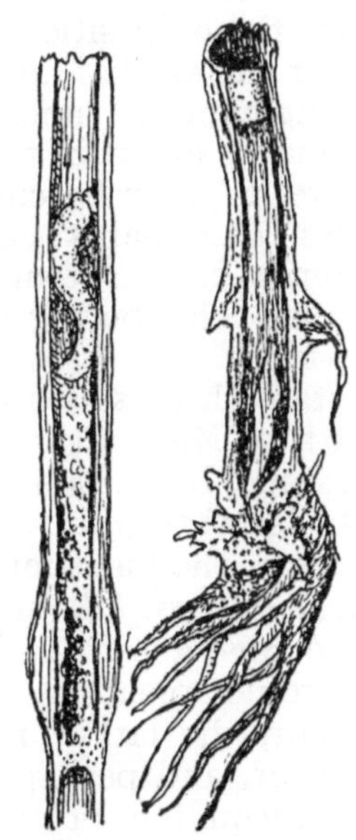

Fig. 54. *Cephus pygmaeus* larva and pupa in wheat stem – after BALACHOWSKY & MESNIL.

The larva attains a length of 12—14 mm and is without legs. The body has a sigmoid curvature and the head is round and yellow. On the ventral side of each thoracic segment there is a constriction which divides it into two sections — posterior and anterior. The first nine abdominal segments are divided in a like manner, while on the tenth there are several lobes and it terminates in a style.

Distribution. This sawfly is found in Europe, the Mediterranean region and North Africa. It has been introduced into North America, where a similar local species is to be found.

Hosts. Hosts of this pest are winter cereals, wheat, barley and rye and also various wild species of this family.

Type of Injury. The larva bores into the stem; when young plants are attacked this may cause the ears to wither and prevent the formation of grain. But damage is generally effected at a later stage when the larvae cause stem breakage near the base, and the grain shrinks without maturing. In Europe this pest causes falling straw in a large portion of the stand; in America damage amounting to 50% is known. In certain fields in Israel about 10% of the plants have been found infested with sawfly larvae.

Life History. The adult sawfly appears when ears are formed (June in Europe, April in the Mediterranean). The adults fly about, visiting flowering weeds in the grain fields or nearby, and feed upon their nectar. After mating, the female lays eggs within the plant stem, the oviposition being beneath the ear. Approximately a week later the egg hatches and the delicate larva bores into the stem, feeding upon the inner tissue. In the course of its development the larva moves about in the stem cavity and fills it with excreta. About 5 weeks after hatching, the larva penetrates through the lower nodes and remains near the stem base. Here it bores a groove in the interior of the stem wall, spins a weak cocoon and goes into diapause. This boring weakens the stem which breaks in the slightest wind.

In infested wheat stems brought to the laboratory at Rehovot on the 14th of May, all larvae were found to be in cocoons in a quiescent stage. In Europe the larva attains this stage in the month of June. In Israel the diapausing larvae pupate during March of the following year, the pupation period lasting about two weeks.

Ecological Factors. Like many other insects, the larvae of *Cephus* in diapause must absorb moisture before they pupate. If the seasonal rains are not plentiful, and the moisture within the pupal cell not sufficient, the larvae may continue their diapause until the following year, and perhaps even later, and resume development only when adequate conditions of moisture prevail.

In the laboratory at Rehovot, larvae remained dormant for two years, and awakened only after moisture was added to the soil in which they were placed (at the end of the second season), while others of the same batch pupated a year earlier because moisture was added at the end of the first season (RIVNEY).

Control. After the harvest the stubble should be ploughed up. In this way the larvae are buried underground, thus preventing their development. In an infested field, a crop rotation excluding all cereals and other host plants of *Cephus pygmaeus* should be applied for at least two years.

LITERATURE

A. BALACHOWSKY & L. MESNIL, 1936. Les Insectes Nuisibles aux Plantes Cultivées. *1076-1090.*

E. RIVNAY. Unpublished Notes.

The Greasy Cutworm — Agrotis ypsilon Rott.

This pest is dealt with more thoroughly in the chapter on polyphagous pests. As a rule in Israel, this cutworm infests legumes, but it often attacks barley and wheat as well. In the autumn of 1942 larvae completely destroyed barley fields in the Gath and Hulda district and wheat fields in the Beth-Hanan and Mikve Israel districts. The attack usually originated in neighbouring fallow fields. In several instances it proved necessary to plough under and plant anew. The farmers were advised to apply bran bait in belts as barriers around the fields.

The Cereal Leaf Miner — Syringopais (= Scytris) temperatella Ld. (Elachistidae, Lep.)

During the months of February and March passers-by may note variable-sized yellow patches in verdant grain fields, where the foliage has completely dried up. Examination of these dried leaves will reveal that the yellowing and desiccation have been caused by a larva gnawing the leaf parenchyma. Often a line of demarcation, looking as if drawn by a ruler, divides the infested and healthy fields.

In cases of severe infestation, the plants of an entire field may be devastated without yielding even the weight of grain that was sown. To farmers in the Eastern Mediterranean region the word "Duda" (larva or worm) has always been synonymous with want and hunger.

Description. The adult female is 5—5.5 mm long. The forewings are generally golden, with dark brown edges and they are covered with yellow and black scales. The hind wings are dark or black, with fringes of hairs, typical of the family, on the wing margins. The abdomen of the female is dark and the tip is yellow. The body length of the male is 6—6.5 mm, its forewings are longer than those of the female and are covered with yellow and gold scales; instead of the brown spot found on the wings of the female the male has bright yellow patches on its wings.

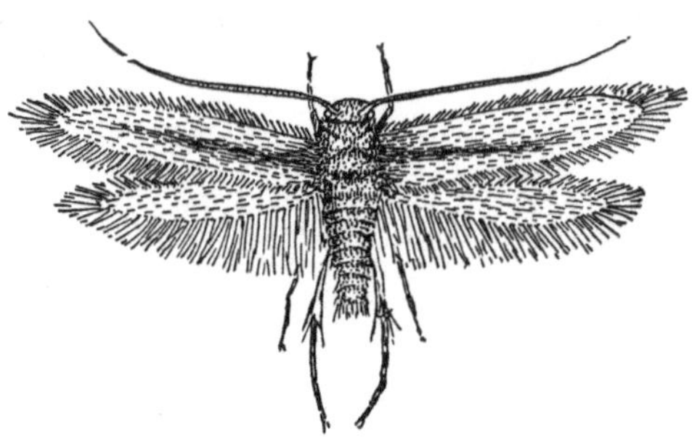

Fig. 55. *Syringopais temperatella* – after RIVNAY.

The Egg. The eggs of the cereal moth are oval and ½ mm long. Their colour is yellowish-gold. Shortly before hatching they turn light grey.

The Larva. The larva upon hatching is 1 mm long and yellow in colour. During the growing period its colour changes to brown and it sometimes attains a length of about 1 cm. At the end of the larval instar, it spins a weak cocoon between clods of earth, in which it pupates.

The Pupa. The colour of the pupa is reddish-brown and its length 5—6 mm.

Distribution. *S. temperatella* is a typical Mediterranean insect and is to be found in Cyprus, Turkey, Syria, Jordan, Sinai and Israel. It may be found in varying soil types — in regions of hills and valleys — in sand, silt and loess.

Hosts. Most of the hosts belong to the grass family Graminaceae. The pest attacks impartially wheat, barley, or oats. When nothing else is available it may also feed on dicotyledons: in Cyprus, for example, it is found on *Convolvulus*, *Plantago* and wild mustard *(Sinapis)*. In Israel too, it is present on various plants. In 1941, serious damage was caused to clover sown in a field in which a large larval population was present.

Type of Injury. The larva mines through the leaf parenchyma. When there is a severe attack and absence of timely rains the plant withers completely. An entire field may sometimes dry up and yield nothing whatsoever.

Life History. The Adult. The first few adult moths appear in February, when large numbers of larvae are still to be found on the leaves. The number of adults gradually increases during March and April. They are capable of mating a few days after emerging from the pupal cocoon. During day-time, the moths may be observed flying and mating. In the month of March, when the average temperature is 17—18° C, the pre-oviposition stage lasts 3—5 days, and in April, when the average temperature is 19—22° C it lasts 2—3 days. Compared to that of other species, the number of eggs produced by one female is small. In laboratory rearings, 30% of females laid less than 50 eggs each, while 60% laid 50—125 and only 10% more than 125. The female lays almost throughout her entire life, but most eggs are produced during her first few days (RIVNAY, 1946). The adult lifespan is 2—3 weeks. In the laboratory, 10% of the moths died before completing two weeks of their life period. Thus most moths have disappeared from the fields by the end of April.

The Egg. The eggs of this moth are adapted to the climatic conditions prevailing in the Mediterranean countries at the time of incubation. They withstand conditions of low humidity, but are sensitive to high humidity. In experiments carried out in the laboratory more than half the eggs died at a humidity below 10%. However, at humidities between 30—75% the mortality rate was less than 10%. The incubation period of the egg lasts about 8 days at a temperature of 25° C and increases to 19 days at a temperature of 18° C. The threshold of development is 9° C (RIVNAY).

The Larva. The larva emerges in spring, when low humidity prevails, and food is scarce because the grain is already ripe, hard and dry. It may be assumed that its immediate entry into a summer dormancy without eating beforehand is a result of a co-adaptation of the larva to these conditions. In order to do so, the larva bores into the soil, finds itself a suitable spot, and spins a slender cocoon, in which it becomes quiescent.

The reaction of the larva to external influences is the determining factor in its choice of the site of dormancy for 9 months of summer and winter. This reaction is not expressed, as might be expected, in a tendency to avoid light (negative phototropism), but rather in a preference for favourable conditions of humidity (positive hydrotaxis). Larvae in a glass tube, with high humidity at one end and low at the other were attracted to the more humid end. When larvae were forced to remain in conditions of high humidity for a few days, they eventually died. Optimal humidity for the larva is 75%. Mortality rates of larva held for 4 days under varying relative humidities are presented in Table XXX.

It may be seen from the table that at a relative humidity above 55% mortality decreased markedly. The optimal relative humidity range for the larvae is therefore very limited. A relative humidity beneath 35% is especially unfavourable. For this reason most larvae are found not far below the surface of the soil. According to obser-

Table XXX.

Mortality Rate of *S. temp.* Larvae under Varying Conditions of Relative Humidity

% Relative Humidity	Number of Larvae Examined	% Mortality
above 90	64	47
84	71	42
76	81	23
55	19	81
32	115	100

vations made on 825 larvae, 27% remained at a depth of 7—16 cm, approximately 65% penetrated as far as 17—25 cm and only 7% reached a depth of 26 cm or more (HUSSEIN, 1954).

As previously mentioned, during all the summer months the larvae are in a state of heterodynamic diapause. They are liable to awaken as early as November and to attack the volunteer growth, but generally they appear in appreciable numbers only at the end of January or the beginning of February. The addition of 2 months to the period of diapause is due to the absence of high humidity needed for the larva to awaken.

Upon emerging from the earth the larva ascends to the tip of the leaf and from there burrows into the parenchymatous tissue between the two epidermal layers. It mines into the leaf and feeds on its tissue. If one leaf is not sufficient, the larva may pass on to a second or even a third. Some 30 larvae may be found within one leaf, and approximately 100 have been found within a single plant (RIVNAY).

The larval stage lasts until the end of February. It extends therefore over a period of about 8 weeks. Upon the termination of its development the larva seeks a sheltered spot, hides between stubble stalks or in the upper layers of the soil and there spins a cocoon and pupates.

The Pupa. The pupal stage lasts about 2 weeks at a temperature of 18—19° C. This means that the adults appear in Israel at the beginning of April. In laboratory rearings, the first moths raised from larvae collected in the field always emerged at the beginning of April (RIVNAY).

Ecological Factors. S. temperatella does not appear every year in the same numbers — there are years when its appearance is of no economic importance, while during others it may cause great damage. In the Mediterranean region there is a saying among farmers: "A year of rain — a year of Duda (worm)". While there is much truth in this saying it becomes apparent that the distribution and not the amount of rain is the main factor. If the beginning of the season is rainy the chances of attack are enhanced, but when there is little rainfall early on, chances are lessened, even if there is considerable precipitation in the late season.

This phenomenon may be explained by the fact that in November, December and the beginning of January, the larva is at the end of its diapause and moistening of the soil may induce activity. This correlation of early rainfall and intensity of infestation may be seen when coastal precipitation figures for the years 1947—1954 are shown together with those recording the intensity of attack in the Coastal Plain of Israel during the same period (Fig. 56). It is noteworthy that during the years 1949—1950 and 1952—1953 the amounts of rainfall were approximately equal. Nevertheless the statistics for the intensity of attack during the 2 years were entirely different. Rainfall distribution explains this fact.

Another factor may be responsible for a heavy infestation: the date of sowing. For example, a portion of a certain field ploughed and sown in November 1949 suffered severe infestation, while the remainder, whose ploughing was delayed because of the rains and which was sown only at the beginning of January 1950 sustained almost no attack.

Control. Cultivation practices. Until recently, the only means of controlling this pest was by thorough ploughing of the soil. Today there are very effective insecticides, which are used at the site of the most severe infestation. SARCOMENOS (1908), HALLAGE (1929) and RIVNAY (1956) deal at length with various cultivation practices applied for this purpose. The best method is basic ploughing and in every case, where its execution is possible, it is to be preferred to other methods.

The weak point in connection with this practice is that the drought-sensitive larva spends the summer in the soil at a depth of 10—25 cm and it can only be killed off by exposing it completely above the soil. This is achieved by thorough plouging. It is clear that one ploughing does not suffice to expose all the larvae and it is therefore necessary to plough a second, third or even a fourth time. The efficacy of these measures as a method of pest control is greatly enhanced when carried out in spring during the Khamseen days. That this type of ploughing is successful may be seen from the results of a survey carried out in Arab fields in 1942. Results of an enquiry conducted among farmers of two villages gave the following picture:

In 12 grain-fields where durra* had been sown the previous year the damage was as follows: — over 95% in 5 fields, 85% in 2, and 40—50% in the remaining 5. Of 7 fields, following chickpeas

* A local *Sorghum* variety.

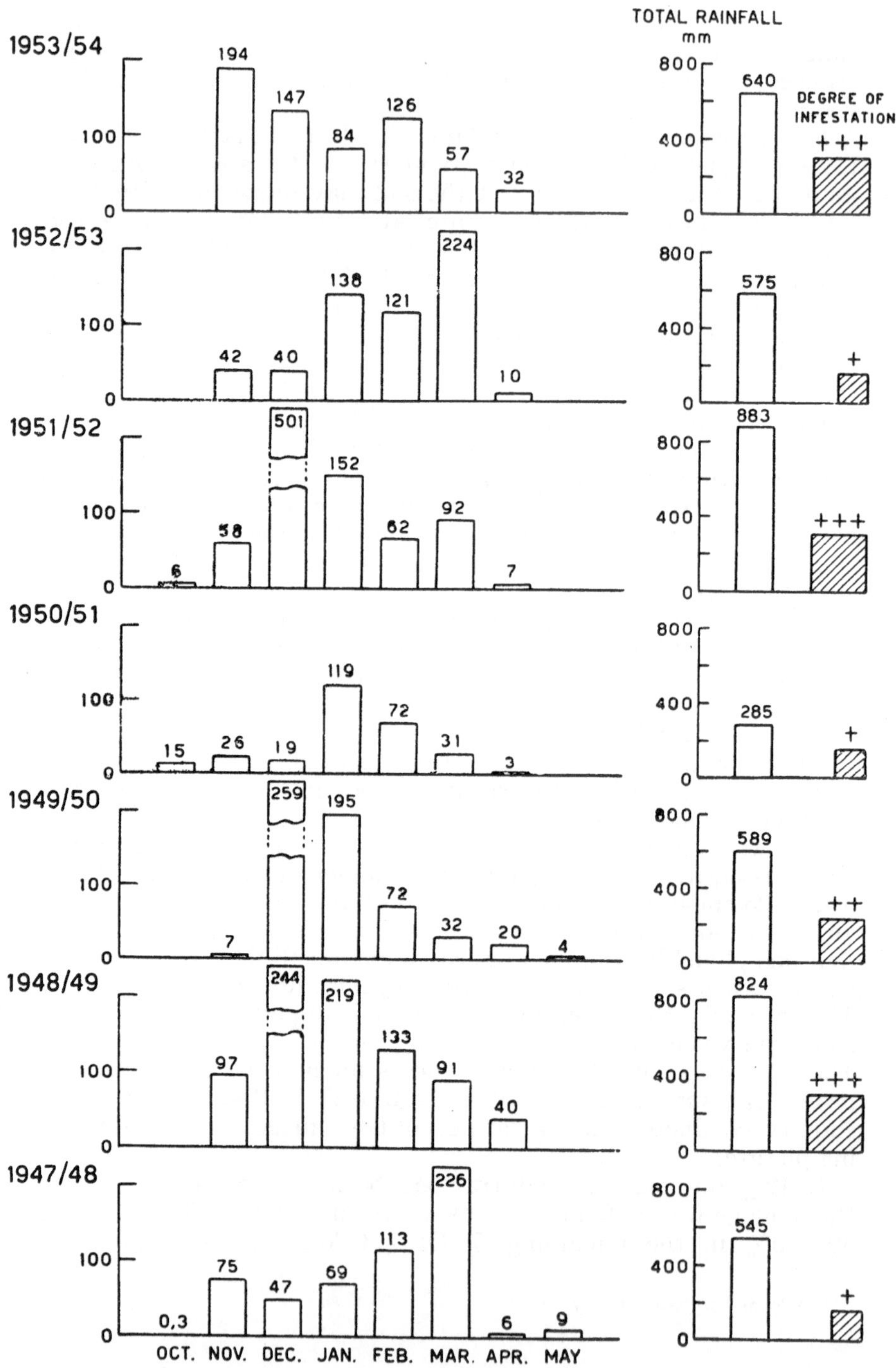

Fig. 56. Effects of Rain on abundance of *Syringopais temperatella* in Israel after RIVNAY.

(Cicer arietinum), one showed a 40% infestation and the remaining six 20—25%. Five fields sown after a sesame crop and one field of watermelons were undamaged. These six fields were pest-free despite severe infestation of neighbouring fields.

There are circumstances under which basic spring ploughing is impossible. It can also happen that despite ploughing the larvae are not eradicated, since they hide in rock crevices which the plough does not reach. This often occurs in hill regions, where only a thin layer of top-soil covers virgin rock. Here synthetic contact poisons may be applied.

Chemical Control. The use of benzene-hexachloride in the form of a 20% Agrocide preparation at a rate of 4 kg/dunam (about 100 g active ingredient per dunam) turned into the soil by the plough in November gave only slightly better results than the untreated control. The reason is that the insecticide was turned under too early and had lost its effectiveness before the pest ascended. Far better results were obtained when the insecticide was turned under a month later.

Treatment with 4% dieldrin at a rate of 4 kg/dunam may also be expected to give good results, since this insecticide, too, kills the larvae, as shown by laboratory experiments.

A more effective method of treatment is applying the poison on infested leaves. Benzene-hexachloride has proved to be the most suitable of synthetic poisons tried for this purpose since its power of penetration into the leaves is greater than that of the others. Dusting with a preparation of 20% Agrocide at a rate of 2 kg/dunam (50 g active ingredient per dunam) gave a mortality rate of above 97%. Similar results were obtained by spraying with an emulsion containing lindane at a rate of 25 g gamma isomer per dunam. Spraying is carried out from the air.

In order to minimise expenses, this preparation may be applied together with 2,4 D which is customarily used for selective weed control in cereals.

LITERATURE

R. Hallage, 1929. *Int. Rev. Agric.* Rome, **I** (1927): *86-89.*

S. Y. Hussein, 1954. *FAO Plant Protection Bull.* **2** (2): *22-3.*

E. Rivnay, 1956. *Ktavim* **7,** *1-27.*

D. Sarcomenos, 1908. Treatise on the Oecophora of wheat. Cyprus Dept. Agric. *1-26.*

The Hessian Fly — Mayetiola destructor Say

(Cecidomyiidae, Dipt.)

The origin of this pest is in Europe, most probably in the Caucasus. However, it is most troublesome in the United States, where it is considered one of the most serious pests of wheat. In Israel, it is unfamiliar to most farmers, and it is possible that it is not recognized even when it does occur in the fields. Occasionally its appearance in the Sharon Plain or in the Northern Negev is noted. It is probable that this fly has been prevalent in the Mediterranean area for some time; it was first noted in Israel only in 1938 when it seriously attacked grain fields in the Karkur region.

Description. The adult is a tiny gall midge, the male being 2.5 mm, and the female 3 mm long. The colour of the head, thorax and legs is dark grey. The abdomen of the young female is yellow, but as the eggs develop within it, it turns red. It tapers posteriorly and terminates in a thin extendable ovipositor. The abdomen of the

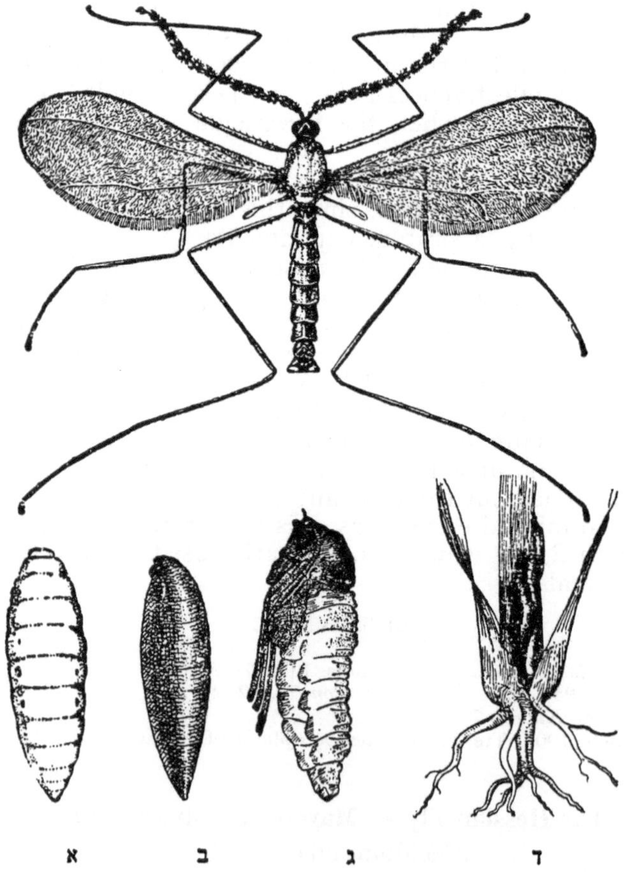

Fig. 57. *Mayetiola destructor* – adult male, larva & pupa – after Walton USDA F.B. 1627.

male is smaller, grey in colour with claspers at the end, that serve to hold the female during copulation.

The egg is 0.5 mm long, spindle-shaped and of reddish or red translucent colour.

The larva, when newly hatched is small and reddish-yellow. At a later stage it turns white and then becomes yellowish-green with white spots. It is round anteriorly and pointed posteriorly. When mature, the skin of the maggot turns brown and its body hardens. At this stage it is 3 mm long and is known as "flax seed", because of its resemblance to this seed in colour and shape.

The pupa develops within the hardened case of the "flax seed" or puparium, which protects it from unfavourable climatic conditions.

Distribution. As stated above, it is believed that the Hessian fly originated in the Caucasus, and migrated from there to other countries. It is believed that the pest was introduced into America in the straw bedding of Hessian soldiers who were brought there during the War of Independence, hence the name. The pest was first found in Long Island in 1780 and since then has become widespread in the United States from coast to coast, except in dry regions. It also occurs in North Africa, Asia and New Zealand.

Type of Injury. The larva attacks the stems of cereals near the base, gnaws the epidermis and feeds on the sap which flows from the injured cells. In case of serious attacks, the main stem stops developing, and in its place secondary ones are produced, which are not always able to develop ears. In lighter attacks, the stem develops, but breaks easily when the ears are being formed. Infected plants are greenish-blue in colour, with broader-than-normal leaves.

Life History. The Adult. The adult flies are not very active, do not feed and are indifferent fliers. Being light, they are carried by the wind to great heights and distances. They have been found 8 m above the ground and 3 km from the nearest wheat-field. The adult lives approximately 3—4 days. The female lays its eggs along the veins of the leaf, between the minute grooves; the eggs are 2—15 in a row and contiguous. On one plant 300 eggs may be found, probably oviposited by several females. In newly-sprouting plants the eggs are laid on the tip of the seedling. On an average one female lays about 2500 eggs.

The Egg. The incubation period lasts several days. Below 10° C it takes up to 10 days, and at the usual spring temperature in the Mediterranean area — about 20° C — it takes from 2—4 days.

The Larva. Immediately after hatching, the larva descends to the base of the blade, enters the sheath and hides near the node. If there are many maggots on the plant, they arrange themselves in a row one behind the other. The larval period lasts 12 days at a favourable temperature, and about a month at the lower one. Before pupating, the maggot changes its position and faces upward. It remains in this "flax seed" state until the adult emerges.

The Pupa. The pupal period lasts 30 days at a temperature of 4.5° C and only 7 days at 21° C. At the tip of its head there is an appendage which aids it in breaking the puparium on emerging.

Ecological Notes. The "flax seed" is the most resistant stage. During this period the pest diapauses, and is able to pass through the winter withstanding the cold, or the summer and its low humidity. The dormant larva in the puparium must absorb a large amount of water in order to pupate. Only after much rain, and at a temperature below 38° C is it able to do so. Under conditions like the above, generations of flies develop without interruption. In certain regions of the United States, five generations are produced throughout the year.

There is no doubt that the low humidity and absence of rain prevailing in California during a certain part of the year lengthen the diapause. In this region the pest raises but one generation a year. It can be assumed that in the Mediterranean area the dryness and high temperatures act as limiting factors inhibiting the development of many generations. In addition it should be noted that both the egg and the adult are sensitive to low relative humidity and therefore the Hessian fly has not become a serious pest in Mediterranean countries. The pest is liable to appear only in rainy years when spring is temperate and "Khamseen" winds are few and mild.

Control. As a rule, agrotechnical preventive measures only are used to control this pest. They are:

a. Turning over the soil after harvest. The Hessian fly in the "flax seed" stage remains attached to the plant near the roots at the base of the stubble after harvest. There it spends the rest of the summer and the winter. The stubble must be plowed under and thus the larvae are destroyed.

b. Destruction of volunteer growth. In very rainy summers the fly produces another generation, which attacks volunteer plants and other cereals. These must be destroyed in order to forestall this additional generation.

c. Adjusting of time of sowing. This method has been found very useful. The time of sowing is adjusted so that the seedlings come up after the adults have emerged from the "flax seeds". In the United States these periods are known exactly and for each state and latitude there are different "border dates", between which wheat should not be sown because of the danger of attack by the Hessian fly.

These methods of control are suitable for North America, but are not appropriate for this region, which, however, as previously mentioned, does not require them in any case.

LITERATURE

J. R. HORTON & HASEMAN, 1944. *Mo. Univ. Res. Bull.* **384,**
C. L. METCALF, W. P. FLINT & R. L. METCALF, 1951. Destructive and Useful Insects. McGraw Hill Co. 3d Edition, *459-462.*
C. M. PACKARD, 1949. *Purdue Univ. Agr. Exp. Sta. Bull.* **440,** *1-15.*
E. RIVNAY, 1948. Field Crop Pests in Israel, *Hassadeh, 1-72* (in Hebrew).
W. R. WALTON & C. M. PACKARD, 1930. *U.S.D.A. F.B.* **1627,** *1-14.*

Oscinidae (= Chloropidae)

In Northern Europe and North America, certain flies of Oscinidae are considered to be among the serious pests of cereals. Only the most important of those found in this or in other regions will be mentioned here.

Most of the adult flies of this family are yellow, brown or reddish in colour, and have a number of dark lines on their thorax. The larvae are soft, transparent maggots, whose bodies are white at the anterior part and green at the posterior. The larvae are cylindrical with straight-sided bodies. The posterior spiracles project from the larva.

The Frit Fly — Oscinella frit L.

In Israel this fly has been raised only once in the laboratory, after being found in wheat sent for examination (BODENHEIMER, 1930). In Egypt its appearance has been noted once.

Description. The adult is a very small fly, 1.5 mm in length; its head is black with large eyes. The antennae are black and hairy;

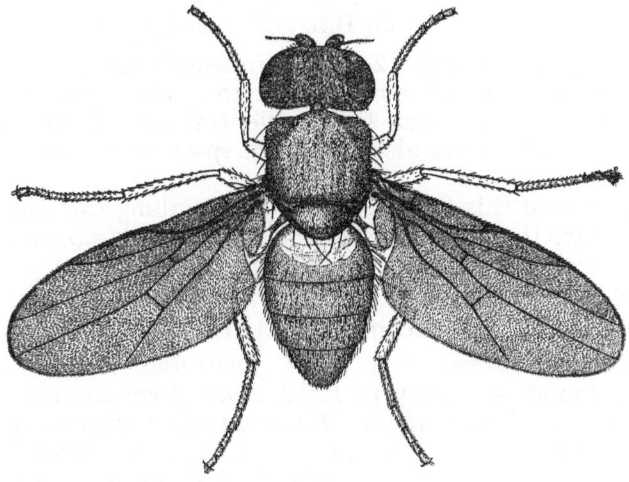

Fig. 58. *Oscinella frit* – after BALACHOWSKY & MESNIL.

the thorax is black and shiny. Its legs are dark and the abdomen is both dark and gleaming. The white, glistening egg is 0.7 mm in length. The larva is white-green in colour and the pupa brown.

Distribution. The frit fly is known as a serious pest in Europe, especially in the north, where much rain falls during the summer. In the Mediterranean area the fly is found in Spain, in Italy, in Portugal, in Egypt and in Israel, but as has been mentioned, in these countries it is rare, and is not considered as a pest. Its chief host plants are wheat and barley.

Type of Injury. The female, which appears in spring, lays its eggs in the sheath of the leaf. The larva, hatching from the egg, penetrates deeper, and feeds on leaf tissue sap, thus interfering with the normal growth of the leaf. The second generation appears at heading, when ears are forming, attacks the tender ear, and causes it to wither. When plants are attacked at a later stage, only a few grains are damaged.

Life History. The female lays 20—60 eggs. The incubation period lasts under European summer conditions 4—5 days, and the larval stage 2—3 weeks. Pupation takes place in the plant tissue, and lasts two weeks.

The fly is therefore able to continue its development, albeit very slowly, under European winter conditions. During the summer, on the other hand, it can produce three generations. The temperature threshold of reproduction is 16° C. The eggs are very sensitive to dryness, and it may be assumed that this is one of the causes of the frit fly's relative rarity in the Mediterranean area.

Chlorops pumilionis Bjerk.

[Syn.: *Ch. lineola* BRL., *Ch. taeniopus* MG.]

The adult is a yellow fly, 2—3 mm in length. On its back are five dark longitudinal lines, the middle one the widest, and those at the sides shorter and narrower. On the ventral side of the thorax are dark patches. There are also two dark spots on the dorsal side of the abdomen.

The maggot of this fly eats out a groove along the stem, underneath the ear. If the plant is young and there is no ear, the pest causes the leaves to wrinkle and wither.

Distribution. The same as that of the frit fly.

Additional Pests belonging to this Family

Camarota curvinervis HAL. (=*C. flavitarsis* MG. =*C. cerealis* ROND), is found on barley in Italy, *Elachiptera carmuta* FALL. is found on grain in North Africa. *Meromyza* sp. attacks the uppermost internode of the stem of cereals. In Europe *M. saltatrix,* and in America *M. americana* are known. The damage to the plant is either at the base of the stem or beneath the ear.

Domomyza ambigua Fall.

(Agromyzidae)

The larva of this fly bores between the two epidermal layers of wheat and barley leaves. The adult is a small fly, black and shiny, and 3—4 mm in length. The larva is white, 5 mm long, and on pupating enters the ground. It appears that the pupa spends the summer in the soil, and awakens only with the coming of rain. The insect is found in Northern Europe and in the Mediterranean area, including Egypt and Israel, but is not considered a pest.

Hydrellia griseola Fall.

(Ephydridae, Dipt.)

During the months of February-March one can sometimes find in grain fields, especially those of barley, leaves or parts thereof whose substance have been eaten, nothing remaining but the dry, brown and empty epidermal covering layers. This damage, which resembles that effected by the attack of the cereal leaf miner is caused by larvae of *Hydrellia*.

Description of the Pest. The adult is a small fly, grey in colour, and 2—2.5 mm in length. Its head is wider than its thorax, and its "face" is yellow. The antennae are black, their third segment being short and round. The arista is long and bears about six small hairs. The legs are black.

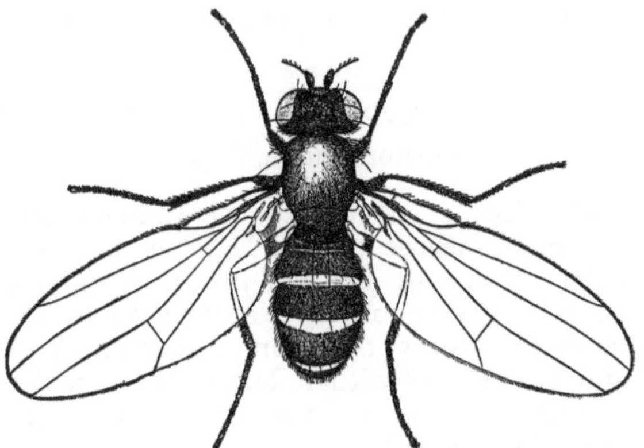

Fig. 59. *Hydrellia griseola* – after BALACHOWSKY & MESNIL.

The larva is a white maggot, 5—6 mm in length. Its head is narrow and at the end of it the black mouth part is visible consisting of a single piece (both pairs of jaws have united). The last segment of the body is long and is rimmed with a wreath of small hooks. The posterior spiracles are each covered by a small cone.

Type of Injury. In older plants the damage from chewing is not great, and they outgrow it. Sometimes, however, the plants are attacked while they are still very young and their leaves tender, and then they are completely destroyed. This type of severe damage is not known in the Mediterranean area, but is familiar in countries of Northern Europe.

Life History. The adult fly oviposits near the base of the leaf. From the egg hatches the larva, which bores into the leaf between the two layers of epidermis. In contrast to the cereal leaf miner larva, the maggot does not leave its tunnel, but terminates its period of development within it.

The larval stage lasts about a week; the pupa, which also remains within the leaf, develops during 8—9 days. The generation developing in February therefore raises a second one, appearing in March. Many details of the larva's life are still lacking, especially scanty is the information on the number of generations which the pest raises during the year.

Ecological Factors. Hydrellia develops mainly in northern countries where there is abundant rain during the summer. The fly has economic importance only in the Baltic countries, in Estonia, in Lithuania, in Denmark etc.

In Israel *Hydrellia* is rare, and appears mainly in rainy years. Even in case of an attack the damage is not great. In a barley field in Mikveh-Yisrael, which was infested with this fly, only that part of the field which was shaded by a row of *Eucalyptus* trees was infested. In other parts of the field very few larvae were found.

The Cereal Root Maggot — Hylemyia flavibasis Stein.

(Anthomyiidae, Dipt.)

In grain fields, during the months of January and February, plants may be found, whose central leaf has withered, while the two or three lateral leaves are still green. A close examination reveals at the base of the plant, near the roots, a small white maggot, which in the process of feeding destroyed the tissue of the middle leaf, and caused it to wither. This is the cereal root maggot of the family Anthomyiidae. In severe attacks more than one-third of the plants may be harmed.

Description. The adult is 4—6 mm in length, and its wing-spread attains 12 mm. The fly is grey-brown in colour, except for its back which is the colour of lead, and its face and dorsal side of its abdomen which have a grey sheen.

The egg, like all eggs in this family, are white and cucumberlike in shape. The larva is white and approximately 10 mm in length. Its head is narrow and the posterior end is wide. The anterior spiracles are near the head of the larva.

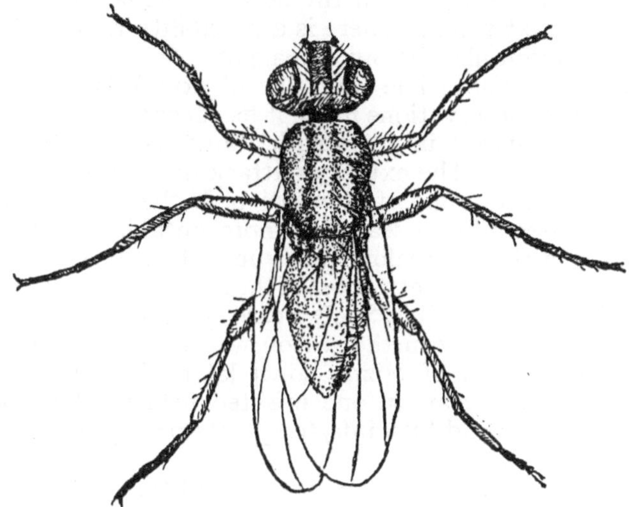

Fig. 60. *Hylemyia flavibasis.*

Type of Injury. The damage consists of the destruction of the plant tissue near the roots, which brings about the withering of the main leaf, so that the entire plant ceases to develop. If the part of the plant below the affected area is strong, and the soil fertile, other lateral shoots develop, which succeed in developing and producing a yield if there is abundant rainfall. The plant is seldom able, however, to overcome an attack by this fly and the damage is serious.

Biology. The female oviposits on the plant or on the ground near them in December or January. A few days afterwards, the egg hatches and the white larva feeds on the juice of the plant cells obtained from its scraping off the leaf tissue near the roots. The plant is completely destroyed at the site of attack. This destruction is not caused by the direct physical action of the larva, but by the excretion of the larva and probably of the bacteria which it carries, as is the case with its relative, the cabbage maggot. The maggot continues eating for about two or three weeks, and afterwards pupates. The pupation takes place among the leaves of the plant near the ground. The pupal stage lasts approximately one week. The flies of this first generation appear as early as February.

The life history of this fly has not yet been studied. However, it can be surmised that its development is similar to that of the onion maggot, which is also monophagous. If this is so, it can be supposed that the new generation oviposits again during the same winter, while some or all of the pupae of this generation enter a summer dormancy from which the adults emerge only at the beginning of the next winter. There is a possibility that it also raises a second generation like the onion maggot.

Ecological Conditions. The intensity of attack is not the same every year. From observations made it has become evident that the pest is most widespread in years with abundant rainfall at the beginning of the season. The extent of attack depends on the state of the grain field, at the time of oviposition. If there are no plants in the field, the insects fly in search of more suitable sites for laying. The situation in the Esdrealon Valley in 1940 will serve as an example. Irrespective of variety, most of the grain-fields were attacked, especially those sown earlier. Such a case occurred in Tel-Adashim, where a field sown early in the season was 40% infested, while in the neighbouring field of Mizra, sown later, the infestation was only 5%. The same year, in one infected field in Heftziba there were eighty healthy and fifty infested plants per sq. metre, and the attack reached 40%.

Methods of Control. In the case of a pest of this kind, preventive measures are preferable. The fact that the pupae remain in the ground throughout the entire summer must be taken into consideration. Constant ploughing of an infested field may greatly reduce the intensity of attack. Attacked plants can be saved if the field is cultivated properly and contains sufficient nutrients. Even a badly attacked field may be saved by top dressing with fertilizers.

LITERATURE

A. BALACHOWSKY & L. MESNIL, 1936. Les Insectes Nuisibles aux Plants Cultivées, *910-1075.*

F. S. BODENHEIMER, 1930. Die Schädlingsfauna Palästinas, Berlin, 438 pp.

E. RIVNAY, 1948. Field Crop Pests in Israel, *Hassadeh, 1-72* (in Hebrew).

SUMMER CEREALS

A Key to the Pests of Summer Cereals

A. The seeds do not germinate, or the sprouting is poor and sparse
   1. When uncovered, the seeds are found to be broken or injured; often beetles are near them . . . . . . : Tenebrionidae page 138
   2. Millipedes are found near the injured seeds . . . . : *Strongylosoma* page 182
   3. White maggots are found in the seed . . . . . . : *H. cilicrura* page 131

B. The plants, after sprouting, are destroyed.
   1. In the ground, near the plant, curled caterpillars may be found . . . . : *Agrotis* spp. page 88
   2. The field is inhabited by grasshoppers of various kinds . . . . : Grasshoppers page 19

C. The entire plant withered.
   1. When the roots are exposed, hard-skinned, brown-coloured worms are feeding upon them or boring into the stem . . . . : Tenebrionidae, page 138; Elateridae page 139
   2. Near the plants, in the ground, curled caterpillars are found . . . . : *Agrotis* page 88

D. Only the central leaf has withered
   1. When the central leaf is pulled out, its base is foul smelling and moist; often a yellow maggot is in the fold . . . . : *Atherigona* page 202
   2. The basic end of the leaf has been chewed by a caterpillar, which may be found amidst faeces in the centre of the plant . . . . : *Sesamia* page 194

E. The leaves have been partially eaten and the stem is perforated
   1. The leaves have been eaten irregularly; often caterpillars are found in the whorl of the leaf . . . . . : *Laphygma*, page 113; *Hyphilare*, page 201; *Pseudaletia* page 199
   2. In the central leaves there are holes arranged in rows of four; in the curl of the leaves, or in the stem, a cream purplish-coloured caterpillar may be found
      a. The head uniformly brown . . . . : *Sesamia* page 194
      b. The head mottled . . . . : *H. loreyi* page 201
   3. In the stem, at the leaf-axis, small holes are found from which faeces are expelled, — in the stem a purplish faintly dotted caterpillar may be found . : *Pyrausta* page 185
   4. In the stem (of rice) a similar caterpillar is found . : *Chilotraea* page 194
   5. The cane close to the ground has small holes, is often girdled; in it cream coloured caterpillars are found: *Chilo* page 379

F. The ear and inflorescence are eaten.
   1. Small holes, less than 3 mm in diameter, are located on the sides of the ear; webs and caterpillars are in the inflorescence . . . . : *Pyrausta* page 185

2. Larger holes, 4 mm and larger, are found in the ear: *Liogryllus* page 42
3. The tip of the ear and the silk are injured; when opening the sheath caterpillars may be found . . . : *Heliothis*, page 119; *Hyphilare* page 201; *Sesamia* page 194

G. Injury by sucking insects
1. The leaves are mottled with white spots . . . . . : Cicadellidae page 183
2. In the whorl of the central leaf, or between the sheaths of the stem, green aphids are found excreting honeydew . . . . : *R. maidis* page 183
3. On the leaves, ears, or inflorescences, large green or brownish bugs are found . . . . : *Nezara*, *Dolycoris* page 49, 51

The Small Syrian Millipede — Strongylosoma syriacus.

As is known, this pest does not belong to the class Insecta but to the class Diplopoda of which very few are injurious to agriculture.

The small Syrian millipede in its general form resembles the wireworm. However, it is simple to distinguish between them by the many small legs which are found along the body of the millipede.

Description. Its body is thin, cylindrical, dark in colour and 20—25 mm in length. It is divided into a large number of narrow segments, each of which bears two pairs of small legs, Its head has a pair of short antennae and two pairs of maxillae.

Distribution. This pest is found in the eastern countries of the Mediterranean area, Israel, Jordan, Syria and probably in their neighbouring countries.

Type of Injury. The millipede, as an adult or larva, gnaws germinating seeds in the soil. This either effects their ability to develop, or destroys them completely. There is no doubt that different kinds of seeds may serve as food. In Israel, however, this species attacks mainly sorghum and maize.

Life History. The millipede lives in the soil, underneath stones, between rotting leaves, and, in general, in damp places. Usually the female lays its eggs in groups in the soil. Complete details of this pest's development are not known; however the Diplopoda's development is holometabolic. From the eggs small larvae hatch which are similar to insect larvae. Their bodies have few segments and only three pairs of legs. With each moulting, both segments and legs are added, until they reach the adult stage.

These creatures live in moist places. A survey of the areas infested with this pest proved definitely that the damage was mainly localized in low areas of the field where the soil moisture is greater.

Control. Soil treatment with BHC or aldrin, as used against wireworms and Tenebrionidae, may also be useful in controlling this pest. The treatment should be limited only to areas known to harbour the pest, and in rows only.

Leaf Hoppers — Cicadellidae

On the leaves of corn small leaf hoppers may be found running obliquely in a lively manner. When these are numerous, the leaves are mottled, the chlorophyll having been emptied from the epidermis cells. These insects belong to the Cicadellidae, the biology of which is little known.

It has been observed that they migrate to the corn from adjacent uncultivated fields.

For the time being farmers do not complain about injury from these insects. Perhaps the injury is too small to effect the crop.

Species known from the Near East are:

Cicadiluna bipunctella MATS — It was first described from Port Said. It is distributed in Africa, Formosa and Japan. Its host is *Zea*.

Erythroneura coacta RIBAUT. — This is reported from Cyprus and Israel.

In view of the fact that DDT and systemic organo-phosphorus compounds controlled leaf hoppers on other plants, it may be surmised they may control these, too.

The Corn Aphid — Rhopalosiphum maidis Fitch.

(Aphididae, Homop.)

Description. The apterous parthenogenetic female is pale green; its eyes, cornicles and cauda are black and its head is a dark greyish-green. The length of the body is 2.20 mm and the length of the antennae is shorter than half the body length — about 0.85 mm. The third antennal segment is as long as the sixth; the length of the cornicles as long as the cauda — 0.14 mm.

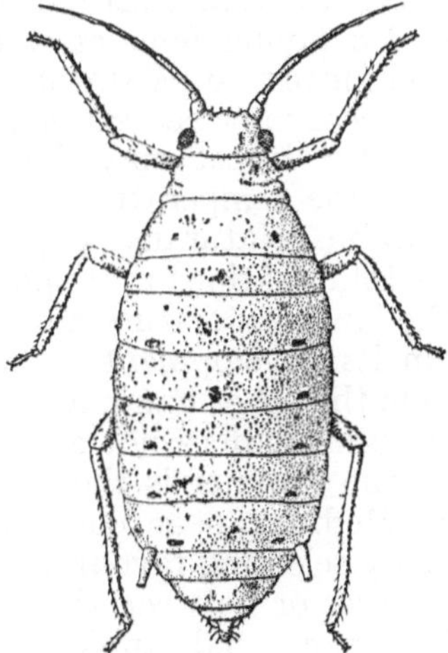

Fig. 61. *Rhopalosiphum maidis* – after SILVESTRI

There is a small form of apterous females whose length is 1.5 mm, and whose colour is dark grey-green, while the abdomen is green.

Distribution. This insect is typical of tropic and subtropic countries. It is found in the United States, in Mexico, South America, Australia, China, Japan, Polynesia, Caucasia, Southern Europe, in the countries around the Mediterranean. In other words, as WILDERMUTH & WALTER (1932) expressed it, between the 40° north and south latitudes.

Hosts. This aphids feeds upon many graminaceous plants: wheat, barley, oats, corn, sugar cane, sorghum, *Panicum, Cynodon, Phalaris* and *Lolium.* The insect was found also on plants such as garlic, *Typha*, etc.

In Israel, since it occurs only during the summer, only summer plants are infested.

Type of Injury. The feeding of the numerous insects weakens the plant. In case of severe infestation the yield from the field may be reduced to as much as a third of the crop. The excretion of honey dew attracts moths of the earworm and infestation of this pest is thus increased. In infested sugar cane the aphid may transmit a virus.

In America, the infestation may be so heavy, that the trousers of a man walking in the field become sticky from the honey dew secreted by the aphids. In the Near East, infestation is not so severe and measures of control are necessary only in extreme cases.

Life History. The Corn Aphid reproduces parthenogenetically throughout the year. Its time of development may last from a week to a month (HARPAZ, 1953). In the Eastern Mediterranean countries it may raise about 35 generations. In South-Eastern United States, according to WILDERMUTH & WALTER (1932), it raised 45 generations. No laying females were discovered in the U.S., nor in Israel. WILDERMUTH & WALTER state that some males were discovered there. These may be vestigial members of the species. In the Near East the insect may develop uninterruptedly throughout the year, as the temperature is favourable and host plants are always available. Naturally, depending upon climatic factors, there are annual fluctuations in its population.

Limiting Factors. The low temperature during the winter months causes a retardation in its development, hence the density of its population decreases. On the other hand, as the insect is of tropical origin, the heat of the summer does not affect it. The only factor which may affect it in the summer is the low humidity, which is typical to the Eastern Mediterranean countries. The insect overcomes these conditions by selecting favourable microclima, such as inside the rolled corn leaves or between the leaf and stem. Should the humidity be higher than usual, aphids may be found also in more exposed places; then its population also increases.

A survey on the fluctuations of the population of this pest in Israel was made by HARPAZ (1953). According to this survey two peaks exist during September and December, and two depressions in March-June and October. The lowest decrease is during the March-June period.

The following are the reasons: The dry desert winds, the so-called Khamseen, occur during these months. Food conditions are not favourable. The winter cereals have matured and begin to dry, while the summer cereals have not yet developed. As mentioned in the case of other aphids, when the host matures and hardens, winged individuals develop which migrate to more succulent plants. The decrease in October is due to the fact that the summer cereals have matured while the decrease after January is due to the low temperature.

LITERATURE

F. S. BODENHEIMER & E. SWIRSKI, 1957. The Aphidoidea of the Middle East. The Weizmann Science Press of Israel, Jerusalem, 378 pp.

I. HARPAZ, 1953. Thesis, Hebrew University.

F. SILVESTRI, 1939. Compendio di Entomologia Applicata, **1**, *472-473*.

E. SWIRSKI, 1959. Notes on Homoptera in Israel (in Hebrew), *Agr. Res. Sta. Rehovot, Bull.* 22, 33 pp.

V. L. WILDERMUTH & E. V. WALTER, 1932. *U.S.D.A. Tech. Bull.* 306.

The European Corn Borer — Pyrausta nubilalis Hb.

(Pyralidae, Lep.)

Little attention had been given to the European corn borer until it was introduced accidentally into the United States at the end of the second decade of this century. In that country it spread, increased and caused injury such as it had never done before.

Many entomologists were employed in the study of its habits, biology and control, and their work has enriched the existing information.

In Europe, and especially in the Near East, this pest was not conspicuous until recently when corn crops increased due to the extension of irrigated fields.

Description. The Adult. The wing span of the adult female is 25 mm. The colours are pale-yellow, yellow, and yellowish-brown in a pattern as in Fig. 62. On the distal third of the fore wing there are two broken lines parallel to the margin. The wings of the male are shorter, its abdomen thinner and longer than that of the female. The wing pattern, too, differs somewhat from that of the female.

The egg is round and flat, about 0.5 mm in diameter. When newly laid, the egg is white and then it turns brown. Eggs are laid usually on the lower side of the leaf in clusters of 15—20 eggs each. They overlap much like roof shingles or fish scales. The appearance and number of egg clusters once served as means of ascertaining the density of the field population.

The mature larva is 18—24 mm long. Its head is dark-brown, and its body cream-white with a purplish tinge. Along the body there

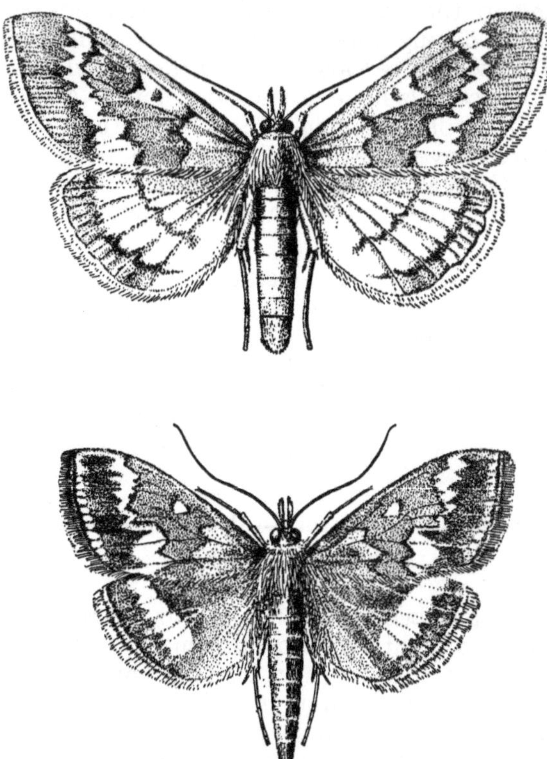

Fig. 62. *Pyrausta nubilalis* female (above) male (below) – after BAKER & BRODLEY U.S.D.A. F.B. 1548.

are faint purplish lines. Each segment of the body has a row of four dark spots and each of these bears a seta. Behind this row two other spots are found in each segment; a brown plate covers the pronotum. The prolegs bear many hooks arranged in a few rows. The pupa is located in a loose cocoon which the larva spins inside a tunnel in the plant. Its colour is brown and it is 12—15 mm long.

Distribution. According to ARBUTHNOT (1949), who gathered data on its distribution in all parts of the world, this pest is found in South and Central Europe, Northwest Asia, Asia Minor and the Caucasus. In America the pest is distributed in the North-eastern and the Great Lake states. To the list of ARBUTHNOT should be added the Southeastern Mediterranean countries. Worth mentioning is the fact that this distribution is not necessarily correlated with distribution of corn. There are countries, usually south of the

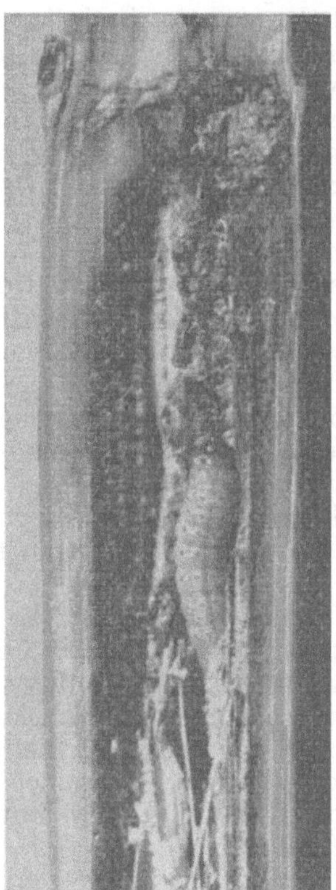

Fig. 63. *Pyrausta nubilalis* – larva in cane of Maize – after Avidov.

Equator, where corn is grown but the corn borer is non-existent. On the other hand, the corn borer occurs in countries where corn is not grown. Thus, in Northwestern Europe the borer lives on plants other than corn.

Hosts. As mentioned above, the corn borer is not limited to corn only, nor does it show any preference to any related group of plants. It seems that the factor controlling host selection is physical rather than olfactory. Plants with thick, succulent stems are preferred for egg laying. Among the cultivated plants which serve as hosts to the borer are, in addition to corn, potatoes, cotton, geranium, zinnia, and gladiola plants.

The larva bores also into the fleshy petioles of beet and rhubarb leaves, as well as into bean pods, tomato and pepper fruit. In wild plants, larvae were found in *Chrysanthemum*, *Artemisia*, *Polygonum*, *Chenopodium*, *Amaranthus* and *Panicum*.

Injury. The main damage is caused to the corn ear. Larvae bore freely into the stem of the plant or its inflorescence, as well as into the midrib of the leaf. Infested parts are easily located by the entrance holes from which faeces are expelled. These drop to the ground or into the leaf axis and expose the infested parts. The boring in the ear renders it unsuitable for market. Boring in the stem weakens it and reduces the yield. Infested inflorescences are easily broken by the wind and they often serve as an indication of the extent of infestation.

The extent of injury depends upon climatic factors and therefore differs every year. Twenty per cent infestation is common. Often 50% of the ears are infested, and on rare occasions the entire crop may be destroyed.

The number of larvae which bore into one ear often varies from 10—20. In a more severe infestation as many as 50 may be found, but there are records where 100 larvae bored into one ear.

Life History. The adults emerge in the spring, or, in cooler territories, in the summer. Thus in the U.S., in the State of Virginia, moths appear in April, in New Jersey in May, in Massachusetts in June, in Maine at the end of June—July. In Israel they appear in April; in Southern France in April, and in Central France in May.

The adults are not active during the day, but hide in debris or the like. At night they fly from plant to plant and lay their egg-clusters — as a rule on the lower side of the leaf.

Egg-laying starts soon after the emergence of the female. On the average one female lays about 400 eggs, but there are records of 2000 eggs laid by one female.

Table XXXI gives the number of eggs laid by 24 females reared in July in Maine. According to HAWKINS & DEVITT (1953), there

Table XXXI.

Length of life and fertility of females of *Pyrausta nubilalis* (after HAWKINS & DEVITT, 1953).

Number of females	Number of eggs laid	Number of egg clusters	Number of females	Length of life in days
5	0	0	3	1— 3
5	1—100	1—12	4	4— 5
3	101—200	3— 6	7	6— 9
5	201—400	13—17	4	10—11
5	401—600	12—18	6	12—14
1	766	33		

is no correlation between number of eggs and egg-clusters laid by one female. It may happen that one female lays only 100 eggs in 12 clusters while another lays 200 eggs in 5 clusters only. On the average each egg cluster contains 15—20 eggs.

Table XXXI presents also the length of life of the female. From this it is evident that in the State of Maine, under the conditions of laboratory breedings, most females died before reaching the age of 10 days. According to others (CAFFREY et al.) female moths lived 10—24 days.

The Egg. The incubation period, depending upon the temperature, may last from 4—14 days. The mortality of the eggs is very great. In Table XXXIII which illustrates the susceptibility of the plant as it ages, it is noticed that under favourable conditions the number of larvae which successfully bore into the plant is only about 25% of the number of eggs. According to this data, over 95% of the eggs and larvae are destroyed. The reasons are: the wind blows many of the crawling larvae off the plant, heat dries the cover of the egg-cluster, and they drop to the ground and are there destroyed.

The Larva and Pupa. Immediately after hatching the larva feeds upon the surface plant tissue. Usually the soft leaves are attacked, and little holes are made. These serve as indication of infestation; when 70% of the plants are thus injured, control application should be made. Afterwards the larva begins to bore. Its length of life lasts 4—8 weeks. During this period it moults 5 or 6 times.

The mode of development after this stage depends upon the race. As will be mentioned below, there are two races of this species — univoltine and multivoltine. In the case of the univoltine race, the larva does not pupate that summer, but enters into a diapause stage which lasts till the following spring when it pupates. On the other hand, the multivoltine larva pupates immediately, develops into an adult and may raise other generations the same year, providing the temperature is suitable. The pupal stage of both races lasts 10—14 days. The multivoltine race continues to develop without an obligatory diapause — only in cold weather its development is retarded — while the univoltine larvae spend part of the summer and winter in diapause; in the spring it awakens, bores a new tunnel into which it pupates, and emerges as adult. In the case of the multivoltine it may happen that larvae late in the autumn do not complete their development but continue this process the following spring.

In Israel mature larvae may be found early in the spring. These are no doubt larvae of the past year, which have resumed their development. Local breedings have shown that the multivoltine race is found in that country.

Number of Generations. The number of generations that the insect raises depends upon the climate of the country. In the U.S., where this pest has been and still is studied extensively, it is known that the multivoltine race raises 2—3 generations annually. Under the same conditions, the univoltine race raises only one generation. ARBUTHNOT (1944) who studied this phase from a genetic point of view claims that the two races are distinct biological races in which

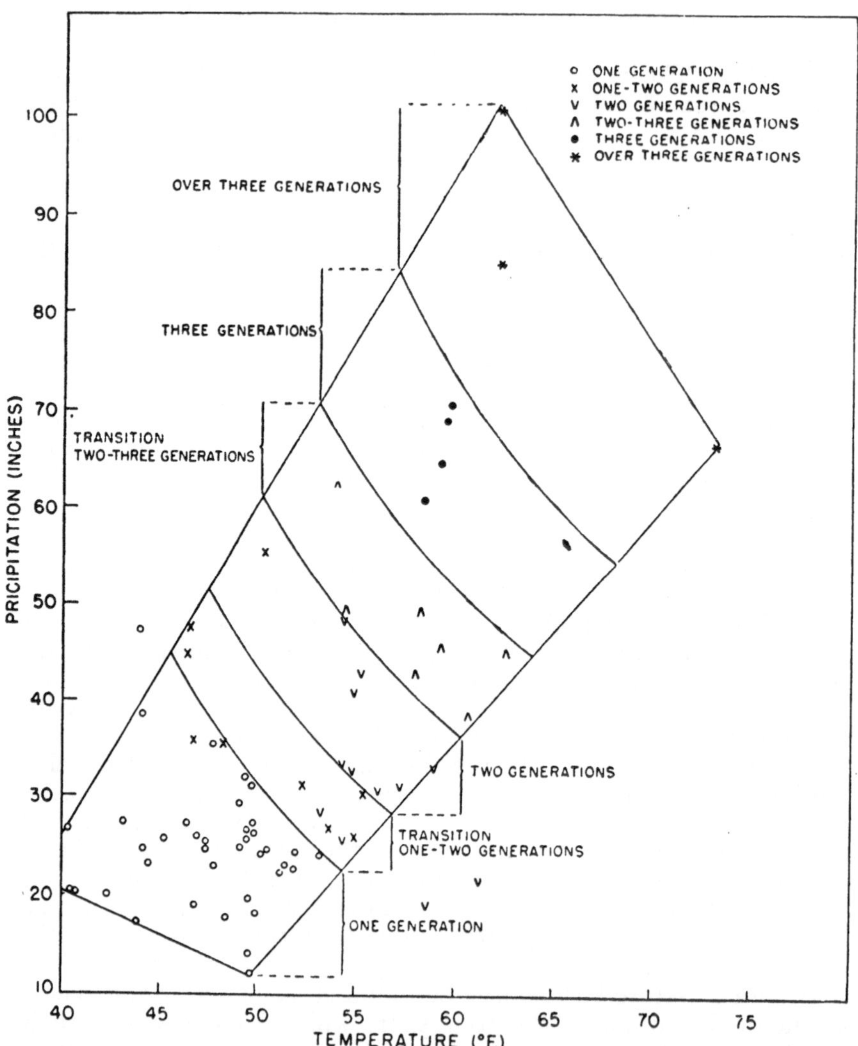

Fig. 64. Effects of Temperature and Rain on the number of generations of *Pyrausta nubilalis* – based on records from various parts of the world – after ARBUTHNOT, 1949.

the main difference lies in the habit of the larvae. In his studies, that author found that some states have a mixed population while other states claim one of the races only. In localities where the mixed population exists he succeeded in isolating only a pure univoltine race. He found, also, that the adults are inclined to mate with the opposite sex of the same race; they do not tend to cross. Especially the males of the univoltine are not inclined to mate with the females of the other race.

In a later work (1949), this author surveys the number of generations in various parts of the world. Having obtained data on rains and the annual average temperature in each locality, he could construct the graph presented in Fig. 64. It is noted therein that the higher the temperature and the more the rainfall, the greater the number of generations.

Limiting Factors. As the pest spread in the United States, efforts were made to find factors which keep the moth populations at low levels in Europe. In an extensive paper on this subject THOMSON & PARKER (1928) published observations which were carried out during six years in France and in the neighbouring countries. These authors speak of three types of factors: *a.* climatic; *b.* agricultural practices; and *c.* biological balance.

a. Climatic factors: According to these authors the high temperature and low relative humidity which prevail during the summer at the time of the moth's flight encourage reproduction. On the other hand, a cool, rainy summer is detrimental to the larvae. The overwintering larvae prefer a cold and dry winter — as a mild and rainy winter is harmful — but in the spring, before the pupation period, rainy weather is preferred. Strong winds also disturb the build-up of a high population. In countries where favourable conditions prevail, 25—50% of the corn plants are attacked.

b. Agricultural practices: In many places in Europe farmers cut the leaves and the inflorescences of the male plants (after the pollination) to feed the livestock. After harvest they also clean the stalks from the fields and use them as fuel. This practice prevents the overwintering of larvae which may be hidden in the stalks.

c. Parasites: In the paper mentioned above parasites of the corn borer are discussed at length. Many parasites of the European corn borer were found in Europe, but not in equal numbers of species and abundance in the various countries. The most important parasites mentioned are as follows: *Masicera senilis* MEIG., *Zenillia roseanae* BEB., and the species of Hymenoptera, *Eulimneria crassifemur* THOMAS, *Dioctes punctaria* ROM., *Apanteles thomsoni* SYLE, and *Microgaster tibialis* NEES. The following facts were established:

1. The abundance of the parasites and their activity depends more upon the climate of the country than upon the abundance of

the European borer, because many of these parasitize other caterpillars.

2. The amount of parasitism is independent of the number of species in that particular territory. For instance, in some places 32% of parasitism had been noted involving a few parasites, while in other places many parasites caused a lesser percentage of parasitism (see Table XXXII).

Table XXXII.

Extent of parasitism of the European corn borer in various countries (after THOMSON & PARKER, 1928).

Locality	Percentage of corn infestation	Number of parasitic species	Percentage of parasitism
France, South	21 —60.00	10	21.66
Rhone, North	1.50— 2.50	6	17.90
Spanish Border	0.08— 0.80	9	16.37
Vausege	8.00	2	17
Belgian Border	24.00—46.00	7	32.65—38.85
Spain	1.7	3	31.9
Italy, Po Valley	21.48	11	17.31—19.53

3. In places which have a great percentage of parasitism, the borer is not considered a serious pest.

Control. In efforts to control the European corn borer direct measures must be considered in order to obtain immediate kill of the pest to save the crop as well as long term measures to prevent a build-up of the pest population.

Direct measures. Before the appearance of the synthetic insecticides, contact insecticides such as nicotine, rotenone and ryania were applied against young neonate larvae crawling on the surface of the leaves. Of the synthetic insecticides DDT 5% dust 5 kg/dunam gave satisfactory control.

The following spray materials were employed: 0.25% nicotine of a 40% solution; 0.1% rotenone; 1% ryania suspension; 0.5% DDT 50% W.P. suspension (250 cc a.i. per dunam).

In view of the fact that the crawling larvae enter the stem near angles or at the axes of the leaves, it is advisable to have the killing agent in larger quantities there. Therefore, application of the insecticides in the form of granules proved to be more efficient. Thus endrin or dieldrin 2% granules at the rate of 100 g a.i. per dunam, or DDT 7.5% at the rate of 350 g per dunam gave better results than when applied in the form of an emulsion (Cox et al., 1956).

The success of the application depends upon the stage of the pest and the stage of growth of the plant. In view of the fact that the application is aimed at the young crawling larvae before they begin to bore, the application should be carried out when the majority of larvae are on the surface of the plant and before they begin to bore into it. Therefore the application should be made when the majority of the eggs have hatched. Because of the fast rate of growth of the plant the application should be repeated 3—4 times.

The Stage of Plant. There is no need to treat a young plant since, at this stage, the mortality of crawling larvae is great even without action of insecticides. The larvae are subject to effects of wind and sun. Furthermore, starting application at such an early date would greatly increase their number.

Applications should begin when the central leaves are rolled and before they open. Table XXXIII shows the mortality of larvae in plants of various ages.

Table XXXIII.

Amount of infestation by the corn borer according to the age of the plant (after HAWKINS & DEVITT 1953).

Stage of development of plant	Height of plant	Survival rate of larvae (percentage) boring into the plant
Before leaves are rolled	10— 30	2.9
Beginning of leaf rolling	15— 50	5— 9
End of leaf rolling	35—100	10
Beginning of inflorescence	70—150	26—28
Appearance of tassel	150—200	26

From this table it is evident that the stage most susceptible to infestation is when the male inflorescence appears. In view of this, the first application in the United States is recommended at the late leaf rolling stage when the entire plant is treated; the second application is given five days later and the stalks only are treated; and two later applications are aimed at the ears only. TURNER (1945) has shown that three applications on the ears alone gave results equal to the four application system as outlined above.

Preventive Measures. These measures are based on the fact that the pest, whether in the stage of a dormant larva or pupa, is located in the stem. It is essential to destroy these immature forms in order to prevent their development into adults. After harvest, stems should be gathered and removed from the field. If these are not used as livestock food, they should be burnt. The field should be plowed in order to turn over the root crowns. BARTLEY & SCOTT (1931) have shown that larvae may emerge from the ground, crawl, seek another stem, and penetrate this. Otherwise they are destroyed by birds or predating insects. Therefore a field must be clean of such remnants in order to prevent the survival of larvae for the next year.

Chilotraea Sp.

(Pyralidae, Lep.)

In the reclaimed area of the Hula Swamps, an attempt was made to grow rice. The crops were very successful but were subject to attack by a species of *Chilotraea*, which, according to TAMS is close to *Ch. bandra* (PLAUT, 1960).

During the first year, according to the growers, the infestation was comparatively low, but in the following years infestation reached 33% (PLAUT, 1960). From casual breedings by PLAUT the following data were obtained:

The adult may live 5—6 days. The pre-oviposition period was 1—2 days; eggs were laid in clusters, each containing many eggs. The maximum number of eggs contained in one cluster was about 200. The average number of eggs laid by one female was 262 (ranging from 96—404). The temperature varied from 17.5—30° C.

The incubation period lasted 4 days at a temperature of 24—34°, while the larva completed its development in the course of 4 weeks. At the end of the summer like other species of this family, the larvae went into hibernation.

LITERATURE

K. D. ARBUTHNOT, 1944. *U.S.D.A. Tech. Bull.* **869,** *1-19.*
K. D. ARBUTHNOT, 1949. *U.S.D.A. Tech. Bull.* **987.**
N. H. BARTLEY & L. B. SCOTT, 1931. *U.S.D.A. Circular* **165,** *1-28.*
D. J. CAFFREY & W. A. BAKER, 1929 and 1941. Later edition by W. A. BAKER & W. G. BRODLEY, *U.S.D.A. Farmers Bulletin* **1548,** *1-45.*
H. C. COX, W. G. LOVELY & T. A. BRINDLEY, 1956. *J. econ. Ent.* **49,** (6), *834-838.*
JOHN H. HAWKINS & J. J. DEVITT, 1953. *Maine Agr. Exp. Sta. Bull.* **522,** *1-26.*
H. N. PLAUT, 1960. *Agr. Res. Sta. Rep.* **288,** *1-4* Rehovoth (in Hebrew).
W. R. THOMSON & H. L. PARKER, 1928. *U.S.D.A. Tech. Bull.* **59,** *1-61.*
NEELY TURNER, 1945. *Conn. Agr. Exp. Sta. Bull.* **495,** *1-43.*

The Sorghum Borer — Sesamia cretica Led.

(Phalaenidae, Lep.)

Until very recently there was confusion when it came to identifying the various species of moths belonging to the genus *Sesamia*, several of which are of great agricultural importance. This lack of certainty gave rise to doubts about their geographical distribution, and there were many divergent opinions as to the species which occur in the Near East. Much light was shed on several problems in this field by the valuable work of TAMS & BOWDEN (1953), who made numerous corrections regarding the identification of species belonging to this and allied genera, and their distribution.

As far as the species found in the Near East is concerned, the revisions may be summarized as follows:

a. *S. vuteria*, on which a great deal has been written in Europe and which causes great damage in the Western Mediterranean basin, is none other than *S. nonagrioides* (the insect's former name) while the true *S. vuteria* is a moth of minor economic importance in Natal Province, South Africa.

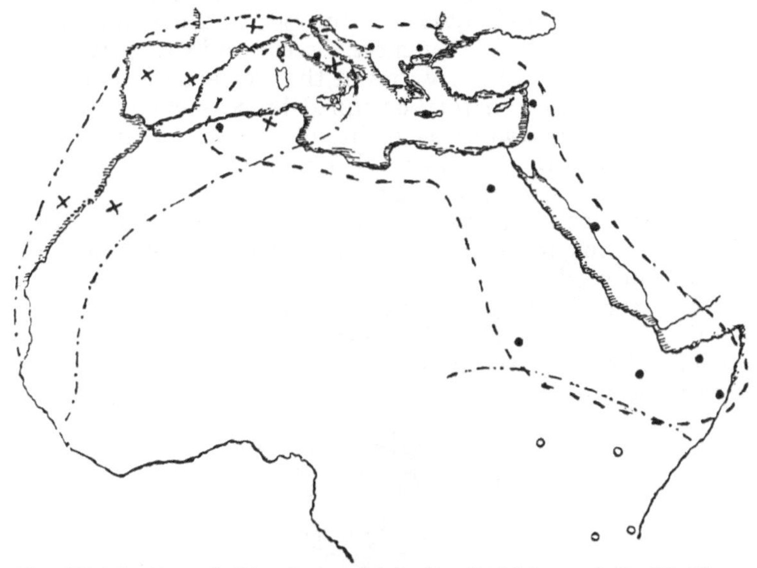

Fig. 65. Distribution of *Sesamia* species in North Africa and the Mediterranean countries – Black dots indicate distribution of *S. cretica*, crosses that of *S. nonagrioides* and circles that of *S. calamistis*.

b. The species prevalent in Mauritius is not *S. vuteria* as mentioned in the literature, but *S. calamistis*, which is common throughout Central and South Africa. A glance at the map proves this opinion to be logical, since Mauritius is close to this part of the continent.

c. *S. cretica* is to be found in East Africa and along the Eastern Mediterranean countries.

d. *S. inferens* is common from India eastwards through the countries of the Indian and Pacific oceans to China and Japan while *S. uniformis*, another Indian species, extends westwards to Afghanistan, Iran, and other West Asian countries.

These two latter species are very similar to *S. cretica* and it is difficult to distinguish between them.

Description. The moth is about 12 mm long, and its wing-span is 26—32 mm. It is entirely cream-white in colour, while the hind wings are white. This species may be distinguished from *S. nona-*

grioides by the fact that the male antenna has a single row of short teeth — while that of the latter species has a double row of longer comb-teeth. The larva is 30—35 mm long. Its colour is pink with longitudinal stripes of a deeper shade. The head is brown, the pronotal plate light like the rest of the body and the spiracles brown.

Distribution. S. cretica is found in Africa. From Somaliland and Ethiopia in the south its distribution extends to Sudan and Egypt in the north, thence westwards to Morocco and north eastwards to Israel, Jordan, Iraq and Syria. In Europe it occurs in Sicily, Corsica and Italy, Yugoslavia, Bulgaria, Greece and Crete. In Asia it is found also in Arabia. In Italy and in the western section of North Africa it invades the territory occupied by *S. nonagrioides*.

The Hosts. The *Sesamia* species in general attack and bore into graminaceous or other monocotyledonous plants. They mostly prefer sorghum, maize and sugar-cane, Sudan grass etc. In those countries where small grains are grown in the summer, these are attacked too. Thus, in the Far East, oats, wheat (chiefly hard wheat) and rice are injured.

S. cretica attacks sorghum, wintersome, maize, sugar-cane, etc. Some think that bamboo and date-palms may also be subject to infestation. However, the major damage is caused to the various sorghum varieties.

Fig. 66. Larva of *S. cretica* in cane of Maize – after AVIDOV

Type of Injury. The degree and type of damage depends upon the age of the plant. In young plants which have only 4—6 blades, the larva of the first generation bores into the centre of the plant and gnaws at the base of the central leaf, causing it to wilt. As a result, small lateral shoots develop into a broom-like growth in the place of a one-stem healthy plant. This brings about a retardation in the formation of the spike. When plants are attacked later by second generation larvae, close to the time of formation of the spike, its growth may be halted altogether. When a plant is large, a larva may bore in the stem without interfering with its development. Such cases do not go unnoticed. The opening of the tunnel is seen in the stem. Later on, second and third generation larvae may prey upon the inflorescences and spikes. Infested plants may be recognized by holes in the leaves. The larvae feed on the folded central leaf. As this is unrolled, rows of four holes are left on the blade. Thus the signs of infestation by *Sesamia* are: 1. a wilted central blade; 2. rows of 4 "windows" in the foliage; 3. a hole in the stem from which sap oozes quite often, and (4) undeveloped spikes or injured inflorescence.

The degree of infestation varies greatly. 5—10% infested plants is common. Severe infestation causes about 40—50% injury, whilst in rare cases, up to 90% of plants may be attacked.

Life Cycle. The Adult appears in Israel at the end of April or beginning of May. The female begins laying the day after its night-emergence. In the laboratory the pre-oviposition period was very brief, lasting one or two days only. The number of eggs laid by one female varies with the different generations. Five females of the first generation which emerged in June 1954 laid an average of 292 eggs, while thirteen of the second generation females emerging in September of the same year laid an average of 141 eggs each (see Table XXXIV).

Table XXXIV.

Oviposition by *S. cretica*

Number of eggs/female	Females of June gener.	Females of Sept. gener.	Number of eggs	June gener.	September generation
50—100	0	3	maximum	350	300
101—200	0	8	minimum	239	53
201—300	3	2	average	292	141
301—350	2	0			

The life-span also varied greatly in the different generations. The five June females lived an average of eight days, whilst the thirteen September females lived an average of five.

During the day, the adults hide among the leaves of the host, while they lay their eggs at night usually in concealed places —

between the leaf-sheath and the stem, for example, or in the leaf-folds.

The Egg: In the laboratory, the eggs were laid in groups upon the sides of the glass containers in which the insects were reared. The incubation period at 25—26% was 4—8 days.

The Larva: In the field, the young larva at first feeds upon the tender folded leaves of the host and only several nights later does it begin to burrow into the plant or stem. In the laboratory, the larva was raised upon folded leaves or upon whole sorghum plants. The period of development lasted several weeks, as follows: larvae which hatched at the end of July or the beginning of August completed their development within 6—7 weeks and pupated in the middle or at the end of September. The average developmental period was 46 days at a temperature of 27—28° C.

Diapause. The larvae which completed their development in the middle of the summer up to the end of September pupated in the same summer. Most of the adults which emerged from these pupae laid the same season as long as the temperature permitted. However, larvae which completed their development from the end of September to November entered a state of diapause. They remained in this state within the stem during most part of the winter, and pupated only in the spring. Before pupating, the larva prepares an opening through which the adult emerges afterwards.

The Pupa. The duration of the pupal period obtained from larvae collected in the field is presented in Table XXXV. It may be observed, that from 1—4 weeks, depending upon the temperature, were necessary to complete development.

Table XXXV.

Pupation period in days of *Sesamia* at various degrees of temperature

Number of pupae	Generation	Date of rearing	Development period	Average period	Temperatures °C
10	Winter	April-May	13—28	21.2	21 —22
7	First(Spring)	June	11—12	11	25.5—26
14	Second (Summer)	Aug.-Nov.	7—11	9.5	26.5—27

Annual Cycle. According to BALACHOWSKY & MESNIL (1936), the species *S. nonagrioides* produces two generations during the year in France. The same is the case in Spain. According to D'HERCULAIS (1896), *S. cretica* produces two generations in Yugoslavia. In Israel, there are three generations as the following survey shows. In Naan in 1955 — early in April — no pupa found; latter half of April —

pupa present; last week of April — empty pupal cases found, which showed that adults had begun to emerge. The offspring of these matured in June-July; the second generation matured in August-September. The offspring of these went into diapause and matured the following April.

Limiting Factors. The eggs are laid in groups of 12 or more, within leaf folds. The fact that within each plant only one or at the most two or three larvae may be found, indicates a high rate of mortality among eggs and larvae. During July and August infestation is at a low level and it increases again in September and October when the third generation larvae become active.

The large larvae which spend the winter in the field are affected by climatic factors. Low temperature apparently affects larvae very much. GENIEYS (1923) states that the larvae of *S. nonagrioides* die at a temperature of 7° C. It may be assumed that *S. cretica*, which originates from a warmer region, is also sensitive to the cold. Therefore it may be assumed that a continually cold winter diminishes its larval population.

Control. Preventive measures: The biology of this moth with reference to its overwintering methods is very much like that of the European corn borer, and the methods recommended for combatting that pest should be applied in this case too.

Chemical Measures. Chemical control methods may be used to kill the adults. For the time being no satisfactory control has been obtained with any of the insecticides, unless several applications are made in succession. Applications directed at the larvae also yielded only partial results.

The Armyworm — Pseudaletia unipuncta Haw.

[Generic syn.: *Leucania; Cirphis*]
[Specific syn.: *extranea*]

This moth is a notorious cosmopolitan pest, distributed in many parts of the world. But not everywhere does it cause as much damage as in the U.S., for instance, and not every year does it appear in outbreaks of large numbers. Noteworthy is the fact that it rarely causes damage in the Near East, although its flights are recorded in some localities of that region.

Description. The adult is a moth 15—18 mm long, with a wing-span of about 35—40 mm. Its colour is brownish-grey and on the center of the front wings there is a white spot, surrounded by an area which is darker than the rest of the wing. A darker, slightly curved line runs from the anterior angle to the hind margin of the wing. The hind wings are darker around the hind borders.

The eggs are globular, slightly flat, over 0.5 mm in diameter, smooth, and resemble small white beads. They are laid in clusters of a few dozen each.

The larva, when mature, is greenish with darker bands, two lateral and one median, running along its dorsum. The latter is divided into two by a broken, light-coloured line. Just below the line of spiracles a cream-coloured band runs on each side along the length of the body. The head is conspicuous and broader than the prothorax. The plate over this segment is shiny yellow, and the head is greenish brown, speckled with black.

The pupa is dark brown, changing to black before the moth emerges. It is 20 mm long and has four minute spines at the end of the abdomen.

Distribution. This is a cosmopolitan insect. In America, probably its native continent, it occurs from Canada to Argentina. In Asia and Australasia it has been recorded in Afghanistan, India, Java, New Guinea, the Philippine Islands, Formosa, China, Japan and Sakhalin. In the Pacific it has been recorded in the Fiji Islands and Hawaii, in Europe in France, Portugal and England, and in the Atlantic, in Madeira and Bermuda. In Africa it occurs only in West Morocco, but it is not found in North Africa, according to BALACHOWSKY & MESNIL (1936), nor in Iraq, according to WILTSHIRE. In view of this, its occurrence in Israel is of interest.

Hosts. This insect feeds upon grasses of all kinds and grasslike grains; millet wheat, barley, oats, rye, rice, corn and sorghum may serve as food. When these are not available it may feed upon other plants: tobacco, tomato, artichoke, flax, beet, etc. have all been attacked.

Type of Injury. The insect feeds upon the foliage and may destroy the entire plant. However, it is not equally abundant every year. In certain years it may appear in outbreaks, as happened in the United States and Canada in 1914, when thousands of acres of grass, grain and corn were entirely exterminated. In the various accounts of such outbreaks we find expressions like "armies of the worm" and "meadows have been rendered as barren as a heath".

In the Near East, however, damage is rare. Although the moth is on the wing in Israel several months during the year, damage is not common. There is one record of damage to a corn field in Kfar Vitkin along the coastal plain in June 1959.

Life History. The life history of this moth is very much like that of *Laphygma exigua*. According to WALTON (1916) it takes from seven to eight weeks for the insect to develop from egg to adult in summer conditions of the middle Atlantic states.

The incubation period is about 4—5 days in the summer, or 9—12 days in the spring or autumn of central New York (KNIGHT, 1916). According to the latter, the larva develops in 30 days in the summer and the pupa in 13 days.

In the autumn the pre-oviposition period lasts 12 days. One female may lay a few hundred eggs.

WILLIAMS considers this moth a "definite migrant", in which case there is no need for further evidence. This view has been expressed as early as 1916 by WALTON, who stated that the moth may fly in large flocks for many miles "in the direction with the prevailing winds" and that this acounts for the fact that outbreaks may occur in regions far removed from any source of infestation. But, he adds, the moth seems to be present in small numbers over most of the areas in which it occurs during a part of every year. This, however, does not account for its sudden outbreaks.

Hyphilare loreyi Dup.

[Gen. Syn. *Leucania, Mythimna*]
(Phalaenidae, Lep.)

This moth attacks summer cereals in the same manner as *Sesamia.* Because of the similarity in the type of damage caused by these two moths, they are often mistaken for one another. It is therefore difficult to evaluate exactly the damage and per cent of infestation caused by each of them.

Description. The adult is 15—17 mm in length and its wingspread is 33—38 mm. It is bright yellow in colour. The forewings are yellowish-brown with a brown triangle on the rim of each and is connected to a brown line which passes along the wing to its base. The hindwings are white and only their border is dark.

The neonate larva is 3 mm long and of a translucent white colour. Its head, pronotum and last abdominal segments are dark; as the larva grows, all these parts become lighter; its head turns greenish-brown and spotted. The larva reaches a length of 25—35 mm and its colour before pupating is yellowish-pink with thin grey-pink stripes along its body. On both sides of its body, above the spiracles, is a faint grey line. Along the middle of the dorsal side is a darker one and on both sides of it is a faint yellow line and a broken brown one. The bases of the setae are brown and it seems as if the body is covered with dark spots.

Distribution. This pest is found in many countries: in Southwest Asia, Israel, Australia, the eastern part of Africa and in Italy.

Hosts. In the Mediterranean countries this moth attacks summer cereals such as corn, sorghum, sugar-cane and rice. It is also found on wheat and its damage to onions was once reported.

Type of Injury. The larva bores into the "heart" of the plant and is usually found in the rolled central leaf. The moist faeces which it leaves in its tunnel generally reveals its presence. In case the plant is attacked during the formation of spikes the larva damages the tender kernels.

Life History. The biology of this moth in this region is still unknown. From field observations, it can be surmised that its population in Israel reaches its peak during July and August.

Limiting Factors. The summer heat is not a limiting factor for this pest. There is, however, a biotic factor which reduces its population. Various countries have reported parasites which destroy the larvae of this insect. In Israel, too, many larvae were found to be parasitized by wasps.

LITERATURE

A. Balachowsky & L. Mesnil, 1936. Les Insectes Nuisibles aux Plantes Cultivées, Paris, *1643-1641.*

Genieys, 1923. *Rev. Zool. Agr. Appl.* **XXII,** *314.*

K. D'Herculais, 1896. *C. R. Acad. Sci. 824.*

H. H. Knight, 1916. *Cornell Univ. Agr. Exp. Sta. Bull.* **376,** *751-765.*

E. Rivnay, 1960. Unpublished Notes.

W. H. T. Tams & J. Bowden, 1953. *Bull. ent. Res.* **43,** *645-648.*

W. R. Walton, 1916. *U.S.D.A. F.B.* **731,** *1-12.*

C. B. Williams, 1958. Insect migration, The New Naturalist Library **36,** 235 pp.

E. P. Wiltshire, 1957. Lepidoptera of Iraq, Govt. of Iraq, 162 pp.

Atherigona excisa Thoms.

(Anthomyiidae, Dipt.)

From the literature dealing with this genus, it is evident that in several countries it appeared as a pest only during the past few years. In Ceylon, for example, it was found to be a pest of rice and in Tanganyika of sorghum. Until 1933—1934, damage due to this pest was not known in Israel. Since then it spread and multiplied greatly and today it is a serious pest of sorghum.

Description. The adult is a small fly, 3—3.5 mm in length. Its thorax is 1 mm wide and its wings are 3 mm long. Its head is silver-grey with a brown frons, black eyes and dark brown antennae.

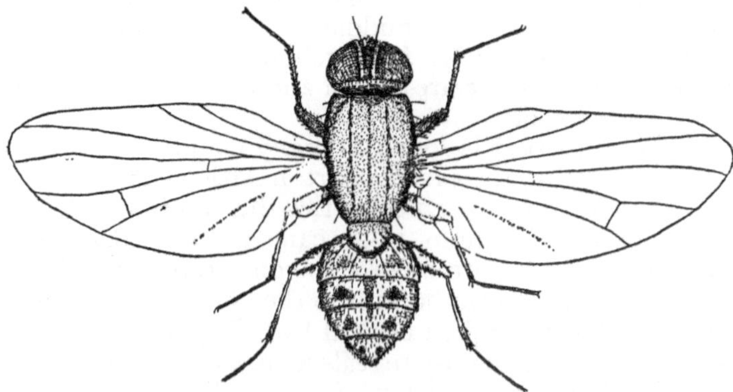

Fig. 67. *Atherigona excisa.* Courtesy Hassadeh.

The thorax is grey, the abdomen is yellow and the tergite of each of its segments bears a pair of dark spots. The spots of the first segment are less conspicuous and sometimes are absent. The fore-legs are brown while the middle and hind legs are yellow with a darker tarsus.

The mature maggot is 7—8 mm long and 1 mm wide. The anterior end of its body is tapered and at its tip is a small and pointed head of which only the black mouth parts are prominent. On both sides of its head are the anterior spiracles which are arranged like a two-lobed fan with four or five "beads" in each lobe. The posterior side is obtuse and on the caudal segment are two small projections bearing the posterior spiracles.

The pupal case is cylindrical, 5 mm long, 2.5 mm wide and reddish-brown in colour.

Distribution. The flies of the genus *Atherigona* are considered pests in India, Southeast Asia, Australia, Hawaii, Tanganyika, North Africa and Israel. In America they were recorded in Brazil.

Hosts. In the laboratory in Rehovot, *Atherigona* were raised on rotting fruits of guava, feijoa, avocado and melon. The flies oviposit freely in these fruits and the larvae develop easily in them. They also attack healthy melons in the fields. The principal damage was caused by their attack on young sorghum plants. From literature it is known that in various countries the larvae damage sorghum, rice, wheat, sugar-cane, *Pennisetum*, *Tapioca*, etc. In Israel *Atherigona* also causes damage to *Pennisilaria*, Sudan grass, corn and lemon grass.

Type of Injury. In pennisilaria, sorghum, corn and Sudan grass the central leaf withers approximately 10—14 days after germination and finally dries up. The leaf can be removed with a light pull and its base is moist and rotten with a foul odour emanating from it. The foul odour indicates that microorganisms share in the destruction of the plant tissue. The maggot responsible for this injury may be removed together with the leaf. It is obvious that a stem, damaged in this manner, does not continue to develop. The death of the main stem causes the sprouting of a dozen or more lateral ones. The many stems form a "broom" and this rarely is able to overcome the others and produce grain. The infestation sometimes reaches great proportions, up to as much as 80—90 percent of the plants.

Life History. For the time being, the details of the life history of this fly and the number of generations it raises in this region in a year are not known.

The female oviposits in the plant and the hatching maggot reaches the crown close to the roots. The way it eats is similar to that of other maggots i.e. the scraping of the plant tissue and the imbibing of its liquids.

Among the various species of *Atherigona*, the biology of *A.*

soccata ROND. was studied in Morocco. According to BLETON & FIEUZET (1943), the incubation period of this species lasts 6—8 days, the larval stage 30—40 days and the pupal period 30 days. In total, the fly raises approximately four generations during the summer. VAN DER LAAN (1951), who studied the species of *A. exigua* in Java claims that the pre-oviposition period lasts 7 days, the incubation period 2 days, the larval period 17, while the pupal period 7 days. In other words, the total development in Java is about 33 days. The pupae of *A. excisa* completed their development in the laboratory in Rehovot during 6—7 days at a temperature of 27—28° C. It can be surmised therefore that *A. excisa* is similar to *A. exigua.* In fact the succession of generations as observed in the field confirms this view.

Encouraging Factors. In Israel, favourable conditions for the development of a large population did not exist until farmers began to grow summer cereals under irrigation in large areas and in a continuity which covered the entire summer. For this reasons the supply of food of the *Atherigona* increased suddenly and allowed this fly to raise generation after generation without pause.

Control. BLETON & FIEUZET (1943) from Morocco suggest plowing the infested field in order to destroy the pupae which are in the soil. HARRIS (1934) from Tanganyika states that treatment with rotenone gave desirable results. VAN DER LAAN (1951) claims to have achieved good results by dusting the field with BHC every day for a whole week.

In experiments conducted in Israel (RIVNAY & ERHLICH, 1957), good results were not achieved from dusting with DDT and dieldrin when applied three times in succession. Spraying with malathion also failed: however, the spraying with dieldrin at approximately 100 g per dunam in each treatment, gave encouraging results.

Further studies of RIVNAY showed that it is essential to treat the plants very early, as the infestation begins immediately after the plants sprout.

LITERATURE

C. A. BLETON & L. FIEUZET, 1943. *Bull. Soc. Hist. Nat. Afr.* **34,** 1-6, *112-117.*

W. V. HARRIS, 1934. Report of the Assistant Ent. Dept. Agr. Tanganyika 1934, *84-89.*

P. A. VAN DER LAAN, 1951. Gen. Agr. Res. Station, Ministry of Agr. Indonesia, **118,** *1-15.*

RAMACHANDRA, 1924. *Rep. Proc. 5th Ent. Meeting Pusa, 19-22.*

E. RIVNAY & S. EHRLICH, 1957. *'Hassadeh'* **37,** *19-22* (in Hebrew).

E. RIVNAY, 1960. Unpublished Notes.

LEGUMINOUS FODDER CROPS

KEY to Insects Injurious to Leguminous Crops

A. Seed injury
1. Seed maggots are found in the seed . . . . . : *H. cilicrura* page 131
2. The seed is eaten entirely . . . . : Tenebrionidae page 138

B. Root injury
1. The roots are gnawed or bored by small white legless grubs . . . . : *Sitona, Apion* page 216, 228
2. Long brown larvae infest the roots . . . . . : Tenebrionidae Elateridae page 138, 139
3. Small globular soft insects are found on the roots . . . . : *Trifidaphis* page 71

C. Root-crown injury
1. The base of the stem is gnawed by caterpillars found near the plant in the ground . . . . . : *Agrotis* page 88
2. The agents of the injury are not found — they may be adult. . . . . : Tenebrionidae page 138
3. In the ground near the plant tunnels of about 15 mm in diameter are made . . . . : *Gryllotalpa* page 44
4. Maggots bore in the stem (in beans) . . . . . : *Melanagromyza* page 239
5. Tiny grubs bore in the stem (in alfalafa, clover): *Apion* page 228

D. Foliage injury by caterpillars
1. Green tapering caterpillars having only 3 pairs of abdominal legs . . . . : *Plusia* group page 125
2. Dark-brown, or dark-olive-coloured larvae with darker longitudinal stripes and black spots near thorax and end of abdomen . . . . : *Prodenia* page 102
3. Green or dark-green larvae with lighter stripes: *Laphygma* page 113
4. Green larvae with dark stripes . . . . : *Heliothis* page 119
5. Dark-green larvae with a velvety appearance, and yellow lines along their sides . . . . . . : *Colias* page 233

E. Leaf injury incurred by insects other than caterpillars
1. The leaves (of clover or vetch) are eaten by small beetles which, as a rule, are found near the plants or upon them . . . . : *Marseulia* page 154
2. Semi-circular sections are cut in the margins of the leaves . . . . : *Sitona* page 216
3. Small holes are made in the leaves (of alfalfa, clover or vetch) . . . . : *Hypera, Apion* page 222, 228
4. In the leaves (alfalfa) holes are made in which the epidermis is left . . . . : *Smynthurus* page 206

F. The leaves are not eaten, but injured by sucking and scratching
1. White silvery spots are seen on the leaves, often concave in shape (in peanuts) . . . . : *Thrips* page 74
2. White spots are formed on the leaves; in a pocket between the two epidermis layers small maggots are found . . . . : Agromyzidae page 238

3. The margins of the leaves (alfalfa and vetch) are folded over to make a pocket in which several pale maggots are found . . . . :	Cecidomyiidae	page 240
4. On the leaves or soft stems many spherical soft-bodied insects are found . . . . :	Aphididae	page 160
5. Leaves which become pale, mottled and covered with spider web, have tiny mites on them . . :	Acarina	page 80
G. Inflorescence and flowers injured		
1. The top leaves (in alfalfa) are attached together with webs enclosing a brown little caterpillar . :	*Cnephasia*	page 236
2. Between the top leaves and inflorescence small green larvae are found with a white median line over the body . . . . :	*Hypera*	page 222
3. Minute white grubs bore into the side buds (in vetch) or into the flower buds (in horse bean) . :	*Apion*	page 228
4. Minute elongate insects are found in the flowers:	*Thrips*	page 74
5. In the flowers (of peas) small flat caterpillars are found — or flowers drop without making pods . . . . :	Lycaenidae	page 234
H. Pods and seed injury		
1. In the seed (of alfalfa and clover) minute grubs of Hymenoptera are found . . . . :	*Bruchophagus*	page 231
2. Small grubs or beetles exist in the seeds . . . :	Bruchidae	page 212
3. In the pod (of beans) a small purple-green caterpillar may be present . . . . :	*Etiella*	page 237
4. In the pod (of peas or beans) a yellowish-white black-headed caterpillar can be found . . . . :	*Laspeyresia*	page 236
5. In the pod of chick peas, peas, or beans larger, light and dark striped caterpillars may be found:	*Heliothis*	page 119
6. Short, flat, greenish caterpillar may be present in the pod . . . . :	Lycaenidae	page 234
7. Large green bugs may be found on the flowers and on green pods . . . . :	*Nezara*	page 49

The Springtails — Collembola
(Smynthuridae, Apterygota)

The springtails or Collembola are, as a rule, very small insects of the most primitive groups included in the Apterygota. They are characterized by a muscular forked appendage at the end of the abdomen which serves as a spring. As a rule members of this group are found in damp places among decaying vegetable matter. A few species, like the one discussed here, are injurious pests.

The Lucern Springtail — Smynthurus viridis L.

Alfalfa and clover are often inhabited by a small springtail which eats holes in the leaves, leaving only the upper transparent epidermis. This species is *Smynthurus viridis*. In badly infested fields in Israel a few dozen individuals of this species may be found on one plant.

Description. The adult is a soft-bodied insect, 2—2.5 mm long, wingless, elliptically shaped and tapering posteriorly. The head is comparatively large, with eyes which are situated posteriorly and four-segmented antennae. Its colour in Europe and Australia is greenish-yellow but in Israel it is yellowish-brown with darker spots scattered over the surface of the body. At the end of the abdomen there is a bi-forked tail, bent ventrad, which aids in jumping. The larvae look like their parents except that they are smaller and their appendages are comparatively thicker.

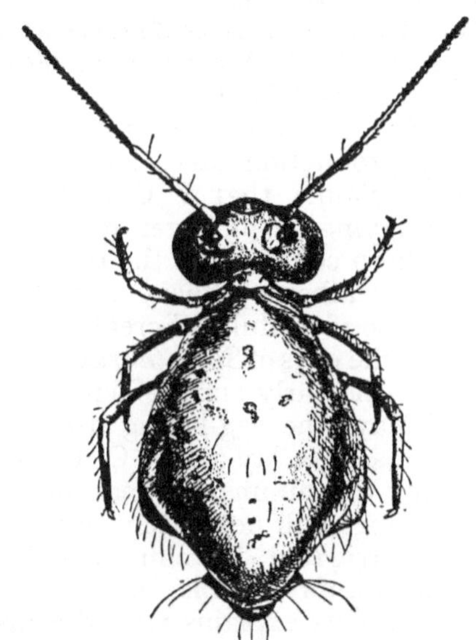

Fig. 68. *Smynthurus viridis* after SWAN. Courtesy Department of Agriculture, Adelaide.

Distribution. This species is distributed in Europe, the Mediterranean region and Australia. In European countries its damage is greatest where summer rains are abundant. In Australia it is most injurious in humid regions. In the Mediterranean countries it is not conspicuous.

Host and Type of Injury. Although the insect may be found on various plants, it favours leguminous crops, particularly clover and alfalfa. Its damage consists of gnawing holes in the leaves and thereby reducing the value of the fodder.

Life History. As in the case with other Apterygota, the development is ametabolic. The female lays her eggs in clusters in the soil and covers them with excretions from the anus mixed with earth particles. This envelope guards the egg against predators and protects them from drying. On the average each cluster contains about 60 eggs and each female lays two such clusters at a 10 day interval.

At uniform and favourable conditions of humidity the incubation period lasts 8 days at 25° C or 11—12 days at 20° C. High relative humidity is essential for egg development. If the egg is removed from a high humidity to a lower level, development ceases. It is resumed when favourable humidity is restored.

The percentage of egg-mortality under these circumstances depends upon the age of the eggs; it is high with both freshly laid eggs and with those near hatching.

The development of the larva lasts 23 days at 20° C, 20 days at 17° C, and 38 days at 13° C. According to McLagan (1932) the threshold of development is 2.65° C.

Factors Affecting the Population. According to Davidson (1934) the highest rate of reproduction takes place at a temperature of 12—15° C. McLagan claims that 7° C is the most favourable temperature for egg-laying. Whichever it is, in the Near East, especially in the southern countries of the region, these temperatures prevail during December, January and February. In these months the rains are also plentiful and create favourable conditions for the development of the eggs of *Smynthurus*. This explains why the species is abundant in the early spring. With the approach of the arid season in April-May, the egg-laying decreases. According to Davidson the insect ceases to lay at 25° C. Therefore no reproduction takes place during the summer. It may be surmised that eggs found in dry surroundings will not hatch until the following rainy season unless the soil is irrigated. The climate of the southern section of the Near East is also unfavourable for the larvae which require high humidity as well. This explains why the insect is not found there in the summer.

This insect is more prevalent in heavier soils since, in sandy soil, moisture conditions are neither stable nor favourable for the eggs.

Control. The application of insecticides such as DDT, lindane, toxaphene or parathion, kills this pest satisfactorily. As a rule, applications against other pests control this insect too. In the Near East, with the onset of the drier weather early in the spring, control measures against this insect are superfluous, as it gradually disappears from the crop.

LITERATURE

J. Davidson, 1934. *Comm. Sci. and Ind. Res. Bull.* No. **79,** *1-66.*

J. Davis, 1928. *Bull. ent. Res.* **XVIII,** *291-6.*

D. S. McLagan, 1932. *Bull. ent. Res.* **23,** *151-190.*

D. C. Swan, *Dept. Agr. So. Aust. Bull.* **353,** *1-12.*

Plant Lice (Aphidoidea)

The Spotted Alfalfa Aphid — Therioaphis maculata Buck.

[Syn.: *Myzocallis ononidis* KALT.]
(Rhynchota)

This aphid is found in alfalfa fields without necessarily causing serious damage. As a rule its population in the Mediterranean countries is sparse: only on certain occasions does it break out in extremely dense populations causing serious trouble.

Description. The agamic apterate female is 1.5—1.7 mm long and elliptical in shape. Its colour is green with a yellow head and thorax, dark antennae and a yellow third segment. The upper side of the abdomen has 6 rows of tubercles, each bearing setae. The tubercles and the areas around them are brownish. The alate is 1.6—2.4 mm long, and the antennae are longer than the body.

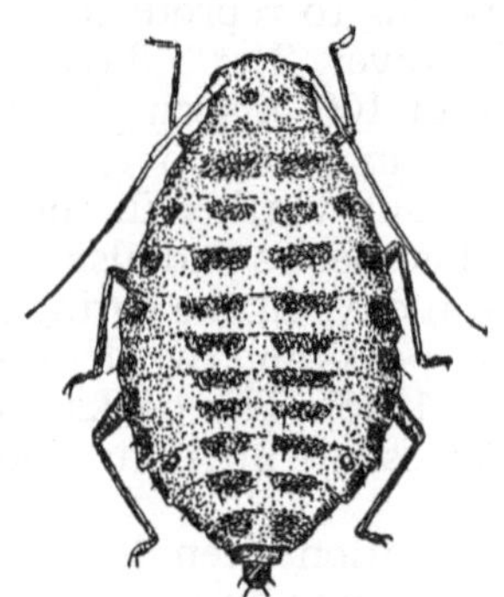

Fig. 69. *Therioaphis maculata.*

Distribution. This species occurs in Europe, North Africa, the Near East and India. It is found also in America, and during the past two decades it has become of extreme economic importance there.

Hosts. The species is limited to leguminous plants. It prefers alfalfa, but not all varieties to an equal extent. It is found also on clover, *Ononis*, and others.

Type of Injury. Like other aphids, the sucking of the juices by thousands of individuals weakens the plant. In the case of this species, the excretions of honey dew are very prolific and troublesome. The plants become sticky, fumagine develops, and the entire crop loses its value.

Life History. BALACHOWSKY & MESNIL (1936) state that during October sexual females lay eggs on the stalks of the plants. These eggs overwinter and hatch towards the spring. In the Mediterranean region, in spite of a careful search (HARPAZ, 1955), no sexual females were found. Only parthenogenetic reproduction goes on all year round, as long as the temperature permits. In California a similar situation exists, although DICKSON et al. (1955) found sexual females. In a later study on the phenology of this species in the U.S. (DICKSON et al. 1958) it was found that *Th. maculata* sexual females lay non-viable eggs. It was found that the egg-laying females are more abundant in the warmer parts of its distribution range. It was found also that the number of eggs laid is very small — 137 eggs per 763 females. The non-viable eggs, and the fact that agamic females do not survive lower temperatures leads to the extermination of this species in the colder range of its distribution. The authors believe that the species is in the process of losing its sexual forms since they are produced erratically, in the wrong location, and the eggs are non-viable. In Europe, however, BALACHOWSKY & MESNIL claim that winter eggs hatch in the spring.

The agamic female begins to reproduce very soon after the last moult. In the winter, however, 2 or 3 days lapse before the female reproduces. An average of 100 offspring may be produced by one female, but fertile females may bear over 150 young.

Ecological Notes. The cold weather in the Mediterranean winter reduces the activity of this species and also its reproduction. On the other hand, the khamseen days in the spring kill off great numbers. Therefore, there are three peaks and two depressions in its population curve in Israel (HARPAZ, 1955). The peaks are in March, June and November, and the depressions in August-September and December-January.

The density of the population often depends upon the prevalence of parasites. Lady beetles feed upon this species and many parasites especially the genus *Praon*, parasitize it. It is possible that the satisfactory situation of this species in the Near East is due to these factors.

Control. This aphid was satisfactorily controlled by applications of parathion and malathion. However, recent studies have shown that this species developed resistance to these insecticides. STERN & REYNOLDS (1958) showed that the systemic insecticides are the most efficient. Outstanding are phosdrin and demeton. The latter is also the least harmful to *Hippodamia* beetles.

The Pea Aphid — Acyrthosiphon pisum (Harris).

[Syn.: *Aphis pisi* KALT., *Acyrthosiphon onobrychis* B. & F., *A. lathyri* MOSLEY, *Illinoia pisi* KALT.]

This is the largest of the aphids which inhabit leguminous crops. It is easily identified by its size, its long legs and slow movements.

Description. The agamic apterate is 4.5—5.5 mm long, its colour is bright green, and the length of its antennae is about 4 mm. The first two segments are light and the third to sixth segments are darker. The legs are long and green, except for the tips of the femora, tibia and tarsi which are darker. The cornicles are narrow and about ⅓ of the length of the body. The cauda is 0.75 mm long and tapers towards the end.

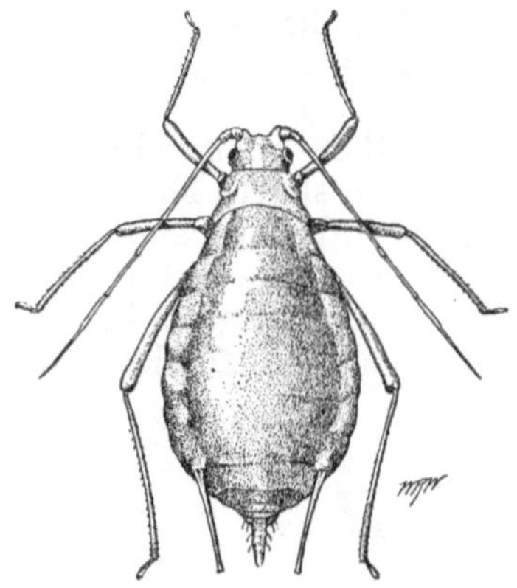

Fig. 70. *Acyrthosiphon pisum*. Courtesy U.S.D.A.

Distribution. This species is found in Central and Southern Europe, North Africa, North America, Central and Eastern Asia. In the Near East it was recorded from Egypt, Jordan, Israel, Lebanon, Syria, Iraq and Turkey.

Host and Injury. This species is limited to leguminous plants only, and feeds in particular on the annual plants of this family. In the Near East its damage is inconspicuous.

Life History. In Central Europe, in the countries of colder climate, sexual forms appear towards the end of the summer and lay eggs on the stems of various leguminous plants. These hatch in the spring, and throughout the summer asexual generations develop. The alates form in May and begin to infest other secondary hosts of the same family.

In countries with a warmer climate, like the southern Mediterranean countries, no sexual forms are found. This species produces continuously and parthenogenetically all year round. The peak of its population is reached in Iraq in Febraury-March, in Israel in March-April, and in Turkey in June-November (Bodenheimer & Swirski, 1958).

LITERATURE

A. Balachowsky & L. Mesnil, 1936. Les Insectes Nuisibles aux Plantes Cultivées, Paris. pp. *1274* and *1286*.

F. S. Bodenheimer & E. Swirski, 1957. The Aphidoidea of the Middle East. The Weizmann Science Press of Israel, Jerusalem, 378 pp.

R. C. Dickson, E. F. Laird & G. R. Pasho, 1955. *Hilgardia* **24** (5), *93-188.*
R. C. Dickson, E. F. Laird & M. M. Johnson, 1958. *Ann. ent. Soc. Amer.* **51,** *346-350.*
I. Harpaz, 1955. *J. econ. Ent.* **48,** *668-671.*
F. Silvestri, 1939, Compendio di Entomologia applicata, Portici. pp. *428* and *524.*
V. M. Stern & H. T. Reynolds, 1958. *J. econ. Ent.* **51,** *312-316.*

Bruchidae (Col.)

The Pea Weevil — Bruchus pisorum L.

Description. The adult is 3.5—5.5 mm long. The prothorax is wider than long and the elytra are short. The colour of the body is dark brown, mottled with grey and white spots. On the prothorax there is a white triangle at its base. The last abdominal segment, the so-called pygidium, is covered with grey hair except for two brown spots. The antennae are black and the legs black except for the tips which are reddish.

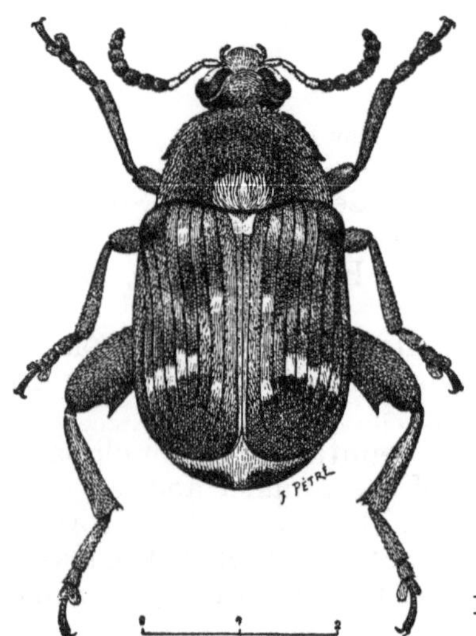

Fig. 72. *Bruchus pisorum* – larva.

Fig. 71. *Bruchus pisorum* – adult after Balachovsky & Mesnil.

The egg is yellowish, oblong, 1.5 mm long and 0.6 mm wide. It is laid on the pod, and is covered with a white translucent mucous, which hardens and serves as a protective cover.

The neonate larva has three pairs of legs, and is capable of free movement. The prothorax bears protrusions which help it to penetrate the pod and the soft kernel. Upon penetration into the pea it moults into the grub-like larva with rudimentary legs.

Distribution. B. pisorum is a cosmopolitan insect and is found wherever peas are grown. It is found from the European Arctic region to South Africa, from Canada to South America, and in Asia.

Hosts. The beetle attacks peas of all varieties, including garden, field peas and wild varieties. The adults feed on the pollen of flowers. They may feed on flowers other than peas, but oviposition is limited to this genus only.

Type of Injury. 1. The larva of the beetle feeds upon the seed, thereby reducing its food potential. Larvae may consume about one fifth of the original weight of the pea; 2. The feeding in the seed may impair the germination. True, in its early stages, the damage to the seed is very little; but as it grows, and consumes more of the seed, the injury increases to such an extent that by the time the larva is ready to pupate the seed may be destroyed. According to LARSON et al. (1938) only 10% of peas containing *Bruchus* pupae germinated. 3. It is an unpleasant experience to find beetles floating in one's soup.

Life History. In cold weather the adults hibernate. Beetles which emerge from the seed before the cold weather sets in spend the winter in the ground, under leaves or in cracks of objects. For this purpose they may fly long distances from their breeding places to woods where they spend the winter under the bark of trees. Should the winter arrive before the beetle emerges from the kernel, it overwinters in it and emerges in the spring. Often beetles remain throughout the summer in the pea only to overwinter in it.

In countries of warmer climate, where peas are grown in the winter or spring, as in the Eastern Mediterranean countries, the weevil does not hibernate but aestivates, and is then active in the winter. The awakening of the beetle in the spring takes place when the temperature reaches 14° C; therefore, the colder the climate in the country, the later the beetle awakens. Thus, in Oregon, where the pest was studied at length, *B. pisorum* begins its activity in late May-June, while in the more southern states, in late April-May. In Israel this weevil is active during the winter months. Upon emerging the adult starts to feed. Egg laying begins about two weeks after the insect begins to feed. One female lays on the average 300—400 eggs. The maximum number laid by one female was 700 eggs. As a rule the egg laying period lasts 6—7 weeks.

The egg is laid on the pod. The female is not selective as to the size of the pod upon which she lays, but usually full pods containing soft peas are selected. One or a few females may lay as many as 100 eggs on one pod, but generally about one dozen eggs are found on a pod.

The incubation period at 13° C is 23 days, and at a temperature of 16—18° C, 14—16 days. As the larva hatches, it bores its way through the pod into the center of the seed. The entrance hole on the pea looks like a small black dot. Often a few such dots may be detected on a single seed, but only one larva matures. At the temperature of 18—20° C the development of the larva lasts 33—45 days. Before it pupates, the larva gnaws a small circle in the peel of the pea which later serves as a lid over the exit hole of the adult. This circle is visible externally and indicates infestation. Then the larva spins silken threads over the cavity walls and pupates.

The pupal stage lasts 8—14 days at 18—20° C. Often the mature adult may remain for a certain period in the seed before it emerges from the pea cavity.

Thus the entire life cycle from egg to adulthood lasts, at a temperature of 18—20° C, seven to eight weeks. The threshold of oviposition of the beetle is 18° C, and the optimal temperature for reproduction is 26° C. Accordingly, oviposition in northern countries may take place throughout the summer to September. In the Mediterranean region the weevil reproduces in the spring, and during the summer it falls into a diapause.

The Bean Weevil — Acanthoscelides obtectus Say

[Syn.: *Bruchus obtectus* SAY]

This *Bruchus* attacks the bean in the field, but unlike *B. pisorum*, it may develop in stored seeds as well.

Description. The adult is smaller than the pea weevil. Its elytra are brown, mottled with white spots. The pronotum is narrower at its anterior border and the pygidium is covered with brownish scales except for three pale spots. The hind femora bear three dents each, one larger and two smaller. The egg is white, 0.6 mm long and 0.25 mm wide.

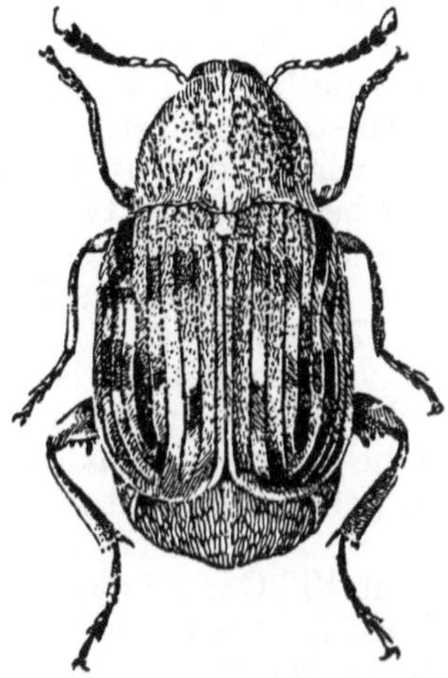

Fig. 73. *Acanthoscelides obtectus*. Courtesy U.S.D.A.

Distribution. This weevil is distributed in all parts of the world. Since it can exist in the store room it depends less upon changes of outdoor climate conditions.

Hosts. Although this weevil is called the "bean weevil" it is not limited to beans only, but attacks all kinds of leguminous seeds.

Type of Injury. This is a more dangerous pest than the pea weevil, for the following reasons: 1. It is polyphagous; 2. It may attack seeds in storage and, if allowed to spread, can infest the entire seed stock; 3. It may raise several generations in succession without a diapause period.

Life History. The eggs are laid in groups of 10—12 each. As a rule they are laid in folds or covered places. After a week or two, depending upon the temperature, the larva hatches and wanders on the plant for some time before it penetrates the pod. Inside the pod it gnaws its way into the seed. The entrance hole is very small and hard to detect. In the store room the eggs are laid on the bare seed.

The development of the larva may last 11—40 days depending upon the temperature. Pupation is similar to that of *B. pisorum,* in the seed.

Upon emerging, and after mating, the female begins to lay as discussed above. Under favourable conditions of temperature the development of one generation may last 6—8 weeks. Unlike the pea weevil, a few individuals may develop in one seed. The larger the seed, the more individuals can develop in it.

A survey of Bruchidae made by CALDRON (1958) in Israel shows that there are about 32 species; the most important of them and their respective hosts are as follows:

Species raising one generation a year

Species	Hosts
Bruchus pisorum L. (The Pea Weevil)	Peas
Bruchus ervi FROL. (The Lentil Weevil)	Lentils.
Bruchus rufimanus BOH. (The Broad Bean Weevil)	Horse beans, Vetch, Peas.
Bruchus dentipes BDI.	Horse beans, Vetch.
Bruchidius trifolii MOTCH (The Clover Weevil)	Clover

Species raising several generations a year

Species	Hosts
Callosobruchus maculatus F.	Chick peas, Cow peas, Lubia.
Callosobruchus chinensis L.	Chick peas, Cow peas, Lubia.
Acanthoscelides obtectus SAY (The Bean Weevil)	Beans

Control. In the Field. Treatment against these pests should be given at the time of blossom. LARSON et al. (1938) claimed that in laboratory tests DDT did not kill the adult weevils, while in the field a very good control was obtained with this material. Apparently DDT is not effective by contact only, but kills the beetle when taken orally. Thus, dusting the field with a 5% DDT dust 2.5 kg/dunam gave satisfactory control. Pollinating insects may be harmed with this substance and it is more advisable to apply a dust containing rotenone 0.75% a.i. which was also effective though not as much as DDT.

In order to prevent the building up of future generations, the field should be turned over soon after harvest, and all plant remains buried. Also volunteer crops should be destroyed.

In the Store-room. In order to insure that seed brought from the field into storage is free from living Bruchidae, one of the following methods should be employed:

1. Retain the kernels for three to four hours at a temperature of 57° C. This will kill the pest in all its stages.

2. Fumigate with hydrocyanic gas 30—40 g/m³ or Carbon bisulphide and carbon tetrachloride at the rate of 200 g/m³. In Israel ethylene dibromide is employed at the rate of 80 g/m³ or a mixture of Ethylene dibromide with carbon tetrachloride 200 g/m³.

In order to prevent infestation in the store-room, the seeds may be covered with a fine dust.

LITERATURE

E. A. Back, 1922. *U.S.D.A. F.B.* **1275,** *1-35.*

J. C. Bridwell & LJ. Bottimer, 1933. *J. agric. Res.* **46,** (8), *739.*

T. A. Brindley, R. Schopp & G. F. Hinman, 1948. *J. econ. Ent.* **41,** *832-833.*

M. Caldron, 1958. The Bruchidae of Israel, Thesis - Hebrew University, Jerusalem. 38 pp.

A. O. Larson, T. A. Brindley & G. F. Hinman, 1938. *U.S.D.A. Tech. Bull.* **599,** *1-48.*

Sitona spp.

(Curculionidae, Col.)

Many species of *Sitona* live on leguminous plants; in the Near East their biology has been little studied.

The characteristics of this genus are as follows: The body is elongate, its sides more or less parallel. The colour is grey, brownish-grey or brown. The head is large, the rostrum short and broad. The second segment of the antenna is elongated, the mandibles strong and large, the prothorax rounded at the side margin and the elytra bear longitudinal dotted lines.

Over three dozen species inhabit leguminous fields in the Near East, but only a few are of economic importance. One of these, *Sitona lineata*, was studied in Europe and its biology will be discussed here.

Sitona lineata L.

Description. The adult is 4—5 mm long, its colour is brown and minute dots are scattered on the pronotum. The elytra bear small scales.

The egg is spherical, about one mm in diameter. Its colour is first yellow, then it turns black.

The larva is 5—6 mm long, grub-like and legless. Its colour is white and the head brown.

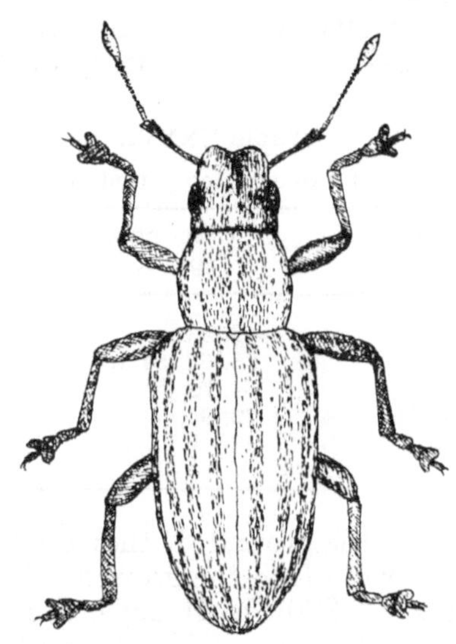

Fig. 74. *Sitona lineata.*

Distribution. The species is distributed throughout Europe. The most northern countries from which it is recorded are Sweden, Finland and the USSR. In the Near East it is recorded from Cyprus and Israel.

Hosts. The species is found in various leguminous fields, but is most prevalent on vetch, *Vicia faba,* and *Pisum.*

Type of Injury. The adult beetles feed on the leaves, and if a young crop is attacked it may be greatly injured or defoliated. The larvae feed upon the nitrogenous nodules, thereby destroying the storage of nitrogen of the plants which depend on bacteria for its supply. The plant weakens and the yield is thereby reduced. The larvae may feed also upon the roots, thus facilitating the entry of decay organisms which may destroy the root and cause the entire plant to wither. This may happen in particular in lower areas of the field where moisture has accumulated.

Life History. The adults emerge from their diapause quarters early in the spring. In the Mediterranean region they appear in November-December. They mate, and after a preoviposition period, egg-laying begins. Reproduction goes on during the entire life-time of the female. According to ANDERSEN (1931) one female lays on the average 1800 eggs. The maximum number laid by one female may be 4500 eggs. In one day, one female may lay, on the average, as many as 40 eggs — while 80 eggs was the maximum daily egg production. In studies at Rehovot the minimum number of eggs was 224, and the maximum 2154.

The egg-laying capacity for 12 females at a room temperature which varied from 15—24° C is given in Table XXXVI (MELAMED unpub. notes).

Table XXXVI.

Number of eggs and oviposition period of *S. lineata* in days

Number of females	Number of eggs laid	Oviposition period	Date of death of females
2	200— 500	42, 43	15.IV;19. V
4	501—1000	56, 62, 54, 67	6.IV;17.IV;17.V;20.V
3	1001—1500	74, 76	5. V;10. V
2	1501—2000	72, 95,	15.IV;26.V
1	2154	95,	13. V

The eggs are laid on the ground. If they are laid while the female is on the plant, they roll off to the ground. The incubation period, according to JACKSON (1920), is 21 days, according to ANDERSEN 28 days at 6.5° C, and 4 days at 36° C. ULASHKEVITCH (1935) found 11—14 days at 18—24° C while, at Rehovot, MELAMED found the values given in table XXXVII.

Table XXXVII.

Incubation period of *S. lineata* (after MELAMED 1962)

Temperature in C°	Period in days
8— 9	32—60
15	14—22
20—21	10—13
24—25	7—11
28—29	6— 9.5

Eggs need contact moisture for developing; otherwise they die.

Larva and Pupa. Immediately upon hatching the neonate larva penetrates deeper into the ground in search of food. After finding a nitrogenous nodule it feeds upon it. After finishing one it may crawl in search of another. As it grows larger and stronger it attacks the roots too. The larval period, according to JACKSON, is 50 days, and 30—42 days according to ULASHKEVITZ. When the larva reaches 4—5 mm it pupates in a cell which it prepares in the soil, in the vicinity of the root. The pupal period lasts 9—21 days depending upon the temperature.

Number of generations. All the writers who discussed the biology of this species state that this species raises but one generation during the year. In Europe this generation is active in the summer and then it hibernates. In the southern Mediterranean countries it is active in the winter and aestivates. It goes into diapause in June and becomes active in November-December.

Ecological notes. Table XXXVIII presents the results of some interesting ecological data based on experiments carried out by ANDERSEN (1934). Beetles were reared at various conditions of relative humidity. In one group the feeding of the beetles was limited to 4 hours only. In the other group the beetles were fed with no restrictions. The average daily number of eggs per female in each of the groups is given in that table.

Table XXXVIII.

Oviposition of *S. lineata* at various conditions of feeding and relative humidity (after ANDERSEN 1934)

R.H. %	Average daily oviposition	
	Group A 4 hours feeding	Group B Feeding with no restriction
100	88	42
70—75	42	28
50		104
32—35	40	—
17—25	—	108

The reason for these results is that, in group A, at the low R.H. the egg production was reduced as the food and moisture intake were limited. With group B the low humidity stimulated the beetles to feed more, thus increasing egg production, too.

The most favourable temperature for egg production is 24.5 °C. Reproduction ceases at 11 ° C which is the threshold. At a temperature above 26° C no egg-laying takes place. The beetle is capable of existing in a temperature range between 20—32° C, as long as the average is 26° C, but the length of life is then shortened. This species can move at the extreme temperatures of 0.7° C and 42.5° C. Its optimal temperature is 25° C, and immediate death occurs at 44° C.

Effects of Moisture. As mentioned above the female is very prolific, but, in comparison, the number of neonate larvae is very limited. Still less is the number of larvae which penetrate into the ground and succeed in finding food. The reason is the need of contact moisture for the development of the eggs and larvae. When this moisture is not available the mortality is extremely high. Eggs hatch only when the soil is moist or after rain, and only larvae in wet soil survive. This may be noticed if we correlate populations with various amounts of yearly rainfall. An illustration may be obtained from a survey taken in Israel during 1957—59. The rainfall in 1956—57 was plentiful and equally distributed, resulting in a dense population of overwintering beetles in 1957—58 (Fig. 75). But this generation produced a far smaller population because of lack of rains during February and more particularly in March 1958. Also 1958—59 was not favourable: low temperature prolonged the development of the larvae till April, when again drought prevailed. Hence a still smaller generation developed during May and June of 1959.

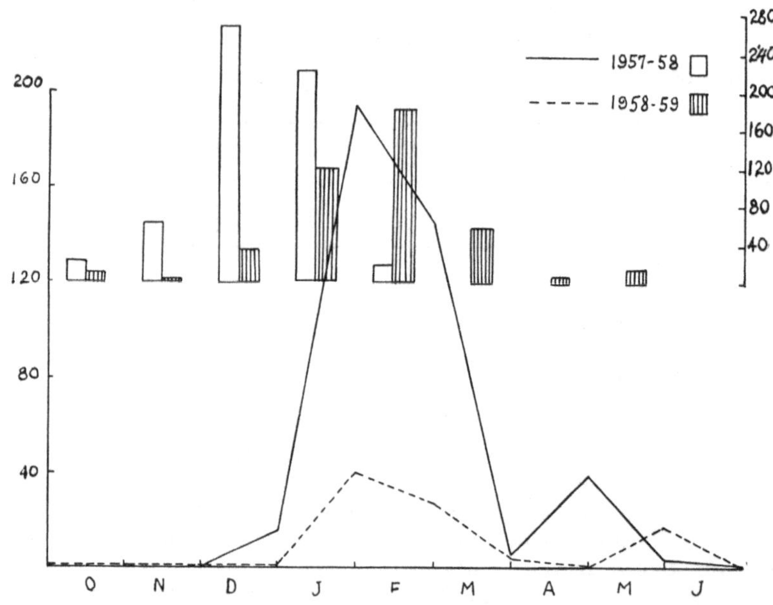

Fig. 75. Effects of rain on the population of *S. lineata* – after RIVNAY & MELAMED. Figures on the left hand side correspond to the curves and represent the number of beetles caught in 100 sweepings each date; Figures on the right hand side correspond to the columns and indicate the quantity of rain each month. For explanations see text.

As mentioned above there are a few more species of this genus in the Near East but their biology will not be discussed. Their description with their hosts in the Near East will be given in tabular form below. A few notes on the phenology of *S. limosa* are also given. The following is taken from a report of observations carried out by PLAUT (1960a) in the northern part of Israel: According to these observations *S. limosa* apparently does not fly, but is distributed by walking. Laboratory studies indicate that the beetles can survive over two weeks without food. The beetles spend the summer aestivating in the same field where they developed during the preceding winter. They began to awake early in December 1958, and their number increased till about the middle of the same month. Subsequently, as no food plants were grown on the field, they left till by the end of the month hardly any beetles remained. In neighbouring fields of leguminous crops, the beetles began to make their appearance by the middle of December, and increased in numbers till the end of the month.

The most prevalent species of *Sitona* in Israel.

Species	Description of adult	Hosts in order of preference
Sitona lineata	Length of body 4—5 mm. Ventral side white. On the pronotum two light straight longitudinal lines. On each elytron 5 grey bands on a brown background.	*Vicia faba* *Vicia sativa* *Pisum* Spp.
S. lividipes	Length of body 4—5 mm. Ventral side dark. A wide band extends laterally from the eye to the end of the body. On the pronotum three light lines. On the elytra pale lines not distinct. White dots are scattered on the pronotum and elytra.	*Trifolium* Spp. *Medicago* Spp.
S. crinita	Length of body 3—4 mm. The entire body is covered with black and white setae. Ventral side entirely grey, on the pronotum three straight longitudinal light lines. On the elytra three rows of dark and white spots.	*Vicia sativa* *Trifolium* Spp. *Medicago* Spp.
S. hispidula	Length of body 4—5 mm. Black and white setae scattered on the entire body. The ventral side of head and thorax is brown and of the abdomen brown with a light lustre. On the pronotum two light lines curved at the anterior ends; a light dot is near each line. On the posterior femura there is a light band.	*Trifolium* Spp. *Medicago* Spp. *Vicia sativa*
S. limosa	Length of body 5—7 mm. Colour of body and vertex brown. On the frons two white dots. Pronotum dark brown and in the middle a light brown band. On each elytron three dark brown bands containing light brown dots.	*Vicia faba* *Vicia sativa* *Trifolium* Spp.

Control. No exact information is available regarding the amount of damage caused by the larvae of *Sitona*. Occasionally bare patches appear in leguminous crops due to injury by the larvae of these species. Such patches are, as a rule, in lower areas where moisture accumulates.

As to the damage by adults PLAUT (1960a) reports severe defoliation by *S. limosa* and *S. crinita*.

Tests at Rehovot, on *S. lineata*, showed that DDT, toxaphene, endrin and dieldrin are all effective against the beetles. Endrin remained effective even after a week of weathering. Tests by PLAUT against *S. limosa* also gave similar results. PLAUT (1960b) found the insecticides to be active even after four weeks of weathering. In later studies this author tested gusathion and sevin against this species and found them to be very effective.

LITERATURE

K. T. ANDERSEN, 1934a. *Biol. Zbl.* **54,** *478-486.*

K. T. ANDERSEN, 1934b. *Vern. dtsch. Ges. angew. Ent.* **IX,** *42-49.*

D. P. JACKSON, 1920. *Ann. appl. Biol.* **VII,** *269-298.*

H. H. JEWITT, 1934. *Kentucky agric. Exp. Sta. Circular* No. **42,** *1-23.*

V. MELAMED, 1961. Unpublished manuscript.

N. PLAUT, 1960a. Observations on the large Sitona *S. limosa* Rossi, *Agr. Res. Sta. Progress Rep.* **279,** *1-6,* Rehovot.

N. PLAUT, 1960b. Effects on Some Insecticides on *Sitona limosa* Rossi, *Agr. Res. Sta. Progress Rep.* **299,** *1-8,* Rehovot.

E. RIVNAY & V. MELAMED, 1959. A Survey on the Sitona spp. on leguminous plants in the winter in Israel - *Agr. Res. Sta. Progress Rep.* **265,** *1-25,* Rehovot.

M. J. ULASHKEVITCH, 1935. *Vinnetza reg. agr. Exp. Sta.* **23,** 75 pp.

F. M. WEBSTER, 1915. *U.S.D.A. F.B.* **649,** *1-8.*

The Alfalfa Weevil — Hypera variabilis Hbst.

[Syn.: *Phytonomus posticus*]

(Curculionidae, Col.)

This is a notorious pest, and a great deal has been written about it, especially in the United States, where it has been studied more extensively than in any other country. Early in this century this pest was first discovered in Utah, and since then it has spread and established itself in other states. In the Near East the amount of damage depends upon the climate. In rainy years with mild temperatures it may develop quite dense populations.

Description. The length of the body is 5 mm. Its colour is light brown. The pronotum bears a wide longitudinal band and a paler line runs along it and divides it into two. The anterior half of the elytra also bears a brown band, wide at its anterior end and tapering posteriorly (see fig. 76). The head and thorax are covered with fine hair, while the elytra bear rows of small black and white setae.

The egg is elliptical, its length is about 1 mm, and its width ½ mm. Its colour is yellow but before hatching it turns grey.

The neonate larva is 1 mm long, its body is bright green and its head black. After it moults the colour becomes a darker shade of green. A whitish dorsal line runs along its entire body. Often two

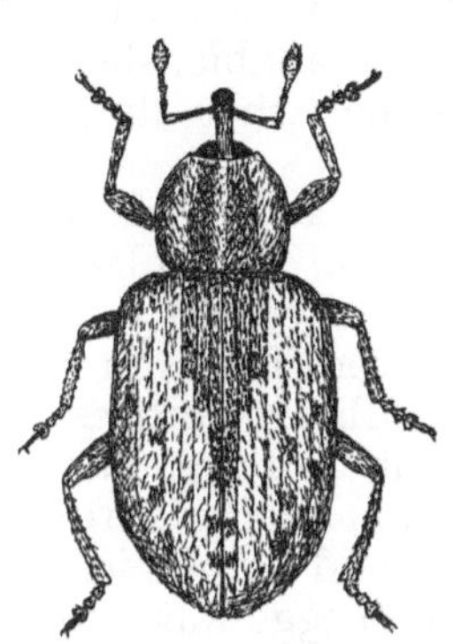

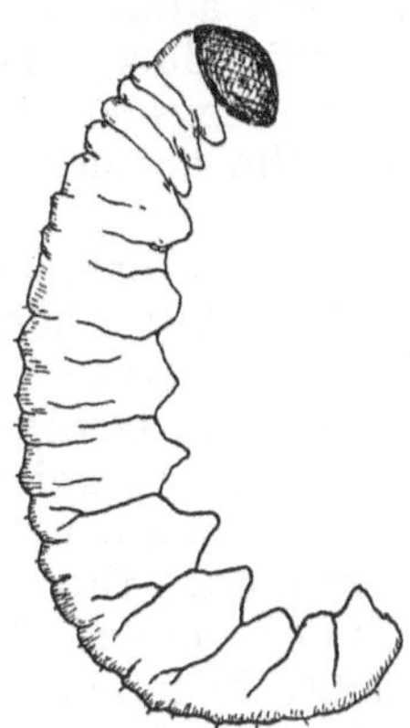

Fig. 76. *Hypera variabilis* – adult and larva

lateral, pale, less distinct lines may run parallel to the median. The body is furrowed laterally by several folds which obliterate the true segmentation. The abdominal segments bear a pair of tubercles which aid the larvae in crawling, thus fulfilling the function of legs which are absent.

Before pupation the larva finds a shelter between two leaves, within a fold of the plant, or on the ground between plant remains where it spins a cocoon made of sparse, rough threads.

It is slightly elliptical, 5—6 mm in diameter and the pupa may be seen through its walls. The pupa is yellowish green with dark wing pads.

Distribution. The origin of the beetle is in the Old World. It is distributed in North- and Central Europe, in the Mediterranean countries, in Asia Minor, Caucasia and the Canarian Islands. As mentioned above, it has been introduced into the United States and is established in many western states.

Hosts. The insect is restricted to plants of the leguminous family, but is most injurious to alfalfa. Sometimes the adult and even eggs may be found on other plants, but, since the larvae do not feed upon them, they should not be considered as hosts. The larvae may feed on peas, beans, vetch, on *Astragalus*, *Melilotus*, *Trigonella* and even on *Robinia*.

Type of Injury. Both adults and larvae feed on the plant and they may both be harmful, particularly the larvae which feed on the soft leaves and inflorescence thus hindering the development of the plant and its seeds. The adults feed upon the foliage and bore into the stem to lay their eggs. In a young crop this activity is detrimental.

Life History. After a period of diapause, towards the end of the summer or in the beginning of the winter, depending upon the country, the adults awaken and, after a short period of feeding, begin to reproduce. The female gnaws a hole in the stem and lays a number of eggs there. If no proper stem is available, she may lay them on the surface of the plant. Eggs may then drop to the ground.

In the laboratory, females laid in the stem as long as these remained fresh but when they wilted oviposition took place on the surface of the plant or even on the surface of the glass jar.

In rearings in Israel as many as 45 eggs were found in each egg cavity. Often one egg only was found. In other countries (TITUS, 1919) similar numbers were found. The number of eggs laid by one female during the three months of reproduction averages from 600—800 eggs per female. The maximum number laid by one female was about 1500. In Israel an average of about 650 eggs was laid by one female; the maximum was 1525 (MELAMED, 1962).

The Egg. According to data available in the literature and summed up by ESSIG & MICHELBACHER (1933), the length of the incubation period may last from 7—20 days, depending upon the temperature. In the rearings at Rehovot, at a temperature of 25° C the incubation period lasted 7—8 days, while at a temperature of 14.5° C, 22—24 days. With additional data, MELAMED calculated the threshold of development of the egg and found it to be 11° C. At higher temperatures, the mortality of the egg increased; while 20% died at 13—14° C, 65% died at 25.5—29° C.

The Larva. The neonate larva feeds first upon the soft internal tissue of the stem. After 2 or 3 days it crawls outside and feeds upon the soft leaves of the top of the plant or flowers. Here it also finds shelter within the folds of the leaves or between them. The more mature larva may also feed upon older leaves which it skeletonizes, leaving only the net of veins.

The length of development of the larva lasts 30—40 days (TITUS, 1913) or 23—28 days (SWEETMAN, 1932). Neither author mentions the temperature, but no doubt this is the factor responsible for these discrepancies.

According to rearings of MELAMED, the larva is capable of completing its development in 12 days at a temperature of 26° C, in 25 days at 18° C, and in about 45 days at 13.5° C.

The threshold of reproduction was found by MELAMED to be 9.5 °C which is quite close to the reports of others (SWEETMAN, 1932, YAKHONTOV, 1934).

The pupal cocoon is built from viscid matter which dries and hardens immediately. According to RIVNAY (1960) this material is squeezed by the larva from its anus with the aid of its mandibles.

From the mouth threads are drawn out and glued together into a loose, basket-like, almost spherical cocoon around its body.

The Pupa. A day or two after the construction of the cocoon the pupa is formed. At optimal temperatures 10 to 14 days are required for the pupa to mature into an adult. At a temperature of 28—29° C it lasts only 5 days, while 25 are necessary at 13° C. The adult remains in the pupal cocoon a few more days before it emerges.

Annual Cycle. The adult alfalfa weevil is active for part of the year, while the other part is spent in a state of diapause.

Since the beetle is widely distributed, the time of its diapause differs in the various types of climate. For comparison, three places may be chosen — namely Germany, Israel and California. The phenology of the beetles in each of these places is illustrated in fig. 77. A glance at this figure shows that while the active period in Europe is in the summer and the diapause is in the winter, in Israel the opposite is the case. The beetle is active in the winter and diapause occurs in the summer. The European summer and Israeli winter are suitable for the activity of this beetle, while the European winter is too cold and the Israeli summer too warm. But in California, in certain regions near the coast, the winter is not too cold and the summer not too warm. So we find the beetle active all year

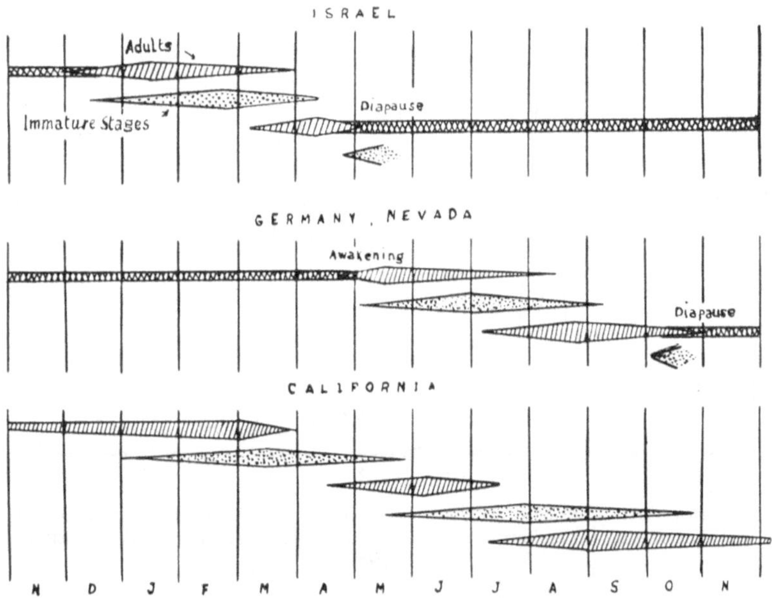

Fig. 77. Phenology of *H. variabilis* in various parts of the world. Note the difference in the periods of activity and diapause in various localities; and the lack of diapause in some localities in California.

around, raising three generations uninterruptedly one after the other.

In Central Europe the insect raises one generation a year. According to some writers, at the end of the summer, eggs and larvae may be found in the field, but they do not complete their development as the cold winter destroys them. In Israel similar conditions exist, except in opposite seasons. Eggs or larvae may be found at the end of the spring, but the heat of the summer kills them before they complete their development. Fig. 78 presents the fluctuations in the population in Israel. In the rearings in the laboratory, too, new females laid in the end of the spring — and it was established that the percentage of such laying females is not more than 10% of their number. This situation reminds us of the phenology of *Lixus junci*. The difference is that the offspring of the few laying females of *L. junci* survives the summer conditions, while the offspring of *Hypera* is destroyed by heat and low relative humidity. After the diapause period moisture is necessary to awaken the beetles. Therefore adults may appear in irrigated fields earlier than in non-irrigated fields where moisture ordinarily results from the rains.

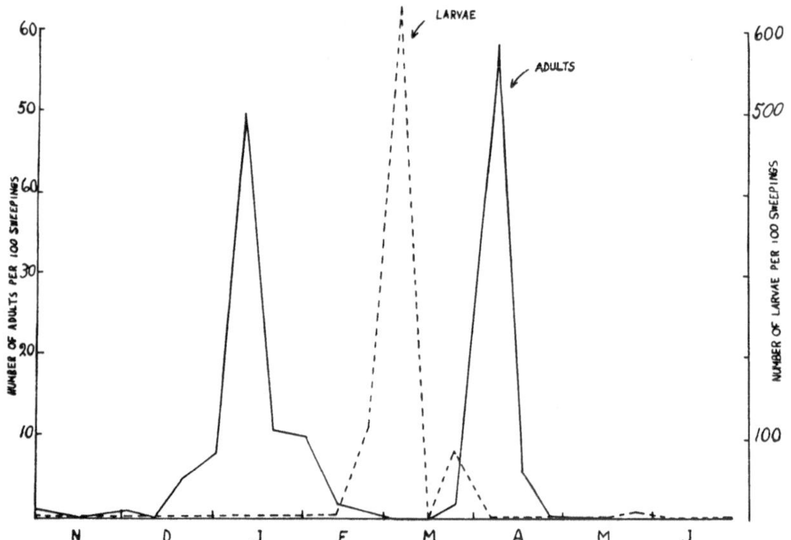

Fig. 78. Population curve of *Hypera variabilis* – adult and larva in Israel. Based on biweekly counts of numbers caught in 100 sweepings each time. After MELAMED, 1959.

Limiting Factors. As mentioned above, the beetle does not reproduce at a temperature below 10° C. The more hours above 10° C during the day, the more oviposition there is in the field. In the

Near East there may be days below 10° C in December and early January. If the beetles begin to oviposit in November and early December there may be an interruption after which egg-laying is resumed around February, when a great part of the day has a temperature above 10° C.

Thus, this egg-laying in early February causes an increase in larvae towards the end of February and early March.

Control. Since applications of insecticides are not desirable on fodder crops, efforts were made to control the pest by methods other than insecticides. The following practices were recommended:

1. Clean fields to prevent beetles hiding in the ground;
2. Healthy crops which can withstand the attack of the pest;
3. Harrowing and sweeping the field after harvest to kill hiding beetles;
4. Burning infested fields after harvest;
5. Pasture;
6. Proper harvesting to kill the pests which are on the plants.

However, with all these practices, control was never complete. Stomach insecticides were employed regardless of the danger to livestock. Care was taken that the harvest should not be made before a lapse of 3 weeks from application. Lead arsenate 100 g per dunam proved to be efficient. Later, dusting took the place of spraying and calcium arsenate mixed with sulphur was employed. In Israel satisfactory control was obtained with synthetic insecticides of the cyclodiene products, of which aldrin proved to be the best. However, this substance is toxic and care should be exercised in times of harvest.

In addition to *Hypera variabilis* the following species were discovered on cultivated crops in Israel:

H. saulcyi Cap.	on	*V. faba*, clover (alfalfa)
H. subvittata Cap.	on	*V. faba*
H. tenuirostris Petr.	on	alfalfa, clover
H. nigrirostris F.	on	clover

These are not as serious pests as *H. variabilis*.

LITERATURE

A. Balachowsky & L. Mesnil, 1936. Les Insectes Nuisibles aux Plantes Cultivées, Paris. pp. *1240-42*.

E. O. Essig & A. E. Michelbacher, 1933. *Univ. of Calif. Bull.* **567,** *1-99.*

H. R. Hagan, 1918. *Utah Agr. Coll. Exp. Sta. Circular* **31,** *1-8.*

J. C. Hamlin, T. W. Liberman, B. W. Bunn, McDuffie W. C., R. C. Newton & L. J. Jones, 1949. *U.S.D.A. Tech. Bull.* **975,** *1-84.*

V. Melamed, 1959. *Hassadeh* **39,** *354-356.*

V. Melamed, 1962. *Israel J. Agric. Res. (Ktavim)* **12** (in Press).

A. E. MICHELBACHER & E. O. ESSIG, 1934. *J. econ. Ent.* **27**, *960-966.*
E. RIVNAY, 1959. *Hassadeh*, **39**, *356-357.*
E. RIVNAY, 1960. *Bull. ent. Res.* **51**, *115-122.*
H. L. SWEETMAN, 1932. *J. econ. Ent.* **25**, *681-693.*
H. L. SWEETMAN & J. WEDEMEYER, 1933. *Ecology*, **14**, *46-60.*
E. G. TITUS, 1913. *Utah Agr. Coll. Exp. Sta. Circular* **10**, *107-120.*
E. G. TITUS, 1919. *Utah Agr. Coll. Exp. Sta. Bull.* **110**, *19-72.*
V. V. YAKHONTOV, 1934. Moscow ob'ed gosud Izd Sredneaz Otd. (Saogiz) 240 pp.

Apion spp.

(Curculionidae, Col.)

The various leguminous crops often harbour dark-blue, long-legged, small snout beetles of the genus *Apion*. Several species of this large genus occur in the Near East, yet their life history in this region has not been studied thoroughly. This discussion is based on European literature and a study carried out in Israel by MELAMED & RIVNAY (1960).

The species which occur in the Eastern Mediterranean countries according to the literature are listed in Table XXXIX, according to the survey of MELAMED & RIVNAY (1960). Over twenty species

Table XXXIX.

Species of *Apion* prevalent in the Near East according to Winkler's Catalogue of Coleoptera.

Name of species	Hosts	Site of oviposition	Distribution Records
A. pisi F.	Alfalfa, vetch, clover	inflorescence	France, Germany, Italy
A. aestivum GERM.	Bean, clover, lettuce, carrot	inflorescence and pods	France, Denmark, U.S.S.R., Israel
A. apricans HBST.	Clover	inflorescence and buds	France, Italy, Poland, Denmark, Czechoslovakia, U.S.S.R.
A. assimile KBY.	Bean, clover, lettuce, carrot	inflorescence and pods	Germany, Denmark, and England.
A. jaffense DSBR.	Alfalfa, clover,	Roots and stem	Israel.
A. virens HBST.	Clover	Roots and stem	Algeria, Italy, Czechoslovakia, Germany, Austria, Denmark.

occur on leguminous plants, but only three are of economic importance. These are *A. aestivum* GERM., the most common in alfalfa fields, the smaller species *A. seniculus* KIRBY and *A. arrogans* WENK., more common in the southern part of the country.

The characteristics of the genus *Apion* are as follows: Its colour is dark metallic blue, dark green or black (one species is red). The snout is very long in comparison to the body. It is thin and bent at the end. The legs are very long. The larvae are white, legless, broad grubs.

The various species may differ from one another in their host selection and habits, as is illustrated in Table XXXIX. However, their life history is similar, and the account of one species will give a description of another. The life history of *A. aestivum* is given because, of all the species found in this region, it was studied more extensively than the others. Notes will be added about *A. arrogans* which is most common in Israel.

Apion aestivum Germ.

Description. The body of the beetle is 2.5—3 mm long, black, very convex and round. The legs are black except for the femora which are brownish-red. The elytra are distinctly striated longitudinally and the pronotum is punctate.

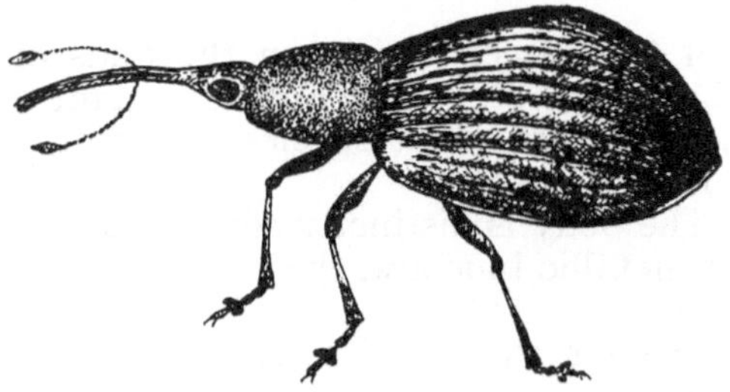

Fig. 79. *Apion aestivum.*

Distribution. This species is recorded as a pest from England, France, Denmark, U.S.S.R., and Israel.

Hosts. Its main host is clover although it is also found on bean. Adults feed also on carrot, parsley and lettuce.

Type of Injury. The adults feed on the leaves and, when plants are small and young, damage may be caused in this way. However, the main injury is caused by the larvae which feed on the ovules or young seeds, destroying as much as 16% of the seed (VASILIEV, 1936). Thus, this species causes very little harm to the fodder crop but in some countries is a limiting factor to seed production.

Life History. In Europe the adult appears in the spring and in June. In Israel it appears in January and oviposition continues during February and March. The egg is laid in the inflorescence between the ovules. According to VASILIEV (1936) the incubation period lasts 8 days at 12° C, and 3 days at 17° C. The hatching grub feeds upon the ovules and may consume 8—10 before it is ready to pupate. At a temperature of 19° C the larval period lasts 26 days, while at 23° C it lasts 22 days. Since the oviposition period is long. larvae are found in the field from June to September, the peak occurring in late July — early August. The larva pupates within the cavity it makes during feeding and the pupal stage lasts 9 days at a temperature of 17° C. Thus, at a favourable temperature of 17—22° C the complete development lasts about 35 days, or even less at a higher temperature.

Number of Generations. Although the life cycle lasts but five weeks, by the time the beetles of the new generation appear, the meteorological conditions are such that the adults enter into hibernation, and begin to reproduce only the following spring. Therefore, in northern countries the beetle raises only one generation.

In warmer countries some authors claim to have followed two generations in the fields as follows: JENKINS (1929) in Wales, VASILIEV (1936) in Russia, and SERVADEI (1940) in Italy.

Apion arrogans Wenk.

The larvae of this insect feed upon the buds of side branches, flowers, and plant tops. Since it is not dependant upon ovules and seeds, it is the more abundant species in Israel, where crops are grown mainly for hay.

Description. The body is distinctly convex, 3—3.5 mm long. The elytra are dark metallic blue and the head, thorax, legs and abdomen are black.

Distribution. So far this species has been reported from Asia Minor, Syria and Israel.

Hosts. This species is found especially on horse beans and vetch. Alfalfa, clover and peas are also infested.

Type of Injury. The injury of this species is particularly conspicuous on vetch. It feeds upon this plant either at the top or in the buds of the side branches. In a young crop such feeding prevents the formation of a branched, healthy plant. Instead, a single weak-stemmed plant develops, if at all. When feeding upon flowers in horse beans it hinders the formation of pods.

During the Israeli winter the adults appear in two peaks — the first, the aestivating beetle; the second, their offspring. The relation between the two depends upon the weather of the particular year.

Control. Laboratory tests showed that DDT, toxaphene, endrin, and dieldrin are effective against this beetle. In Israel dusting with DDT and dieldrin in infested fields of vetch reduced the population to a great extent.

LITERATURE

P. BOVIEN & M. JORGESEN, 1936. *Tidsskr. Planteavl.* **41,** *337-353.*
E. EGOROVA, 1940. *Bull. Plant. Prot., 35-36.*
J. R. W. JENKINS, 1929. *Welsh J. Agric.,* **V,** *176-186.*
V. MELAMED & E. RIVNAY, 1960. A Survey of the Apion spp. on Leguminous Plants in Israel, *Agr. Res. Sta. Progress Rep.* **275,** *1-19,* Rehovot.
A. SERVADEI, 1940. *Redia,* **26,** *177-212.*
N. S. SHCHERBINOVSKII, 1939. *Plant. Prot.* **18,** *136-141.*
TH. SHTCHERBAKOV, 1916. *Rev. Russe Ent.,* **XV,** (4), *529-557.*
K. A. VASILIEV, 1936. *Vsesoyuzu Akad. S.Kh. Nauk. Denny* **8,** *1-96.*

The Clover Seed Wasp — Bruchophagus gibbus Boh.

(Chalcididae, Hym.)

Alfalfa or clover pods may contain empty, perforated seed shells, or even remains of seeds. Among the intact seeds some may lack the natural lustre typical to these seeds. When these are opened a tiny grub may be found: this is the larva of the clover seed wasp, which is responsible for the above-mentioned damage to the seed.

Description. The adult is a tiny wasp whose colour is shiny black. The length of the female is 1.9 mm, and the wing span is 2.7 mm. The head is wider than the prothorax. In the male, which is slightly smaller than the female, the abdomen is shorter than the thorax, while in the female the abdomen is longer than the thorax. The

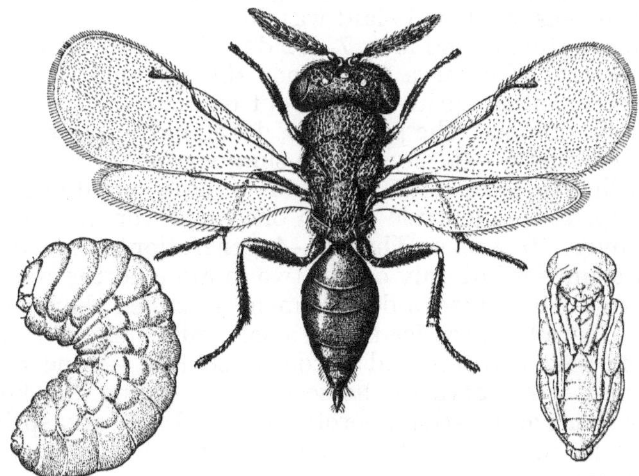

Fig. 80. *Bruchophagus gibbus* after URBAHNS – U.S.D.A., F.B. 812.

egg is small, elongate, 0.2 mm long and 0.08 mm wide and at the end, it has a pedicel 0.35 mm long.

The larva, like others of the same family, is legless and grublike. Its length is 1.9 mm and its width 0.9 mm. It is entirely white, only the mandibles are brown. The pupa is 1.8 mm long, white except for the eyes which are brownish, and as it matures, it becomes entirely black.

Distribution. Apparently this wasp originates in America and, through infested seeds, it has spread and established itself in other countries. According to our knowledge today it occurs in Germany, Turkey, Siberia, Israel, South Africa, and Chile.

Hosts. This wasp attacks seeds of alfalfa and clover, and all varieties of these crops are subject to damage. Leguminous seeds of larger size are not attacked.

Type of Injury. The seed is the only part of the plant which is infested. No other part of the plant is injured. The infested seed is destroyed entirely, and the amount of seed crop damaged by this pest may be as great as 60% — this may render the cultivation of crops for seed unprofitable. As stated above, infested seeds lose their lustre and become perforated by the adult when it emerges.

Life History. The adults appear in the spring. In southern countries, like in the States of Utah, Southern California and Arizona, and probably also in the southern countries of the Near East, the adults appear in March; in more northern countries, in April and May.

For oviposition the female choses pods in which the seeds are still in the milky stage. With the aid of the ovipositor she bores a hole through the pod and deposits the egg in the soft seed. The highest number of eggs one female laid was 20.

The incubation period lasts 7—12 days in the spring or 4—5 days in the summer. The larva feeds upon the contents of the seed. Its development in the spring lasts about one month, but only twelve days in the summer. The pupa, too, develops during 30 days in the spring, but only 6—10 days in the summer.

According to these data, under favourable conditions the insect may complete its life cycle within one month or, at cooler temperatures, over 10 weeks. Thus 3—4 generations may develop in warmer countries but only one or two in more northern regions.

Diapause. The course of development described above takes place when the larva has been feeding in a soft, milk-stage seed. However, when the seed matures and hardens too fast — due to climatic conditions — the larva in the seed stops feeding and falls into a diapause. This aestivation is prolonged until the early spring, when the larva resumes its feeding, completes its development, pupates and emerges as adult.

Control. It is difficult to control this pest with chemicals. Its increase should be prevented by agrotechnical methods. The following methods are proposed:

1. Should there be a high percentage of infestation in the blossoms, it is advisable to cut the crop early and postpone the seed crop for a later date;
2. Eradication of volunteer crops;
3. Destroying plant remnants in the field and threshing grounds;
4. Immediate ploughing after harvest in order to bury the plant remnants;

The farmer often relies on the possible changes of a biological balance. Quite often parasitic wasps take a heavy toll of the population of this pest. According to HARPAZ (1954) who reared the insect in Israel, the percentage of parasitism is quite high. *Habrocyctus medicaginis* GAN. and *Tetrastichus* sp. were reared.

LITERATURE

I. HARPAZ, 1954. *Hassadeh,* **34,** *509-510.*
T. D. URBAHNS, 1920. *U.S.D.A. Bull.* **812,** *1-9.*
V. L. WILDERMUTH, 1931. *U.S.D.A. Farm Bull.* **1642,** *1-14.*

The Clouded Yellow — Colias croceus Fourc.

(Pieridae, Lep.)

The larvae of this species are hardly noticeable on their food plant, but the yellow adults are quite conspicuous as they fly over the alfalfa fields.

Description. The length of the butterfly is 20 mm, its wing span 40—50 mm. The forewings are yellow with a brown band running parallel to the side margin. At the anterior wing apex it widens somewhat and on the disc there is a round black spot. The hind wings are sulphur yellow, and olive green at the base. In the middle of the wing there is an orange spot, and a brown band runs parallel to the side margin. The underside of the wing is sulphur yellow with a pale spot in the middle.

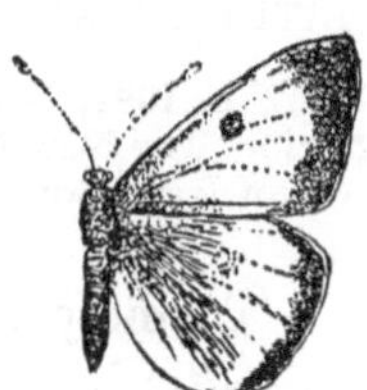

Fig. 81. *Colias croceus.*

The neonate larva is dark brown, but turns green; when mature it reaches 30 mm in length. The body is covered with short dense hair giving it a velvety appearance. On each side of the body along the spiracles there is a yellow line, sections of which are orange. Under this line a black spot is found on each abdominal segment. The head is green and also covered with hair.

The pupa is 25—28 mm long. Its colour is apple-green above and dark green below. A yellow line decorates each side of the thorax. The pupa, like others of the family, is fastened to the plant by means of a silken thread which is wound around the back.

Distribution. This butterfly is distributed throughout Europe, Turkey, Iraq, Iran, Lebanon, Syria, Israel, and Jordan.

Hosts. Although the adults may be seen flying over various crops, the larvae feed on alfalfa or wild clovers.

Life History. The butterflies may be found throughout the year, but they are more numerous over the alfalfa fields during the early spring months. Mature caterpillars may be collected in the field as early as November.

The adult female lays her eggs singly on the plant. One female may lay a few hundred eggs. The incubation period lasts a few days only, and judging from the life history of its related *Colias eurytheme* in California, its larval period is probably two or three weeks, and the pupal period one week only.

Migration. Colias croceus is a migrant. According to WILLIAMS (1958) it spends the winter months in the Mediterranean countries and migrates northwards during the summer. It reaches England in June where butterflies are seen flying northwards. They produce a generation and then, in October, butterflies are seen flying south. In Iraq, WILTSHIRE (1957) states that the butterflies spend the summer in the mountains.

The population of adults, therefore, diminishes in the Mediterranean countries because of migration. The larval population is subject to attack by various parasites. A great percentage of larvae collected in Israel were parasitized by *Apanteles* spp. and others.

Control measures are not necessary, as this insect has never been of a serious economic importance in the Near East.

LITERATURE

V. L. WILDERMUTH, 1920. *U.S.D.A. F.B.* **1094,** *1-16.*
C. B. WILLIAMS, 1958. Insect Migration, The New Naturalist Library **36,** 235 pp.
E. P. WILTSHIRE, 1957. Lepidoptera of Iraq, Govt. of Iraq, 162 pp.

The Pea Blue — Cosmolyce boeticus L.

[Syn.: *Lampides baeticus* L.]

(Lycaenidae, Lep.)

The pods of various kinds of peas may be infested with green, flat caterpillars which feed upon the soft kernels. These are the larvae of the pea blue.

Description. The butterfly is 12—14 mm long with a wing span of 25—30 mm. In the female, the forewings are bright-blue with brown lines running along the side margin. Near the hind margin there are two black spots surrounded by golden rims. A small tail projects from this place. The underside of the wing is grey with several pale brown, wavy lines running across it. In the male, the fore wing is black with a blue glitter. At the end of the hind wing there are black dots with golden borders.

Distribution. This is a subtropic insect and occurs throughout the Mediterranean, yet reaching as far as France and England. It is found also in Africa, Central and South Asia and in Australia.

Host. The caterpillars feed on peas of several varieties, and on vetch.

Type of Injury. The caterpillars feed upon and destroy the green kernels in the pod. Either a few or all of the seeds in the pod may be destroyed.

Life History. The biology of this insect is only slightly known. It may be summarized that the egg is laid on the young pod, and the hatching larva penetrates the pod. When mature, it leaves the pod and pupates outside.

Polyommatus icarus lucia Cul.

(Lycaenidae, Lep.)

Fields of cow peas of many varieties may blossom well, but produce no pods. This may happen also to plants of *Digitalis.* Upon examination of the blossoms of such plants it is revealed that a flat, green caterpillar destroys the blossom and is responsible for the absence of pods. This caterpillar is the larva of *Polyommatus icarus lucia* Cul.

Description. In the male the upper side of the wings is blue, the underside greyish-brown with a row of black and orange spots encircled with white. In the female, the upper side of the wings is brown and near the margin of the forewings there are five orange spots. On the hind wings, near each orange spot, there is a black dot. The underside of the wing looks like that of the male. This butterfly is distributed throughout Euro-Siberia, but was never recorded as a pest. Recently it caused almost total damage by destroying the young pods in various fields in Israel.

Applications with DDT and dieldrin gave complete control of this pest.

Tarucus rosaela Aust. (= *mediterraneae* B. & B.) was reared from caterpillars collected in alfalfa fields in Israel.

LITERATURE

A. Balachowsky & L. Mesnil, 1936. Les Insectes Nuisibles aux Plantes Cultivées pp. 1274-5.

E. Rivnay. Unpublished Notes.

Laspeyresia dorsana F.
(Tortricidae, Lep.)

Description. The adult has a wing span of 14—17 mm; the forewings are narrow and their colour is dark brown. At the outer side margin there is a row of white spots with a pale spot located on the disc of the forewing.

The larva is cream-coloured, the head and pronotum black and when mature it may reach the length of 12—18 mm.

Distribution. The moth is found in Europe and in the Mediterranean countries.

Host and Injury. The larva feeds upon and destroys the soft pea kernels. Often only a few kernels in the pea are eaten but the rest becomes worthless as the pod dries before it matures.

Life History. The adult emerges in the early spring when the plant begins to form pods. The female deposits the eggs upon the pod, and after an incubation period of about 10 days the larva hatches. It is uncertain whether it feeds on the leaves first or bores into the pod immediately. Another species of this genus has the habit of mining into the leaves for sometime before it enters the pod. The larva feeds on the soft beans in the pod and the faeces are pushed out through the entrance hole. When mature, the larva leaves the pod and lowers itself on a silken thread to the ground. It builds a silken cocoon between debris and remains in a state of obligatory diapause until it is ready to pupate the following spring. In other words, it raises but one generation a year.

Control. No control tests were made against this insect. It may be surmised that synthetic insecticides which kill other species of this genus may kill this pest, too. But, in view of the fact that application of control measures should be made in time of blossoming, it should be done only when absolutely necessary.

Cnephasia incertana Tr.
(Tortricidae, Lep.)

In the alfalfa fields in Israel, and probably in other countries too, one may often find top leaves spun together with silken threads forming a "knot". Upon unrolling the leaves one may find a dark-brown caterpillar of the Tortricid family; this is the larva of *Cnephasia incertana* Tr. It is distributed in Europe and in the Mediterranean countries.

The life history of the species has not been studied in this region because it has never rated as a pest of importance.

LITERATURE

A. Balachowsky & L. Mesnil, 1936. Les Insectes Nuisibles aux Plantes Cultivées. p. 1272.

The Legume-Pod Moth — Etiella zinckenella Treit.
(Pyralidae, Lep.)

In various varieties of bean pods green larvae may be found which feed upon the soft seeds. These are the larvae of *E. zinckenella*. In the Near East their damage is hardly noticeable, but in some places they may be troublesome.

Description. The length of the body is 10—12 mm, the wing span, 22—28 mm. The fore wings are brown-grey with a white anterior margin and light-brown-grey hind wings.

The larva is 10—12 mm long. Its colour is greenish purple, its head is brown and a brown line runs over the body.

Distribution. This species is found in all countries of temperate climate. In Europe it was recorded from Russia, Hungary, Austria, Italy and France. In America it is troublesome especially in California, Washington, and also in Puerto-Rico.

Hosts. E. zinckenella is restricted to leguminous hosts. Of the cultivated plants, beans are particularly infested, but the list of food plants of the larva is large. Species of the following genera were found infested with larvae: *Crotalaria, Coronilla, Calycotome, Cytisus, Lathyrus, Lotus, Lupinus, Phaseolus, Robinia, Spartium, Vicia.*

Type of Injury. The larva feeds upon the soft seeds in the pod. The number of seeds it destroys depends upon their size. In case of very small seeds it consumes the entire contents of the pod and may wander to another pod. A large seed may supply the entire food requirements of the larva. However, the remaining kernels in the pod are rendered useless because of mould which develops due to the moist excrement in the pod.

In case of ordinary infestation, 10—15% of the seeds may be destroyed; in severe infestation over 70% of damage was recorded.

Life History. In Central Europe the moths appear in May. Eggs are laid on the pod, as a rule along the rib or under the calyx. The incubation period at 25° C lasts 5 days. Immediately upon hatching the larva bores into the pod and feeds upon the seeds. Several eggs may be laid upon a single pod but, as a rule, one larva remains in it. The larval period lasts 12—14 days during the summer in Central Europe, at 25° C only 7—9 days are required for the larva to mature (Parker, 1951). Then the larva leaves the pod and crawls to the ground. The pupa requires about 9—10 days at 25° C to complete its development. Thus the insect may complete its cycle in about a month.

In Central Europe or in France the moth may raise 3—4 generations a year, depending upon the climate of the locality. Being a polyphagous insect it finds hosts upon which to lay its eggs throughout the summer. A certain percentage of the larvae of the second generation do not complete the cycle the same summer but fall into a state of diapause and pupate, emerging as adults only the following spring. The percentage of such larvae is greater in the following generation and the larvae of October all remain in diapause before pupating.

Biological Control.

In Central Europe and in France several parasites keep the insect in balance. In California efforts were made to establish a biological control by introducing parasites from Europe. Among others the following were shipped from Europe: *Microbracon piger* (Wsm.), *Microbracon pectoralis* (Wsm.), *Phanerotoma planifrons* Nees.

LITERATURE

S. Abul–Nasr & A.M. Awadalla, 1957. *Bull. Soc. Ent. d'Egypte,* **41,** *591–620.*
Cheu Seh-pong, 1943. *Kwangsi Agr.* **III** (6), *351-370.*
F. H. Chittenden, 1909. *U.S.D.A. Bull.* **82,** (III), *25-28.*
A. Couturier, 1943. *C. R. Acad. Agr. Fr.* **29,** *292-293.*
H. L. Parker, 1951. *U.S.D.A. Tech. Bull.* **1036,** *1-28.*
J. Suire, 1943. *C. R. Acad. Agr. Fr.* **29,** *293-294.*

Agromyzidae

Leaves of pea, alfalfa, clover and other leguminous plants may be mined by tiny maggots. As the mesophyll is consumed a pocket-like space is made in which the maggot may be seen through the transparent epidermis. A few species of flies are involved and will be treated in short.

(a) Phytomyza atricornis Mg.

Description. This is a small fly, 1.75 mm long, having a wing span of 5 mm. Its colour is as follows: The thorax is entirely grey, the abdomen, occiput, antennae and legs are dark brown, the frons, face and kness yellowish.

This is one of the most widely distributed species in Egypt and Israel and is probably so in neighbouring countries as well. The species is very polyphagous, and the list of plants in which maggots were found is very large. Many species of Compositae and Cruciferae serve as hosts; it was also found in many leguminous plants including pea, alfalfa, clover, *Lupinus* and others. As a rule the infestation is very slight, so control measures are superfluous.

Life History. The female lays the egg in the tissue beneath the epidermis. The egg hatches after two or three days and the maggot bores into and feeds upon the mesophyll, thus creating a mine which is very narrow at the beginning and which becomes wide and pocket-like in the end. After two weeks, the maggot pupates within the mine, and two weeks later the adult emerges. Thus the fly is capable of raising a few generations during the year, the peak in its population occurring in March-April. Many parasites attack this species and prevent it from building up dense populations.

(b) Liriomyza congesta Beck.

This species also mines into the leaves of peas. The length of the fly is 1.25—1.50 mm. The colours are as follows: The frons (except a brown spot), face, pleura, ventral side of the thorax and the legs are lemon-yellow; notum and abdomen brown.

This fly is rare and its injury is of little importance.

(c) Liriomyza sp.

This species resembles *L. congesta,* but has not been identified as yet.

It attacks the small leaves of chick peas. Until about 20 years ago, the species was hardly known in Israel but today it infests this crop to a great extent. Quite often, in the case of heavy infestation, not a single leaflet may be found which does not contain a maggot, and the number of maggots per plant may be several dozens. Such conditions call for control. Spraying with parathion and phosphamidon gave adequate control of this pest.

(d) Melanagromyza phaseoli Coq.

The maggots of this pest are found in or near the root crown, and its feeding often cuts off the communication between the roots and the plant above the ground.

Description. The female is 2 mm long, with a wing span of 4.5 mm; the male is 1.7 mm long. The colour of both is shining black except for the eyes which are dark brown.

The egg is oblong, has the form of a banana and its dimensions are 0.3×0.13 mm.

The larva is 3 mm long, with conspicuous spiracles.

The puparium is yellowish-brown and its dimensions are 1×2.3 mm.

Distribution. M. phaseoli is a tropical and subtropical species. It occurs in Indonesia, Australia, Africa and in Israel, and attacks beans of all varieties as well as other leguminous plants related to beans.

The damage of the maggot is caused by its feeding around the root crown and thereby weakening the plant or destroying it entirely. According to VAN DER LAAN (1949), 30—40% of the plants may be infested, and in some cases 65—80% may be injured in Indonesia.

Life History. With the aid of the ovipositor the female lays the egg into the tissue of the cotyledon; as a rule it selects for this purpose a place near the petiole on the upper surface. The incubation period lasts about two days. In the tissue the maggot makes a snaky narrow mine whose course runs all over the cotyledon and returns to the petiole. From here it mines, some two days after hatching, into the stem of the plant reaching the root crown. After ten days of feeding there it pupates close to the surface of the stem. The pupal stage lasts 9 days; thus, total development of the insect takes place in 21 days.

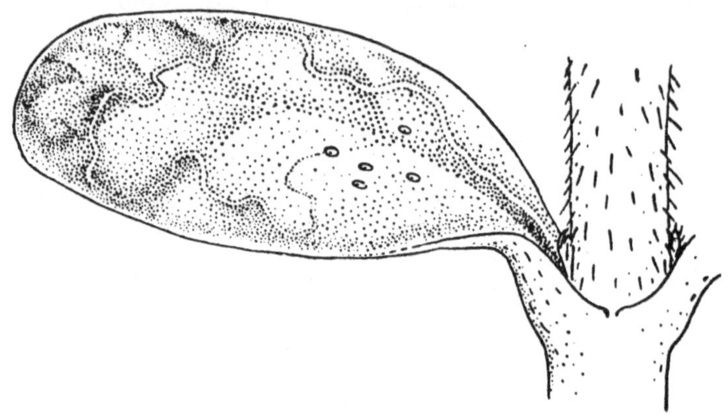

Fig. 82. A Cotyledon of a bean infested with a larva of *Melanagromyza*. Note path of mining before boring into the stem. After v. d. Laan.

One female lays on the average of 94 eggs (the maximum number of eggs was 183) (Van der Laan).

The insect is killed easily with BHC, DDT, or toxaphene.

Application of 0.1% of active ingredient should be applied immediately after the plants have sprouted. If necessary, the application may be repeated. From 100—200 grams of active ingredients are applied in two or three applications.

The Vetch Gall Midge — Perrisia viciae Kief.

(Cecidomyiidae, Dipt.)

Leaves of vetch or alfalfa often have pocket-like folds which contain a few or many reddish-yellow, flat maggots. These are the larvae of a gall midge, *Perrisia viciae,* a species which is found in Europe, and occurs in Israel too.

It attacks vetch, alfalfa, clover and other leguminous plants. The infestation and the resulting damage depend upon the crop. In small alfalfa plants containing few leaves, even a few maggots disturb the plant. On the other hand, in spite of heavy infestation, vetch, about 60 cm high and with a rich growth, was not affected.

The life history of this insect is unknown to us. The small amount of information we have is that crops may become infested both in

November and in May, that the larva completes its development within three weeks, and that the peak of larval population is in March-April.

Since the insect is not found in the summer, it may be surmised that it aestivates from June to November.

Control. Recent trials in Israel have shown that diazinon, dipterex and phosphamidon killed the pest satisfactorily.

LITERATURE

P. A. van der Laan, 1949. *Meded. alg. Proefst. Landb. Buitenzorg*, Java, No. 98. 28 pp.

E. Rivnay. Unpublished Notes.

IV. PESTS OF STORED GRAIN

After the farmer has ploughed, sown, reaped and brought home the harvest, he still has another problem to overcome, namely how to store and keep his harvest free from household pests. He can keep it free from mice (rodents) if the store is well constructed and tightly shut — but how is he to control the many insects?

Accordingly, a brief description will be given of the insects that are found in stores, their life history and distribution together with control measures.

The Rice Weevil — Sitophilus oryzae L.

[Syn.: *Calandra oryzae* L.]
(Curculionidae, Col.)

This beetle is the most destructive pest of stored grains especially in the Mediterranean area, and it is necessary to deal with it at some length.

Description. The adult is a dark brown beetle, 3—3.5 mm in length. It has 4 yellow spots on its elytra, and its body is narrow

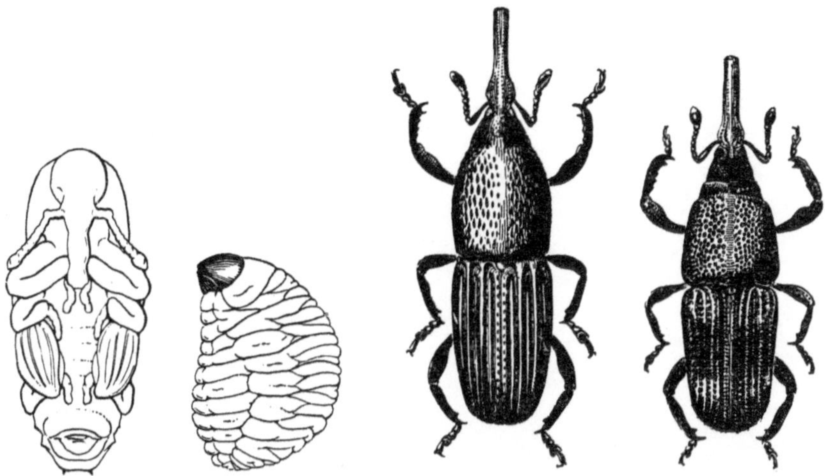

Fig. 83. *Sitophilus* – S. *oryzae* (right) *S. granarius* (left). Larva and Pupa. after Bull. 676 Calif. Agr. Exp. Stat.

with its head prolonged into a long snout. It has sunken dots on its prothorax, and elevated lines on its elytra run from one end to the other.

The larva is grub-like, legless, whitish, with a dark head. Its length is up to 3 mm.

The pupa is similar in shape to the adult, its head is bent downwards, its legs folded on its thorax, and its wingpads pressed to its body with the ends at the tip of the abdomen. At first the pupa is white; later it becomes brown.

Distribution. Although LINNAEUS first described this beetle from rice grains brought from Surinam in South America, it seems that the original home of the beetle was India, and that it was known in South Africa for a long time. Today it is found in all countries with warm climates, both in the Old and New World. In the U.S.A. it causes great damage, especially in the Southern States.

Food. The adults and larvae feed on all kinds of grain — wheat, barley, maize, durrah, sorghum, etc. The adults feed on seeds or flour, but the larvae cannot survive in the free state and live only when enclosed in grain or in dough which has dried and hardened.

Type of Injury. The female bores a small hole in the grain in which she lays an egg. The larva that hatches destroys the contents of the grain. If neglected, the greater part of the harvest may be destroyed in the store. The hole made by the egg-laying beetle serves as an entrance for other pests which, by themselves, are unable to penetrate the hard coat of the grain.

Life History. The female bores into the grain and lays a single white egg into the hole and seals the entry with a gelatinous fluid. The egg hatches and the larva bores into the kernel and feeds upon it. After a number of weeks the larva pupates, the adult appears and, after the mating and pre-oviposition period, starts the cycle again.

The rice weevil is a powerful flier and leaves the storehouse in the spring for the grain fields where it lays eggs in the grain ears. The following generations lay their eggs on grain in the stores.

According to BIRCH (1945a), the incubation period lasts 3½ days at a temperature of 32° C, and 8 days at 20° C. The optimum temperature for larval development is 29° C and, at this temperature, both the larval and the pupal periods together last 3 weeks. The same author found that when the egg, larval and pupal periods as well as the few days in which the adults remain inside the grain are taken into account, then the period from egg to adult lasts about 4 weeks at 29° C or more than 7 months at a temperature of 15° C. BIRCH also stated that the pupal stage comprises about 25% of the total developmental period of the generation; that is, about 7 days at 29° C and about 7 weeks at 15° C.

The data given by Back & Cotton as quoted by Bodenheimer (1927) are somewhat different (Table XL).

The female begins to lay eggs a few days after mating, and the egg-laying period may extend to 3 months. The number of eggs and rate of laying depend upon two factors — the temperature and percentage of moisture in the grain. This will be discussed more fully later. It must, however, be pointed out that in good conditions at a temperature between 24—30° C and at a moisture content of the grain of 13—14%, the female may lay about 300 eggs (maximum 384).

The life-span of the beetle also depends upon these factors. They may live many weeks resulting in overlapping generations.

Number of Generations. The fact that the period of development during the Mediterranean winter lasts such a long time — about half a year — to a great extent lessens the number of generations produced under these conditions.

Bodenheimer studied this and found that in Cairo the beetle may produce 10 generations a year; in Israel in the Jordan Valley 9 generations and in the Coastal Plain 6, while in Naples only 4 generations. As already stated, the above observations hold true only under good nutritional conditions; i.e., when the moisture in the grain reaches 14%.

Ecological Notes. Temperature and composition of the grain are of primary importance to the rice weevil. Problems of grain storage and attempts to produce conditions of storage that will avoid or lessen damage by the rice weevil have resulted in investigations on the influence of the quality of the crop in relation to the survival of the weevils.

In an extensive investigation by Robinson (1926) the following conclusions were reached: When the grain is rich in moisture, the water content of the body of the beetle is smaller, while in "dry" grain, relatively speaking, the percentage of water in the beetle's body is increased. Nevertheless, beetles feeding upon grain high in moisture are bigger than those living on dry grain. Beetles feeding on grain high in moisture have a lower freezing point — that is to say, that they are more resistant than those beetles which feed upon dry grains. The beetles themselves are attracted to high-moisture grains. The optimum water content of grains for these beetles is, as stated, 14%.

Birch (1945b) stated that beetles are unable to develop in grain containing less than 10.5% moisture, and that with the decline in water content of the grain below 13.5% the number of eggs laid by one female decreased. In Fig. 84 the amount of egg-laying under different conditions of temperature and humidity is shown. It will be observed that the maximum number of eggs was laid between temperatures of 24.5° C—29° C and at a moisture content of 13.5—

14% in the grain. If the temperature or humidity was greater or less than these values the number of eggs decreased, until none at all were laid at a temperature above 32° C and below 18° C, or at a moisture content in the grain of less than 10%.

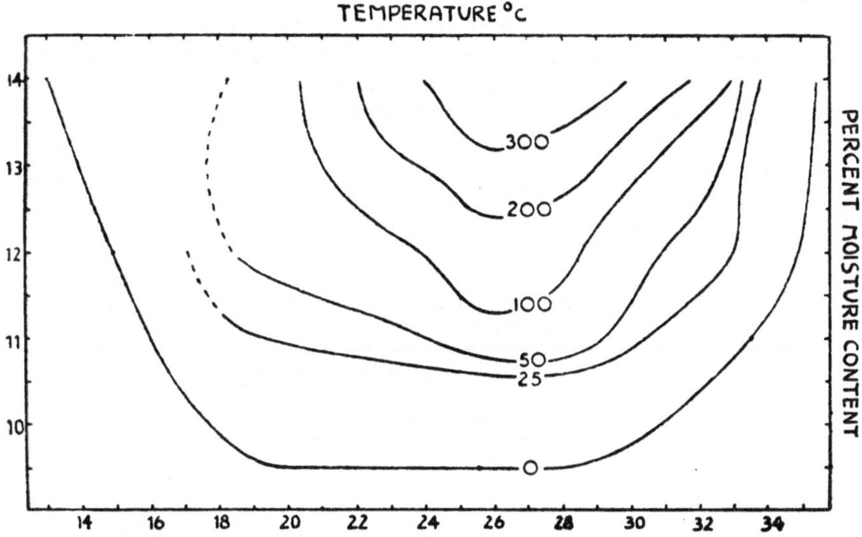

Fig. 84. Effects of Temperature and Moisture content of the grain, upon reproduction of *Sitophylus oryzae*. After BIRCH. Courtesy Australian Journal of Experimental Biology and Medical Science.

BIRCH (1945b) also stated that 34° C is the highest temperature at which the beetle can survive and that it is able to develop in temperatures up to 32.2° C. The threshold of reproduction is 18° C, the minimum temperature for survival, 15° C. The lowest percentage of beetle mortality was at 30° C.

According to ROBINSON, the beetles do not survive longer than 17 days at a temperature of 7° C.

The influence of these factors on the beetle population will be discussed in comparison to the data available on the granary weevil.

The Granary Weevil — Sitophilus granarius L.

[Syn.: *Calandra granaria* L.]
(Curculionidae, Col.)

From the point of view of food and type of damage, this weevil is very similar to the rice weevil. It is more prevalent in the northern countries.

Description. The adult weevil is 3—4.5 mm long, somewhat bigger than the rice weevil. Its shape is similar to that of the rice weevil but differs from it only in the following morphological details: The dots on the prothorax are elongated and not round as they are on the rice weevil; there are no markings on its elytra and it lacks hind wings — in contrast to the rice weevil which flies well.

The egg, larval and pupal stage are all similar to those of the rice weevil.

Distribution. This weevil is found chiefly in Central and Northern Europe, in the northern states of the U.S.A., and in Canada, etc. Further details on differences in distribution between the two beetles are dealt with in the discussion below.

Life Cycle. Its life cycle, too, is similar to that of the rice weevil. There is, however, a difference in length of time of development at different temperatures. Data on this subject were assembled by BODENHEIMER (1927) and are as follows:

Table XL.

Length of Developmental Period of Grain Weevils (after BODENHEIMER (1927) quoting BACK & COTTON)

Mean Temperature (in C°)	Development in Days	
	Rice Weevil	Granary Weevil
12	—	209
15	94	95
18	73	61
21	45	45
24	33	36
27	25	29
30	21	25

It will be observed that at low temperatures the granary weevil develops more rapidly than the rice weevil, whereas at high temperatures the rice weevil develops more speedily. At 21° C they both develop at the same rate.

Other physiological differences, as given in Table XLI, explain to some extent how the granary weevil successfully survives the unfavourable winter conditions in northern lands, and the reasons why the rice weevil cannot survive in cold countries.

The granary weevil cannot survive in a country where the mean temperature in any one of the months of the year is above 25° C. The rice weevil cannot survive when the mean temperature of any month falls to 3° C; therefore, in Israel and in Egypt only the rice weevil is present. In Naples, Italy, both weevils are found, but the population of granary weevils is larger than that of the rice weevil, while in N. Europe the rice weevil is not found at all.

As regards the factor of moisture in the grain, BODENHEIMER stated that in the Jordan Valley, where the harvest is cut dry and

Table XLI.

Physiological Differences between the two Storage Weevils [After BODENHEIMER (1927) (first 3 lines), and ROBINSON (1926) (last 2 lines)]

	Rice Weevil	Granary Weevil
Higer temperature limit for survival (°C) . .	+30	+25
Lower temperature limit for survival (°C) .	+ 3	— 5
Threshold of development (°C)	+13.1	+ 9.5
Temperatures at which beetles become inactive (°C)	+ 7	+ 1.6
Length of life at this temperatures (in days) .	17	38

contains less than 10.5% moisture, the beetles do not, as a rule, develop.

The beetles differ, too, in their sensitivity to poisonous vapours, as shown in Table XLV. These data were taken from SHEPARD et al. (1937). The grain weevil is more resistant than the rice weevil to all the substances mentioned in the table, with the exception of sulphur dioxide, to which the rice weevil is more resistant.

The Lesser Grain Borer — Rhizopertha dominica Fab.
(Bostrichidae, Col.)

Many species of the Bostrichidae are known as wood borers. The lesser grain borer is one of the smallest species in this family and one of the smallest of the pests of stored products.

Description. The adult beetle is about 3 mm long. It has a cylin-

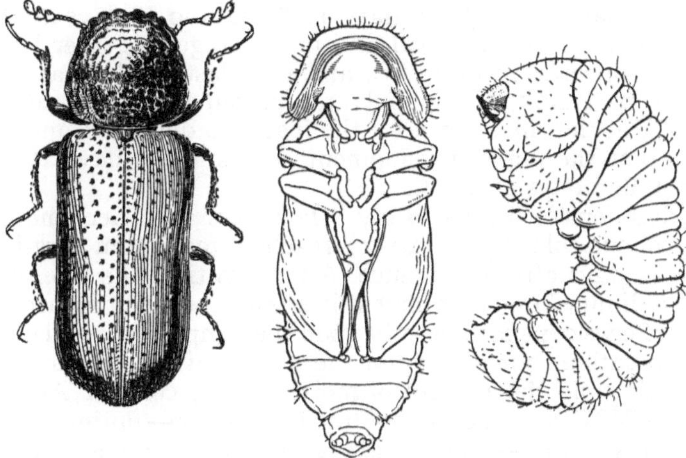

Fig. 85. *Rhizopertha dominica* after Bull. 676 Calif. Agr. Exp. Stat.

drical body and, characteristic of the family, its head is bent downwards so that, from above, it appears as if hidden under the prothorax. The body is covered with sunken dots which on the elytra are arranged in rows.

The length of the larva may reach 2.5 mm. It is white, with a round and fleshy body somewhat curved, similar to the grubs of Scarabaeid beetles. Its thorax is wider than other parts of its body and its mouth-parts are strong and projecting. Its legs are small and degenerate. The pupa is similar to the adult, with wings and legs close to the body.

Distribution. It is believed that this beetle is of Indian origin. Today, however, it is widely spread in many countries with warm or temperate climate. It is found especially in Australia, South America, the Southern States of the U.S.A., the Mediterranean region, etc.

Food and Damage. The beetle bores into grains of wheat and other cereals, damages them and opens the way for attack by other pests. As a rule it prefers damaged grain. It is also known to feed on other substances, such as flour, skins and wooden implements. The pests in the store bore into the walls of wooden bins and destroy them.

Life Cycle. The adult female lays its eggs in groups on the surface of the grain, generally in cracks or crevices. The newly hatched larva seeks a convenient place to gnaw and penetrates the grain. Relatively few manage to penetrate whole grains. The larva (at the end of its development period) pupates in its burrow and eventually appears as an adult.

The biology of this pest has been investigated intensively by Birch (1945a) in Australia, and the following data are taken from his work.

The incubation period of the egg lasts 5 days at a temperature of 36° C, 10 days at 25° C, 16 days at 22° C, and 50 days at 18° C. The development of the larval and pupal stages together lasts 20 days at a temperature of 34° C, and may last 70 days at 22° C. According to Birch, the pupal and prepupal periods comprise about 15% of the time taken by the insect to develop from egg to adult, so that the pupal development lasts about 4 days at 34° C and about 12.5 days at 22° C.

The adult female begins to lay a few days after her emergence and, at a suitable temperature, reaches maximum egg-laying capacity 14 days after emergence. After this time the intensity of egg-laying diminishes but it may still extend another 4 months. The quantity of eggs depends not only upon the temperature and density of the population, but also upon the ability of the female to find broken grains and the amount of moisture they contain. At a temperature of 34° C, and in grain holding 14% water — optimum conditions for this beetle — the female is capable of laying more than 400 eggs.

Ecological Notes. According to BIRCH (1945b), of all the developmental stages, the eggs are the most resistant to low atmospheric humidity. Eggs of this beetle hatched and developed under conditions where moisture was lacking. It appears that the shell of the egg is impermeable and protects it from drought. The larvae, however, are very sensitive to even small fluctuations in humidity, and many of them die before finding a suitable grain into which to bore.

The most desirable amount of moisture in the grain is between 12—14%. If the water content of the grain falls below 11% mortality increases, and if the moisture content falls to 10% there is twice the death rate of the larvae than in grains containing 11% water. The lowest amount of moisture at which the larvae can develop is 9%.

The tropical origin of this beetle explains the temperature range at which it is able to survive. The upper limit both for normal egg-development and for oviposition is 39° C. At a temperature of 38.5° C the insect can undergo all the stages of development — from incubation to adulthood — although the percentage of death will be very high — 83%.

At a temperature below 20° C the individual develops very slowly, while at 15—18° C egg-laying is not possible. This may be considered the threshold of reproduction.

The Common Flour Beetles — Tribolium castaneum Fab. and Tribolium confusum Duv.

[Syn.: *T. castaneum* = *ferrugineum* FAB.]

(Tenebrionidae, Col.)

These beetles are known not only to the farmer and miller but also to the housewife who, when buying pearl barley or semolina, may find small brown beetles in the produce. Both in the kitchen and the larder; farinaceous foods are attacked by these beetles.

Other beetles may be found in this sort of food but these two beetles are the most common.

Description. The adult is an elongate beetle 3.5 mm long, whose sides are parallel; the large flat head narrows towards the front. The antennae are thin, clavate, and as long as the head and half the prothorax. The prothorax is rectangular, wider than long, and equal in width to the elytra. The legs are short.

The two beetles are very similar, and it is possible to differentiate between them only after careful examination, as follows:

Differences in Shape Between the Two Flour Beetles of the Genus *Tribolium*

T. confusum	*T. castaneum*
The space between the eyes on the lower part of the head is 3 times or more the width of an eye.	The space between the eyes on the lower part of the head is about the width of an eye.
The expansion of the club of the antenna is gradual.	The expansion of the club of the antenna is sharp and abrupt.
Margins of frons extend over the eyes.	Margins of frons do not extend sidewards.

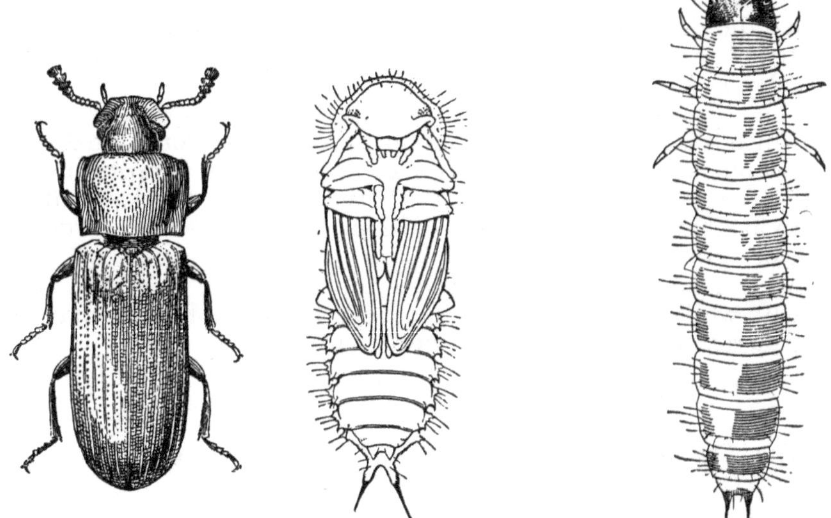

Fig. 86. *Tribolium confusum* after Bull. 676 Calif. Agr. Exp. Stat.

The egg is white, translucent, 0.60 mm long and 0.35 mm wide. Its surface is sticky, so that particles of flour adhere to it. It is always covered with foreign matter.

The larva is shaped like a wireworm. It may reach 6—7 mm in length. The general colour is yellowish-white, while the upper part of the head, the prolegs and the rings of the body are dark. The ninth segment of the abdomen (last body segment) has two projections at its tip. The body is covered with sensory hairs.

In the pupa the wing-pads reach the fifth abdominal segment. The prothorax bears setae. Two-lobed projections extend on both sides of the abdominal segment while the last one bears two spines.

Distribution. Both beetles are found in all tropical and subtropical areas. *T. castaneum* is more common in warmer countries, so, as a rule, it is not found in northern lands. *T. confusum* is more common in cold countries.

Food. Grain products such as flour, pearl barley, bran, etc. are the principal sources of food for both the larvae and adults of these beetles. They like high protein foods. The insects may also be found on other foods, such as chocolate, nuts, and raisins.

Life Cycle. In cold climates the adult hibernates and only awakens in spring or summer. Its reproduction period may last more than one year and, if the winter interrupts, it will continue into the next year. The adult can, therefore, live 2 years. The female lays her eggs on the food, and the larva upon hatching conceals itself, tending to shun the light. It only rises to the top layers of the food supply at the time of pupation.

Details of the length of the developmental periods of the beetles under different temperatures and nutritional conditions are given in Table XLII, and a comparison is made of specific characteristics of the adult beetles in Table XLIII.

Table XLII.

Duration of Development (in days) of *Tribolium* Spp. (The numbers in brackets are averages)

Temperature in C°	Nutritional conditions of larvae	*Tribolium confusum*			*Tribolium castaneum*		
		Egg	Larva	Pupa	Egg	Larva	Pupa
21		12.8					
22					8.8		
25					6		8.8
27	Bran	6.8	33— 64(45)	7.9	5.2	26— 36(30)	7.1
27	Flour		65—112(90)	—	—	73—103(90)	
30							5

It can be seen from these tables that the beetles may live more than 2 years and continue to lay for a long period.

Table XLIII.

Physiological Differences between the 2 Species of *Tribolium* (after GOOD, 1936)

		T. castaneum	*T. confusum*
Length of life of adults (in days)	Males	634	547
	Females	447	226
Length of egglaying period (in days)	Maximum	432	308
	Mean	240	165
Number of eggs per female	Maximum	976	956
	Mean	458	327

A marked characteristic of these beetles is their ability to starve — this may be seen in Table XLIV. The larvae may starve for almost 2 months at 15° C, and at higher temperatures for 3 weeks.

The adults are less able to live without food than the larvae, but even they are relatively resistant to starvation when compared to other insects.

Table XLIV.

Number of Days which *Tribolium* Spp. may live without Food.

Mean Temperature in C°	Adults	Larvae
30	18	23
22	23	46
15	27	54
10	51	—

Ecological Notes. OOSTHUIZEN (1935) carried out extensive investigations on the influence of high temperature on the life of *T. confusum.* He found that, when the eggs are laid and kept at a temperature of 32° C, 88.5% of them hatched. If, however, the eggs were transferred to a temperature of 37.5—38.5° C only 64% hatched, and if oviposition occurred at 37.5—38.5° C only 40% of the eggs hatched. When the larvae were held at a temperature of 37.5—38.5° C, the adult females that developed from them were completely sterile, although the males were fertile. If the larvae and pupae were transferred for a number of hours to a temperature of 40.5—41° C, then the females, which developed under

Table XLV.

Amount of Poison (mg/litre) required to kill 50% (left column,) or 99% (right column) of Grain Storage Beetles. (25° C, duration of exposure 5 hours)

Substance	Boiling point of poison	*T. confusum*		*S. oryzae*		*S. granarius*	
Hydrocyanic acid	26	0.6	1.1	—	—	5.8	11.4
Chloropicrin	112	4.6	7	2	15.2	5	21.0
Sulphur dioxide	10	5.7	10.7	17	46.9	5.7	11.3
Carbon disulphide	46	61	91	26	40	40	66
Methyl bromide	5	11.2	14.4	4	6.2	7.4	8.4
Ethylene dichloride	84	37.5	73	31	137	138	246
Carbon tetrachloride	76	185.0	405	160	559	360	859

these conditions, laid infertile eggs. The fertility of the eggs diminished as the length of the period at high temperature increased.

Control. Data on the sensitivity of the common flour beetle, *Tribolium confusum*, and of the granary weevil to various poisonous vapours are given in Table XLV.

Tenebrio Spp.

Two other beetles of the Tenebrionidae family, *Tenebrio molitor* L. and *T. obscurus* F. are very important as storage pests, especially of flour.

Description. The adults are black, flattened, and 18—25 mm long. The egg is whitish-purple in colour and shaped like a bean seed. The larva is shaped like a wireworm with a hard and smooth skin and it is 25—30 mm long. The colour of *T. molitor* is yellow, and of *T. obscurus* brown.

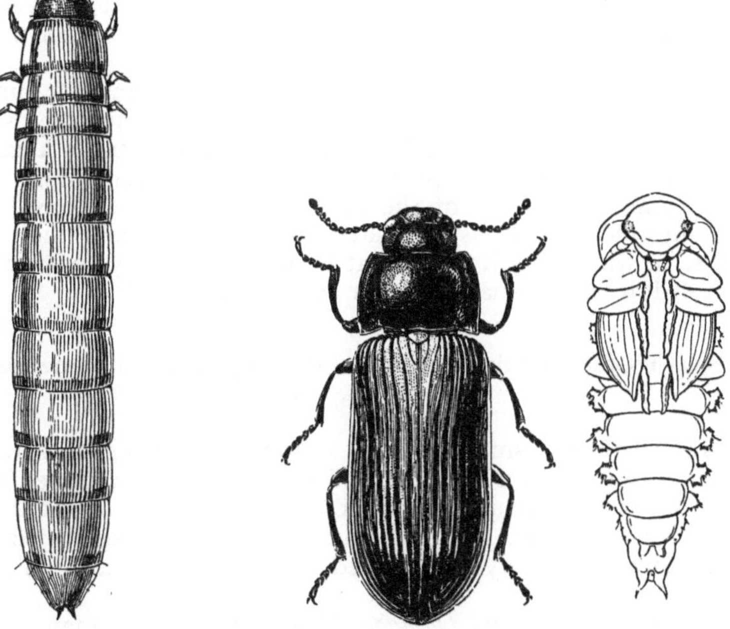

Fig. 87. *Tenebrio molitor* after Bull. 676 Calif. Agr. Exp. Stat.

Distribution and Damage. These beetles are found all over the world and feed chiefly on flour. In damp places they may be found on poultry food in chicken runs, on feathers, etc.

Life Cycle. The female lays several hundred eggs which hatch after 4 days in summer and after 2 weeks in winter. The larval period is very long, from 6 to 9 months. These beetles raise a generation in one or two years and hibernate as larvae.

The Cadelle — Tenebroides mauritanicus L.

(Ostomidae [Syn: Temnochilidae], Col.)

This beetle, from a family closely related to the Tenebrionidae, is very common in mills and flour storehouses. Of all the grain storage pests, only *Tenebrio* spp. are larger in size than this species.

Description. The adult beetle has a flat black body about 8—12 mm long. Its head is separated from the thorax by a narrow "neck", a typical characteristic of this family.

The larva is whitish-grey with a round fleshy body about 18 mm long. Its head is black, and there are two patches on the upper side of the thoracic segments — on the first segment the spots are larger. At the end of its body there are two horn-like projections.

Distribution, Food, and Damage. This beetle is found all over the world. The adults and larvae feed on grain, flour, and other milled products.

In addition to destroying grain and flour, these insects burrow into the woodwork of grain bins and ruin them. The holes serve as cover for these insects and for other storage pests — a factor which makes it difficult to preserve cleanliness in the store.

Life Cycle. Incubation of the eggs lasts from 7—10 days and, under suitable conditions, the larva develops in more than two months. In winter its development is slower and may continue all season. Adult beetles may live 2 or even 3 years.

The Saw-Toothed Grain Beetle — Oryzaephilus surinamensis L.

(Cucujidae, Col.)

This is one of the smallest beetles found in stored grains. It can penetrate into places which are well sealed, even into closed tins.

Description. The adult is a dark coloured beetle, about 2 mm long, with a flat body. Its prothorax bears six saw-like projections on either side which may be observed clearly only with a magnifying glass.

Distribution, Food, and Damage. This beetle attacks many types of stored products — grain, flour, flour products, dried fruits, sugar, nuts, tobacco, dried meat — in short, any food that man eats. This pest is widespread and its damage consists of destructions and spoilage of food.

Life Cycle. Eggs are laid on the food and hatch after a week or two.

Under favourable conditions 4 weeks may elapse from the egg to the adult stage. Depending on the temperature, the beetles can raise a number of generations a year.

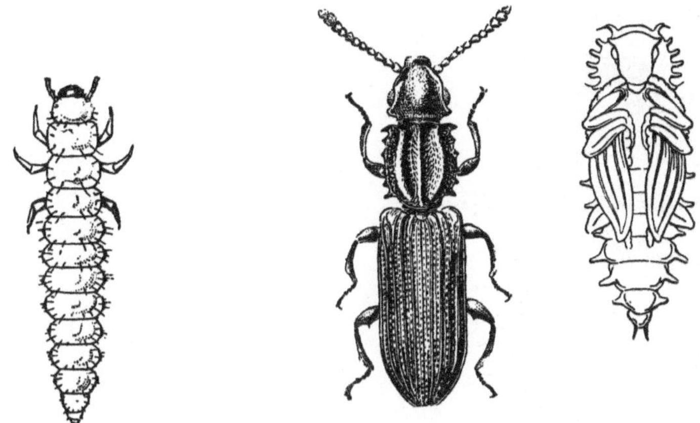

Fig. 88. *Oryzaephilus surinamensis* after Bull. 676 Calif. Agr. Exp. Stat.

The Khapra Beetle — Trogoderma granarium Evts.

[Syn.: *T. afrum* PR.; *T. khapra* ARROW]
(Dermestidae, Col.)

T. granarium has recently spread a great deal; apparently it reached the U.S. in the early forties of this century and since then has spread into many states. Also in the Near East its invasion is recent. PRIESNER described specimens which differed somewhat from the accepted *T. granarium* as a new species which he called *T. afrum;* recent studies by HOWE & BURGESS (1956) have shown this to be synonymous with *T. granarium.*

Description. According to HADAWAY (1956), the adult is a small oblong-oval beetle, 1.6—3.0 mm long, and 0.9—1.7 mm broad. The male averages 2 mm in length. In the antennae the first and second segments are larger, while the third to sixth are smaller. Segments 7—11 form the club. The terminal segment is twice as long as broad. The body is brownish-black, while the elytra and legs are lighter in colour. The body is covered with fine hair which gives it a velvety appearance; a fringe of fine hair decorates the tip of the abdomen.

The egg is cylindrical, 0.7 mm long and 0.25 mm broad. One end is rounded, the other more pointed, and it bears spine-like projections. When newly laid it is milky white, later becoming yellowish.

The first-stage larva is 1.6—1.8 mm long. Half of it consists of a long tail, made of a few hairs borne on the last abdominal segment.

The body is yellowish-white except for the head and tail which are brown. Later on it assumes a reddish-brown appearance and when mature it is 6 mm long.

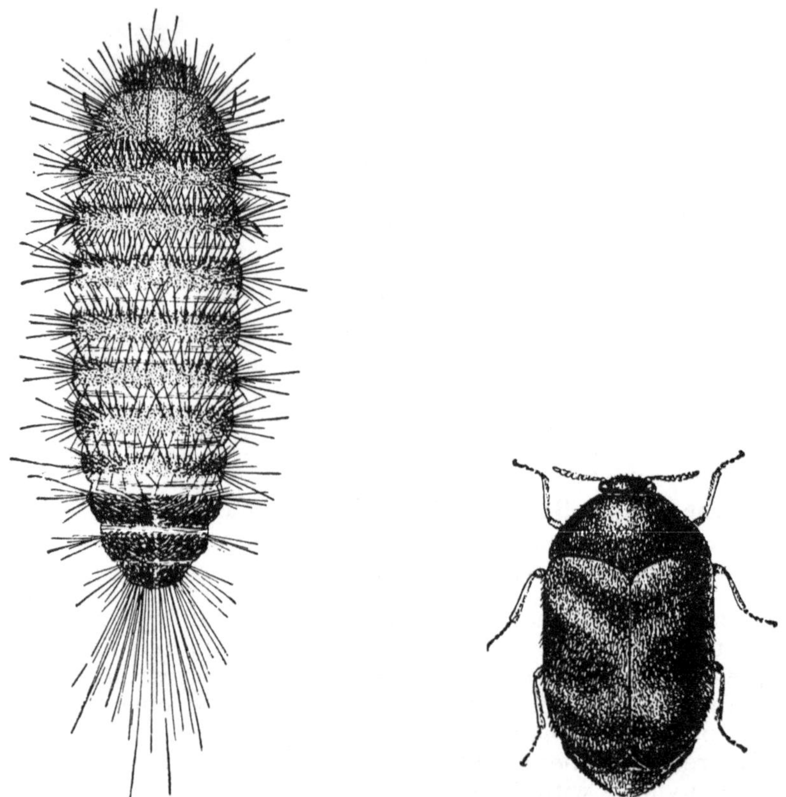

Fig. 89. *Trogoderma granarium* after U.S.D.A. F.B. 1260.

The pupa remains wrapped in its last larval skin. The thorax and basic part of the abdomen are broader and the dorsal surface is densely clothed with hair. The length of the pupa is 5 mm for the female and 3.5 mm for the male. The pupae are usually found near the surface of the grain.

Distribution. This beetle is found in all the continents of the Old World. In North America it was recently introduced and in South America it is not found at all. Records of damage come from England and Germany in Europe; in Asia it is reported from Japan, China, Korea, the Philippine Islands, Ceylon, India, Malaya, Cyprus. In Africa it is reported from Egypt, Nigeria, Madagascar. It is found also in Australia, USSR, as well as in Israel.

Food and Damage. This Dermestid is confined to vegetable products and thrives on malt, wheat, barley, kibbled rice, flour, peas, beans, groundnuts, etc. In England serious damage is reported to moltings. Unlike other dermestids it prefers grains.

The damage is caused by the feeding of all stages on the grain. Young larvae feed on damaged kernels, older larvae attack whole grain. They may thus reduce the stored material to a powdery mess. Infestation by the khapra beetle is characterized by the presence of a large number of larvae and their cast-off skins.

Life History. The eggs are laid singly and are not glued to the substratum. They may be laid in grooves or cracks, or loosely over the grain.

The incubation period at a temperature of 40° C lasts 4—5 days or 9—12 days at 25° C (HADAWAY, 1956, SHULOV, 1955). SHULOV has also demonstrated that the relative humidity hardly affects the length of the incubation period.

The larval development lasts 16—38 days at a temperature of 40° C and a relative humidity of 25—75%, and 27—64 days at 25° C at the same relative humidity; when the humidity is reduced the rate of development is retarded (HADAWAY).

The records of SHULOV differ from the above mentioned. According to him the larval development lasts 5 days at 40° C and 10 days at 25° C. He found the threshold at 14.8° C, and the thermal constant 108 days degrees C°. The records of LINDGREN et al. (1955) are closer to those of HADAWAY.

The pupal development lasts 3—5 days at 4° C and 8—15 days at 25° C.

Ecological Notes. Experiments by LINDGREN et al. show that the beetle can survive certain periods when exposed to low temperatures. Some eggs hatched when exposed for two hours to a temperature of 20° C below freezing point. Eighty-five percent of the larvae of fourth instar died after an exposure of 30 minutes to this temperature while all the pupae succumbed under these conditions, which shows that the larvae are more hardy than the pupae.

Larvae are also hardier at higher temperatures. It took 2:15 hours to kill 95% of the larvae exposed to a temperature of 51° C, while 95% of the pupae died after an exposure of 1:30 hours. At 48° C larvae survived 19 and pupae 12 hours. The data refer to conditions of relative humidity of 75%.

Control. The insect may be killed by various synthetic contact insecticides. LINDGREN et al. have established the LD 90 dosages of various insecticides for this beetle.

According to these authors the dosage of parathion is 2.7, malathion 3.6, aldrin 30, lindane 50, DDT, dieldrin and others over 100 micromilligrams per gram of beetle weight.

Insects may also be killed by fumigation of the grain.

According to LINDGREN et al., in order to kill 95% of the pupae at an exposure of 24 hours it is necessary to apply 1.8 mg of HCN, 7 mg of ethylene dibromide and 38 mg of carbon bisulphide per liter of volume. When calcium cyanide was applied to infested barley at the rate of about half a kg of actual hydrocyanic acid to one ton of grain, no live beetles were found after several days.

The Tobacco Beetle — Lasioderma serricorne F.

(Anobiidae, Col.)

This beetle is primarly a pest of dried tobacco but it may also infest dried leaves or grasses of all kinds. Occasionally it may enter grain stores and cause damage there.

Description. The adult is a brownish oblong beetle about 1.5—2.5 mm long; it may be recognized by the characteristic position of the head which is bent downwards, and almost entirely covered by the pronotum. The larva is hairy and bent like a scarabaeid grub.

Distribution. Its distribution is world wide. The entire life cycle may last from one to two months depending on the temperature; thus several generations may occur annually.

Control methods against other storage pests may be successfully applied against this beetle.

The Angoumois Grain Moth — Sitotroga cerealella Oliv.

(Gelechiidae, Lep.)

Sometimes moths may be seen fluttering to and fro on the walls of grain bins, or flying over the grain itself. These are the adults of the Angoumois grain moth (named after a province in France), one of the most important pests of stored grains.

Description. The adult is small, brownish-yellow, with a wing-span of 12—16 mm. Dark spots are present on the forewings and the hind wings are light grey.

The side margins of the hind wings continue into a kind of finger which is useful in recognizing the moth and distinguishes it from other similar moths.

The egg is 0.4—0.5 mm in size and whitish-red in colour. The larva is 5—7 mm long, white with a yellow-brown head and vestigial abdominal feet.

The pupa is reddish-brown in colour and may be found in the grain.

Distribution, Food and Damage. This pest was first found in France in 1736 in the province of Angoumois. From that time it has been reported in many countries, and it causes severe damage especially in areas of temperate climate. All types of grains are attacked — both summer and winter cereals. The larva bores into the grain and eats out a large cavity. If neglected, up to 25% of the stored crop may be destroyed.

Life Cycle. The female either lays eggs singly or in groups of up

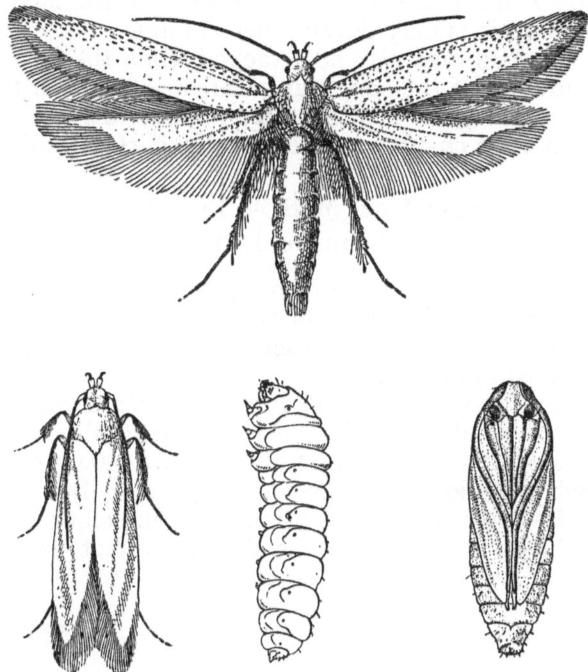

Fig. 90. *Sitotroga cerealella* after Bull. 676 Calif. Agr. Exp. Stat.

to 20 eggs. A female lays on the average about 40 eggs (maximum 389). Under suitable conditions the eggs hatch within four days, and the larva bores into the grain. The larva spins fine threads around its body, which serve as protection, support, and which facilitate its penetration. Only one larva develops in a grain, although the amount of food available in the grain is enough to supply a number of larvae.

The larva feeds on the contents of the grain and grows along its tunnel. Its development lasts 20—24 days. Before pupation, the larva eats a channel to the outside of the seed, leaving a thin layer of the seed-coat intact. When the moth has sufficiently matured, it emerges by breaking through with its head, thus opening an exit for itself.

The pupal period lasts 9—12 days, making a total of 35 days from the egg stage to maturity. At lower temperatures (20° C) the developmental period lasts 2 months.

Number of Generations. The above data hold true for suitable conditions at high temperatures. At lower temperatures, the development is considerably longer. The egg period, for example, may last a number of weeks. Thus, in the more northern countries the pest has only 1 or 2 generations a year while in the Mediterranean area 4—6 generations may develop per annum.

Grain in the field may also be attacked when the female lays its eggs in the glumes of the ears, but severest infestation takes place in the store. Since the moth is not able to penetrate deeply into the grain, infestation is chiefly restricted to the surface grain.

The Mediterranean Flour Moth — Ephestia kühniella Zell.

(Pyralidae, Lep.)

When flour is stored in unprotected places webs containing small worms may be seen. Similarly, in the vicinity of stores and flour mills light grey moths may be observed — these are specimens of the Mediterranean flour moths.

Description. The adult is 6—12 mm long with a wingspan of 25 mm. At rest, its head and abdomen are slightly raised — a specific characteristic of this insect. The forewings are pale grey in colour, with somewhat darker transverse dotted lines. Before pupation the larva is 12—14 mm long, purplish-white in colour, with 5 pairs of abdominal legs. The hooks on the abdominal feet are arranged in

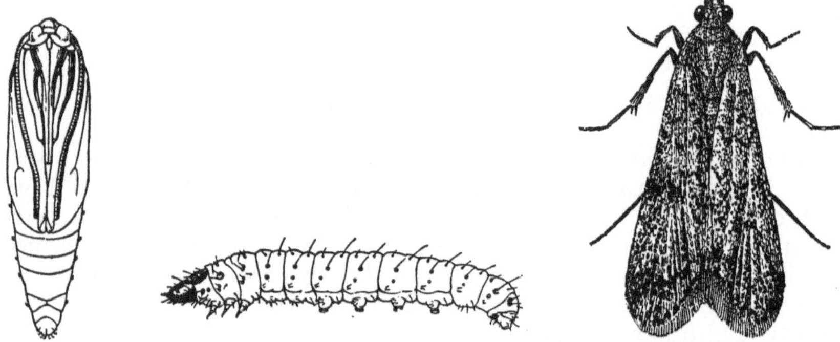

Fig. 91. *Ephestia kühniella* after Calif. Agr. Exp. Stat.

three rows. There are some hairs on the body with white spots at their base.

The pupa is brown, and is found in the web of the larvae.

Distribution, Food and Damage. This pest, it seems, is of European origin. Today it may be found in all the temperate zones both in the Old and New World.

The larvae feed chiefly on flour, but they attack whole grain as well. They readily feed on bran, cornflour, and other substances.

Apart from destroying the flour, the larvae mat together the flour particles forming large lumps which cause clogging and interfere with proper utilization of the food.

Life Cycle. Eggs are laid in groups in the flour or in cracks and crevices of the walls of the bins, etc. After a period of 3—10 days, depending on the temperature, the eggs hatch, and the larvae immediately begin to spin threads and form webs in which they live and feed. The size of the web increases with their growth. The larval period, in the summer, may extend to a month or more, and in the winter up to 80 days. Before pupating the larva spins a cocoon.

The pupal period lasts about 8 days in summer and about 16 days in winter. Therefore, the developmental period of one generation lasts about 6—8 weeks in the Mediterranean summer and about 12 weeks in the winter.

Not all the females are fertile, and some of them do not lay eggs at all. The pre-oviposition period lasts 2—3 days, and a female may lay 150—200 eggs.

The length of the adult period lasts an average of 8—9 days, but some females may live for a fortnight or more.

The Indian Meal Moth — Plodia interpunctella Hbn.

(Pyralidae, Lep.)

Besides the Mediterranean flour moth, other moths of the same family are found in similar circumstances; e.g. the Indian meal moth.

Description. The adult moth has a wingspan of about 16 mm. The forewings are whitish-grey at the base and reddish-brown. The hindwings are grey with a fringe of hair at their hind margins. The greyish-white egg is oval in shape and 0.3—0.5 mm long.

The larva, before pupating, is 10—17 mm long, but usually averages 13 mm. Its colour depends upon the kind of food it eats and

Fig. 92. *Plodia interpunctella* after Bull 676. Calif. Agr. Exp. Stat.

may be whitish, greyish-white, whitish-green, or whitish-red. The tergites on the prothorax and the last abdominal segment are brown. The body bears setae but, in contrast to similar moths, there are no dark spots at the base of these setae.

The pupa is shiny brown with the posterior portion darker than the anterior. It is 6—11 mm long.

Distribution, Food and Damage. This moth was incorrectly named the Indian meal moth since its origin is in Europe. Today it is found in all parts of the world as is its relative, the Mediterranean flour moth.

The moth is not selective in its feeding habits and feeds on all kinds of food in stores — flour, nuts, dried fruit, sesame, dead insects, milk powder, and even pollen gathered by bees.

Apart from destroying the food the presence of the moth and its silken web spoils the food which it has not eaten.

Life Cycle. Hibernating larvae pupate at the end of April. Adults appear in May and begin to lay eggs. Eggs are laid in groups on or close to food. The incubation period at 18° C lasts 10—12 days, and only 3 days at 25° C.

The length of the larval period varies greatly even among those individuals from the same parents which hatched at the same date and were raised under the same conditions. The developmental period of the slow ones may last 4 times that of the early developers. At a temperature of 15° C — typical of the Israeli winter — the development of the larva, according to HAMLIN et al. (1931), lasts 160 days. At a temperature of 23—25° C, it lasts 40 days, and at a temperature of 29—30° C, an average of 36 days.

The pupa develops in 26 days at a temperature of 16° C, or in 4 days at 31° C.

The Adult. The pre-oviposition period depends also upon the weather conditions. In summer it lasts only 2—3 days. The egg-laying period itself lasts only a few days (2—5), but in winter it may continue for a fortnight or more. The majority of the eggs are laid in the first 2—3 days, and a female lays on an average of 50 eggs. According to HAMLIN, the mean egg-laying figures are higher in the summer than in the autumn. In the summer a mean of 178 eggs were laid as compared with 127 eggs in the autumm. The maximum number of eggs was 275 to a female as reported by the same author.

Number of Generations. If the length of development of the moth in its various stages is considered, then 6—7 generations occur in the Mediterranean area. According to data by HAMLIN, who raised only 5 generations in California, all the larvae of the first and second generations pupate and emerge in the same summer. A small proportion of the larvae of the third generation hibernate and mature the following spring.

A larger proportion of the fourth generation larvae hibernate while those of the fifth generation all enter diapause.

However, the results of Tosi (1929) for Bologna in N. Italy differ. According to this author the adults that originated from hibernating larvae appear at the end of May, and their descendants at the middle of August; larvae of the next generation hibernate in the pre-pupal stage.

The Meal Snout Moth — Pyralis farinalis L.

(Pyralidae, Lep.)

This moth is found in badly ventilated stores where the air is stagnant and of high humidity.

Description. The wing-span of the adult moth is about 25 mm. The forewings are brownish-red, with the middle third paler than the rest and bound by 2 white wavy lines.

The larva is white with its first body segment black. Before pupating it may reach 25 mm in length.

Distribution, Food and Damage. Like related members of the family this moth is also found in stores throughout the world.

Flour and grains are attacked as well as other foods that have been in a damp store for prolonged periods of time — such as old potato tubers.

Life Cycle. The development of the insect from egg to adult may last, in summer, 7 weeks. The female lays about 250 eggs.

The Itch Tick — Pediculoides ventricosus Newp.

(Pediculoididae, Acarina)

Farm labourers are sometimes attacked by the "Itch", which either interferes greatly with their work or causes a quite serious illness lasting for a number of days. The cause is a small arthropod known as the itch tick, which is not a pest in the usual sense but, as stated, whose presence is totally undesirable.

Description. The neonate female is about 0.2 mm in length, and 0.1 mm wide at the middle of her body. She has 4 pairs of legs; 2 pairs at the front, and 2 behind. After the female has been fertilized and fed, her abdomen becomes distended and round in shape. The abdomen may reach 1 mm in diameter when it greatly interferes with the movement of the tick. The male is similar in shape to the female, but its back is more convex, and its legs, especially the hind ones, are stronger than those of the female. It is 0.2 mm long.

Distribution. The tick is widespread in all places where a warm and suitable climate prevails. Plagues of the pest are known in America, Central Europe, the Mediterranean area and Australia.

Life Cycle. These creatures are carnivorous, feeding on the body fluids of different insects especially those with soft skins which remain motionless when concealed within their food supply, such as grain- or wood-borers.

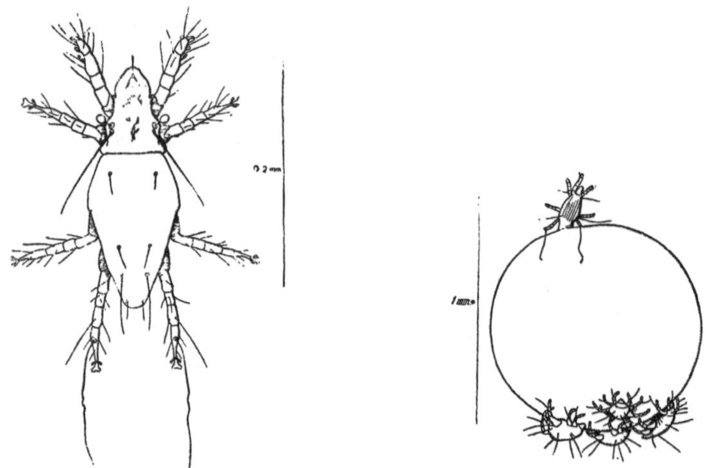

Fig. 93. *Pediculoides ventricosus* after SWAN. Courtesy Department of Agriculture, Adelaide.

The female bears living young which are, in fact, adult ticks. The young males emerge first, remain near the sexual aperture of the mother, and immediately fertilize the newly born females.

The young female immediately begins to feed and her stomach swells according to the amount of food and development of the embryos within her. A well-fed female is able to produce 200—300 offspring — the majority of them females and only about 3% males. The gravid period lasts for a week or more, depending on the prevailing temperatures.

Factors influencing the size of the population. The factors that cause an increase in the population of these ticks are hıgh temperature — about 27° C — and high humidity. Under laboratory conditions insect cultures in closed thermostats suffered chiefly from attacks by ticks. Another factor is abundance of food. If the cereal store or hay in the field is affected by itch ticks, then it is usually a sign that insects are present, especially cereal moths. weevils or other seed-feeding larvae.

In Israel, different degrees of infestation were noted in bundles of hay from various fields. This was due to the fact that in some of the fields the hay was mown late. Seed had already developed and had been attacked in the field by the cereal moth, resulting in a later infestation by the ticks.

Early mowing — at the time of flowering — before the seeds have formed (which is also better from the point of view of food value of the hay) prevents the development of moth larvae and, therefore, the development of ticks in the hay.

Control of Store Pests.

A. Factors favouring development of pests in store:

1. *Amount of moisture in the grain.* As stated, some insects attack only the soft grain with a high water content. The rice weevil cannot exist let alone develop in grain where the moisture content is less than 9%, and it generally selects grain whose moisture is above 11%.

Therefore grain should be dried to a point which does not permit the development of these pests. When the cereals are well dried, the development of moulds in the stores is also avoided. It is desirable to dry the cereals in the sun and to turn-over the layers from time to time so that the sun's rays kill any eggs and larvae that may have entered the grain while in the field.

The methods of storage adopted in many places allow numerous insects to attack the grain almost without interference. The storage of scattered grain in a store, or even in a house, makes possible a considerable increase of moths and beetles. It is customary for the Arabs to collect the grain into a pit in which the prevailing humidity allows the insects to continue their development and reproduction. Farmers also store their grain between partitions built inside the store. These compartments are not covered, and small insects often enter freely through cracks in the walls. The storage of grain in sacks does not solve the problem as insects can easily penetrate through the mesh.

The safest way of storage is to collect the grain into a storehouse that neither mice nor small insects can enter. It is desirable that such a storehouse be covered, built of tin or cement, but not of wood.

It is especially important that no moisture penetrates into the storehouse.

2. *Clean Storage.* The farmer often puts his newly harvested grain into a storehouse that has not been properly cleaned. If grain from the preceding years crop remains, then a greater attack of pests may be expected, especially if the storage conditions are unsuitable.

Therefore, the store or granary must be well cleaned of all remains of grain, straw, flour or dust and well aerated. For safety it is worthwhile to fumigate the store for some days. For such a purpose ½ of the amount of substance is required than in a store filled with grain.

B. Preventive measures in the control of store pests.

In ancient times it was customary to cover the heaps of grain with fine earth, wood ashes or other substances to avoid attack by pests. Sometimes these substances were mixed with the cereal grains and this is still the custom today in certain places. More recently fine pure quartz-sand has been used at the rate of 0.1% of the weight of the cereals. This substance is non-poisonous but sticks to the body of the pest, rubs off its defense layers, and withdraws the moisture from its body.

The substances which were tried included bentonite, magnesium carbonate, crystalline silica, the finely ground silica aerogel and others.

As stated, these materials abrade the cuticle causing rapid loss of water and death, therefore they are more effective at higher temperatures and low relative humidity (Chiu, 1939 a and b). Furthermore, the smaller the particles the greater their efficiency.

The comparative effectiveness of the substances is not the same in all cases. Thus Chiu (1939 a and b) finds that in the case of the bean weevil bentonite was the best in a series of six dusts and crystalline silica the third, while against the rice weevil crystalline silica was the first of the series of the same dusts and bentonite the fourth.

At times poisonous repellents such as naphthalene, or stomach poisons like barium fluosilicate were spread on the heaps or mixed inside. Such methods can be used only for grains intended for sowing and not for human or stock food. Great caution should be exercised since certain poisons, such as paradichlorobenzene, which have been mixed for a long time with the seeds, can impair their quality.

A 5% DDT dust at the rate of 1 kg per ton of seeds is effective for a whole season. The DDT may be mixed with fungicides as well.

C. Disinfectation of used sack and stores.

Old sacks are a source of infection by insects and they should be disinfected before use. They should be put into a barrel which can be hermetically sealed and disinfected by scattering over the sack 80—100 g of ethylene dibromide to 1 m³. The barrel should be then closed immediately and left this way for 3-5 days.

In the empty store various insecticides may be sprayed or dusted in order to control pests found in corners. Their effect will continue for an additional period. These substances are 2—3 kg of wettable DDT (50%) to 100 litres of water, 3—5 kg lindane emulsion to 100 litres of water, or DDT or lindane dusts at a concentration of 5%.

D. Direct control of store pests.

1. *Heating*. Temperatures above 60° C for a 10-minute period

will kill all insects which may be found on and in the grain. Experiments have shown that wheat seeds are not harmed by this or even a higher temperature as regards their germination or the quality of their flour. In such cases where the storage space is heated to a high temperature for a short period it is important that the temperature does not go too high and that excessive moisture is not present in the store.

In Israel, due to technical difficulties, this method is not feasible.

2. *Fumigation.* The most usual method of control is by the fumigation of different poisonous vapours through the infested grain. The fumigation is carried out as follows:

a. At the time of transfer of grain from one storehouse to another (for storage, or movement from place to place, or for aeration) the poisonous vapours are introduced into the stream of grain (this is usual under the conditions where large quantities of cereals are sent or stored in very big storehouses which are difficult to fumigate satisfactorily).

b. It is possible to fumigate any store containing cereals. The store must be hermetically sealed.

c. It is possible to fumigate sacks when covered with an impenetrable meshed cover. The cover must close and fit well, especially at the ends. Between the sacks and the covering a space must be left for movement of air.

d. It is possible to fumigate cereals when they are enclosed in an impenetrable containers. (Since penetration of vapour to depth takes time, the container should not be too deep; however, it also should not be too wide lest much vapour escape. The area of the surface of the cereal should be at least 30 cm lower than the sides of the container. If the cereals fill the container to the very top — the gas is apt to "flow-over" its sides).

The majority of fumigants used are heavier than air; they penetrate the spaces between the grains and kill the pests there or inside the granules.

As is well known, action of the fumigant is quicker when the temperature is higher. Accordingly, if the fumigation is carried out in winter, the vapour should remain in the spaces for 2 days longer than in summer.

The substance is poured over the grains, or put in wide open vessels resting on the surface of the cereals. If the container is very wide, the substance is put into a few saucers and the container is then covered.

The penetration of gas into heaps of different types of grain is not equal. The penetration depends to a large extent upon the space between each grain, and also on the power of absorption by the grain. CALDRON (1960), for example, found that with durrah the gas does not penetrate more than 6 metres, so that in the case of deep containers, they should be filled only to a depth of 5 metres, fumigated, refilled, and fumigated again, and so on until the container is full.

In Israel, a compound which was found very satisfactory for killing cereal pests is known as "Calendrex". It contains 20% carbon disulphide, 10% carbon tetrachloride, and 70% ethylene trichloride of boiling point 97° C, and specific weight 1.47 at a temperature of 15° C.

Persons fumigating should be careful not to inhale the vapours (it is desirable to use respirators).

The following are the quantities of the various substances required for a volume of one m^3 of cereals:

Carbon disulphide + carbon tetrachloride (1 + 4) . . 200 g
Three parts ethylene dibromide + one part carbon tetrachloride . . . . 200 g
Ethylene dibromide . . . . 80 g
Hydrocyanic gas . . . . 20–80 g

LITERATURE

E. A. Back, 1929. *U.S.D.A. F.B.* **1156,** *1-17.*
E. A. Back & R. T. Cotton, 1924. *J. agric. Res.* **28,** *1043–1044.*
E. A. Back & R. T. Cotton, 1929. *U.S.D.A. F.B.* **1483,** *1-35.*
L. C. Birch, 1945a. *Austr. J. exp. Biol. med. Sci.* **23,** *29.*
L. C. Birch, 1945b. *Trans. Roy. Soc. S. Austr.* **69,** (1), *140-149.*
F. S. Bodenheimer, 1927. *Z. wiss. Inskbiol.* **XXII,** *65-73.*
M. Caldron, 1960. Verbal communication.
S. F. Chiu, 1939a. *J. econ. Ent.* **32,** *240-8.*
S. F. Chiu, 1939b. *J. econ. Ent.* **32,** *810-21.*
R. T. Cotten & J. C. Frankenfeld, 1949. *J. econ. Ent.* **42,** *553*
R. T. Cotton, 1943. Insect Pests of Stored Grain and Grain Products. Burgess Publishing Co., Minneapolis, *1-242.*
N. E. Good, 1933. *J. agric. Res.* **46,** *327-334.*
N. E. Good, 1936. *U.S.D.A. Tech. Bull.* **498,** *1-58.*
A. B. Hadaway, 1956. *Bull. ent. Res.* **46,** *781-796.*
J. C. Hamlin, W. D. Reed & M. E. Phillips, 1931. *U.S.D.A. Tech. Bull.* **242,** *1-26.*
R. W. Howe & H. D. Burges, 1956. *Bull. ent. Res.* **46,** *773-780.*
D. L. Lindgren, L. E. Vincent & H. E. Krohne, 1955. *Hilgardia* **24,** 1, *1-36.* Univ. of Calif., Berkeley Col.
M. J. Oosthuizen, 1935. *Univ. Minn. Agr. Exp. Sta. Tech. Bull.* **107,** *1-44.*
H. Priesner, 1951. *Bull. Soc. Fouad 1 Ent.* **35,** *133-135*
W. Robinson, 1926. *Univ. Minn. Agr. Exp. Sta. Tech. Bull.* **41,** *1-43.*
H. H. Shepard, D. L. Lindgren & E. L. Thomas, 1937. The Relative Toxicity of Insecticides and Fumigants. *Univ. Minn. Agr. Exp. Sta. Tech. Bull.* **120.** *1-44.*
A. Shulov, 1955. *Proc. Indian Acad. Sci.* **XLII,** 1, Sec. B. Rep. *1-13.*
P. Simmons & G. W. Elsington, 1933. *U.S.D.A. Tech. Bull.* **351,** *1-34.*
D. C. Swan, 1934. *J. Agr. S. Aust.* **37,** *1289-1299.*
R. Tosi, 1929. *Boll. Lab. Ent. Bologne* **II,** *292-300.*

V. VEGETABLE PESTS

CUCURBITACEAE

KEY to Insects Injurious to Cucurbitaceae

A. The seeds do not sprout		
1. Upon examination, small white maggots are found within the seed coat . . . . :	*H. cilicrura*	page 131
2. Only remains of the seed are left . . . . :	Tenebrionidae	page 138
B. The roots are infested; as a result the fruit-bearing plant may suddenly wilt		
1. Upon examination smooth yellow larvae stick out of the root . . . . :	*Rhaphidopalpa*	page 273
2. The roots are cut and gnawed. Large fleshy white grubs are found in the ground . . . . :	Scarabaeidae	page 273
C. The stem is injured		
1. Newly sprouted plants wilt; maggots bore into the stem . . . . :	*H. cilicrura*	page 131
2. the stem of young plants has been cut at the soil surface; curled grey-green caterpillars are found in the soil . . . . :	*Agrotis*	page 88
3. Tunnels 15 mm in diameter may be found in the ground . . . . :	*Gryllotalpa*	page 44
4. Black beetles inhabit the area . . . . :	Tenebrionidae	page 138
5. The outer surface of root and stem is gnawed-off. Ants are in the ground . . . . :	Ants	page 282
D. The leaves are injured		
(I) Portions of the leaf have been consumed.		
1. Caterpillars or their faeces may be found on the leaves or in the neighbouring ground. . . . . . :	*Prodenia*	page 102
2. Red beetles are on the plant or fly about (on melons) . . . . :	*Rhaphidopalpa*	page 273
3. Red spotted beetles, or spiny yellowish larvae are on the leaves (watermelons) . . . . :	*Epilachna*	page 270
(II) Leaves harbour small insects which cause injury by sucking.		
1. Leaves are curled and many soft-bodied small insects inhabit the leaf; the presence of ants is common . . . . :	Aphididae	page 160
2. Small green elongate insects crawl quickly on the leaves . . . . :	*Empoasca*	page 53
3. On the underside of the leaves are pale spots from which chlorophyll has been removed; minute yellowish or grey elongated creatures are present . :	*Thrips*	page 74
4. On the underside of the leaves minute flat-bodied insects are found. Small white flies hover over the plant when disturbed . . . . :	*Bemisia*	page 56
5. Leaves become chlorotic; on the surface spider webs may be seen under which minute red round-bodied creatures abound . . . . :	Acarina	page 80

E. Blossom is injured

1. Red beetles feed upon the blossom . . . . . . :	*Rhaphidopalpa*	page 273

F. Fruit is injured

1. Very young fruit (watermelons) wilts and drops:	*Myopardalis*	page 282
2. The young fruit (watermelons) is partly denuded of its layer of chlorophyll . . . . :	*Prodenia*	page 102
	Trichoplusia	page 128
3. In young fruit (melons or cucumbers) the peel is gnawed, tunnelled or penetrated . . . . :	*Rhaphidopalpa*	page 273
	Agrotis	page 88
	Prodenia	page 102
4. Young fruit shows many punctures; the stem end is girdled, fruit is wilted or shrunken and upon opening many white grubs are within . . . . . :	*Baris*	page 278
5. Upon opening fruit, maggots are found within		
a. Maggots tunnel into the fleshy part of the fruit; their posterior spiracles are not elevated . . :	*Myopardalis*	page 282
b. Maggots are mixed with the seeds; their posterior spiracles are elevated on black conspicuous buttons . . . . :	*Atherigona* and others*	page 202

The 12-Spotted Melon Beetle — Epilachna chrysomelina F.

(Coccinellidae, Col.)

In the watermelon fields "lady bugs" are quite conspicuous. There are two species which superficially look very much alike since both are red with black spots. They differ, however, entirely in their habits. One is *Coccinella 7-punctata,* which feeds upon aphids. the other is *Epilachna chrysomelina,* which is phytophagous and injurious.

Description. The body of the adult is hemispherical and 6—8 mm

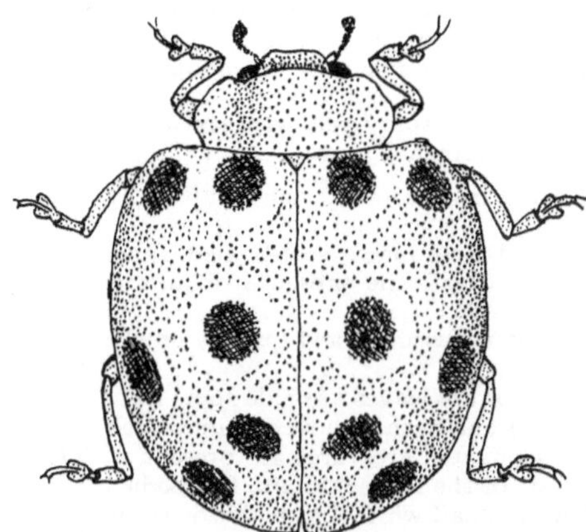

Fig. 94. *Epilachna chrysomelina* – adult.

* Various Muscoidea may be found on melon. From Cyprus and Israel for instance, *Lonchea aurea* was reported.

long. Its colour is brick red, decorated with 12 gold rimmed black spots. The distance between the two antennae is smaller than between the antenna and the eye. The elytra are covered with a golden pubescence. The egg is orange-yellow, elongate, 1.7—2 mm long, tapers at both ends, and is laid in clusters.

The body of the larva is lepismoid, it tapers posteriorly and is 8—9 mm long. Its colour is yellowish. The head is round and 6 ocelli decorate each side. The body is covered with ramified spines, six on each body segment except for the prothorax and the last abdominal segment which have only four.

When ready to pupate the larva attaches the tip of its abdomen to the object upon which it is found, head-downward. The skin at the pronotum splits and the pupa is thus partially exposed. It remains in the old larval skin until it matures.

Distribution. This beetle is typical to the Mediterranean countries and reaches as far as the shores of the Black Sea.

Hosts. All cucurbit plants may be attacked. Preference is shown, however, to watermelon and melon. Among the uncultivated plants it feeds upon *Bryona* and *Ecballium.*

Type of Injury. Both larvae and adults feed upon the foliage. The neonate larvae feed upon the soft tissue, leaving the veins. In case of a heavy attack the plant may be entirely defoliated. Attacked plants do not yield fruits.

Life History. The adults awaken towards the end of the spring when young cucurbits are already in the field. If the temperature is favourable, copulation and egg-laying soon begin. The eggs are laid in clusters on the underside of the leaf and are arranged in rows one next to the other. From a few to 50 eggs may be laid in one cluster. The incubation period is, according to MELAMED (1956) 3—6 days at 27—30° C, and 9—13 days at 19° C. Thus the threshold of development was found to be 14° C. In fact eggs which were retained at this temperature as well as at 16° C failed to hatch (MELAMED, 1956).

The larva moults three times before it is ready to pupate. The complete development of the larva lasts 14 days at a temperature of 28° C, and 27 days at 22° C. The threshold of development is thus 14.5° C.

Often the larva wanders away in order to pupate on some object other than the host plant. The development of the pupa lasts 7 days at 26° C, and about 10 days at 23° C. The threshold of development is 13° C.

The optimal temperature for the pupal development is 22—23° C; any deviation from this temperature increases the mortality of the pupae.

The preoviposition period may last a few days only at 25—26° C, but 15 days passed before females laid when reared at 23—24° C.

One female of the first generation may lay on the average of 650 eggs, the maximum being 1500 eggs. Females of later generations laid on the average of 300—400 eggs. A few dozen eggs may be laid in one day. The adults are long-lived. Non-diapausing beetles may live 100—200 days, hibernating beetles 250—300 days.

Limiting Factors. As the temperature rises above the optimum, the mortality of larvae and pupae increases. From the accompanying table XLVI it seems that the pupae are more susceptible to such changes than the larvae.

Table XLVI.

Mortality of larvae and pupae of *E. chrysomelina* at various degrees of temperature.

Insect stage	Mortality at 24—25.5° C (favourable temperature)	Mortality at 28—29.5° C (unfavourable temp.)	Increase in mortality
Larva	70%	87%	26%
Pupa	20%	77%	285%

The longer the incubation period the greater the mortality of the egg. Hence the smallest mortality of eggs occurs at 26—29° C, while at 21—24° C eggs died in great proportions. In fig. 95 it is noted that the eggs are susceptible to low humidities as well as to

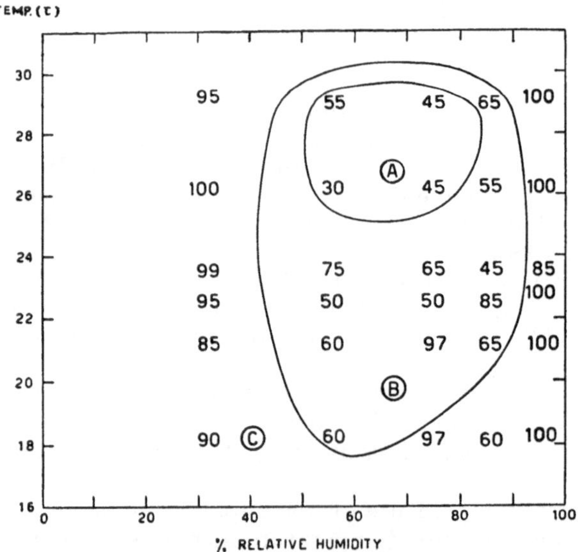

Fig. 95. Effects of temperature and humidity on the mortality of *Epilachna* eggs. Numbers indicate % of mortality of eggs at the particular temperature-humidity-combination after MELAMED.

the relative humidities above 87%. The most favourable humidity was that of Zone A., e.g. a 60—80% R.H. and 26—29° C.

In addition to these factors the beetle has a very high threshold of reproduction. No egg-laying took place at a temperature below 23° C. In view of this no reproduction takes place early in the spring or late in the autumn although the beetles may be active in the field. Fewer generations are raised, therefore, during the summer.

Control. In laboratory tests in Rehovot it was found that of the five insecticides tested dieldrin was hardly effective, DDT had very little effect, but methoxichlor, sevin and parathion were satisfactory.

May Beetles — Scarabaeidae, Col.

Scarabaeid grubs occasionally cause damage to young cucurbit plants. There may be a few species involved, but not all have been identified. From Syria *Polyphylla fullo* L. is reported as a pest in the vegetable garden while in Israel and Egypt one of the species which was reared from such larvae was *Pentodon* sp.

Description. The species belonging to this genus are large beetles about 20 mm long. They are shiny dark brown and the ventrum is hairy. The prothorax has a distinct concavity, the anterior tibiae are dentate and the elytra punctate. The egg is white, elliptical and 2.5 mm long. The larva when mature is about 30 mm long, curved into a U-shape, with a brown head and strong mandibles. The pupa is purplish-white and is found in the ground in a cell made of glued soil particles.

Hosts. The larvae may attack plants of various kinds including grasses and even shrubs. In the vegetable garden they often destroy cucurbit plants.

Type of Injury. The larvae feed upon the roots of the plants and chewing off the main root of a young plant is sufficient to destroy the entire plant.

The life history and its larval habits are very similar to those of *Phyllopertha nazarena,* and the reader is referred to the full account of that species.

One difference in the habits of the adults should be pointed out. *Pentodon* sp. feeds upon the foliage of various trees, while the adults of *Phyllopertha* do not feed at all.

Control. Dusting the ground around the soil with lindane or aldrin yielded satisfactory control and may serve also as a preventive measure.

The Red Pumpkin Beetle — Rhaphidopalpa foveicollis Luc.

(Chrysomelidae, Col.)

During April and May, when the cucurbit plants are still very young, red elongate beetles may appear in the field and remain there throughout the summer. These belong to the species *Rhaphidopalpa foveicollis*.

Description. The adult beetle is 6.5—7 mm long. Its colour is red except for the underside of the meso- and metathorax and abdomen which are black. The eyes and mandibles are also black. The antennae are eleven-segmented and are inserted in the middle between the eyes. The first segment is longer and thicker than the others. The species *R. foveicollis* has a transverse furrow on the pronotum, and the last abdominal segment of the male has three pointed lobes.

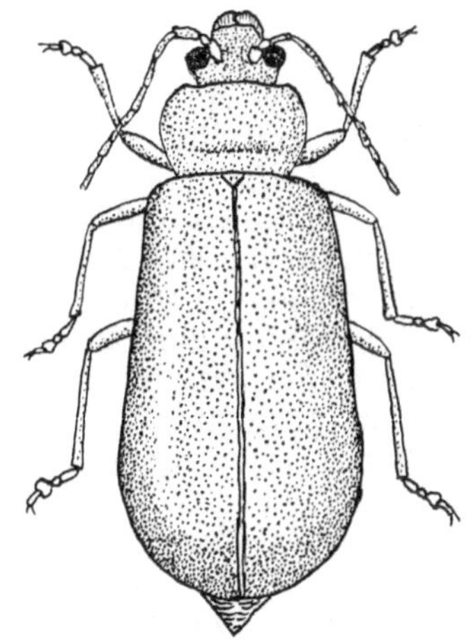

Fig. 96a. *Rhaphidopalpa foveicollis* adult.

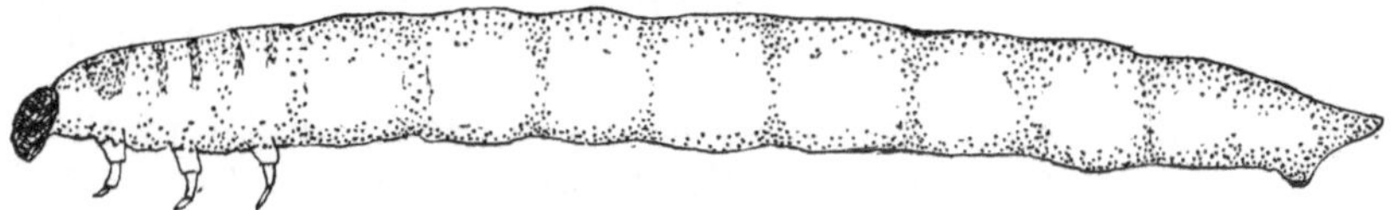

Fig. 96b. *Rhaphidopalpa foveicollis* larva.

The egg is elliptical, yellow and 0.6 mm long. The larva is cylindrical, long, and lemon yellow. When mature it is 15 mm long and 1.5 mm broad. The head is black dorsally and yellow ventrally. The ninth abdominal segment is hardened. Beneath it the margins of the anus protrude into a sucker, with the aid of which the larva fastens itself to objects, thus serving as an auxiliary leg.

Distribution. This beetle is distributed in all countries around the Mediterranean basin except in Southern France. It is found also in

Portugal and Southern U.S.S.R. Damage was reported from North Africa, Italy, Greece, Turkey and Israel.

Hosts. All Cucurbitaceae are subject to attack by this beetle. Preference is shown, however, to cucumbers and melons. Watermelons and squashes attract the beetle to a far lesser extent. The adult may be found in clover fields and, according to PAVLAKOS (1943), adults may feed on tomatoes and maize. Larvae, however, develop on Cucurbitaceae only.

Type of Injury. The adults feed upon the soft leaves, buds and flowers. Early in the season, when the plants are young and the awakening beetles are voraceous, they may destroy the entire plant. In larger plants buds and flowers are eaten. The larvae are troublesome in a different way. These feed upon the roots of the host plant and bore into them. In addition to direct injury bacterial rot of the roots develops and the entire plant may succumb suddenly. Entire fields may thus be destroyed. The damage is aggravated because it often occurs a week or two before the harvest. Injury of a third type may be caused by the larvae which may emerge from the ground and damage the fruit which lies closely. The peel of the fruit is gnawed and the larvae penetrate into it (RIVNAY, 1954. Fig. 97). Such injury was reported also from Australia as caused by *Rh. abdominalis* (MAY, 1946).

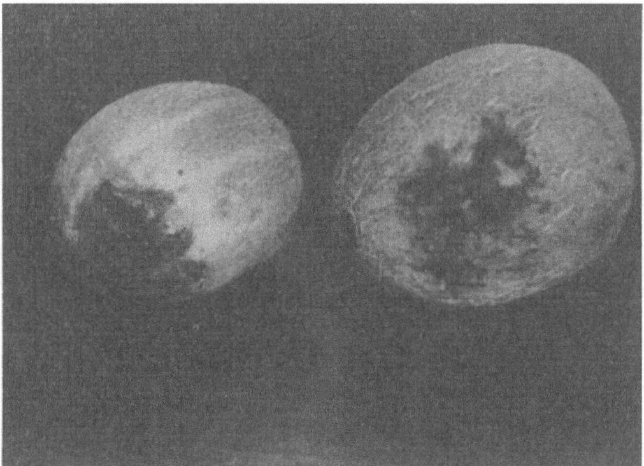

Fig. 97. Cantaloups injured by larvae of *Rhaphidopalpa*. After RIVNAY, Courtesy Hassadeh.

Life History. The adults spend the winter in a diapause. They awaken in the spring with the rise of temperature. In Europe they may occur in April and in Israel even in March. If cucurbit plants are not available they feed upon plants of other kinds. Copulation takes place and as soon as melon plants are available, egg-laying begins.

The eggs are laid singly or in small groups in the ground under earth particles close to the stem of the plant. Under favourable temperatures the egg hatches after 9—14 days (PAVLAKOS). At Rehovot, the incubation period was 9—12 days at 27—30° C, and 15 days at 21—23.5° C. Eggs never developed at a temperature below 20° C which shows that this is near the threshold of incubation (MELAMED, 1960). Upon hatching the larvae bore deeper into the ground and feed upon the roots into which they penetrate. Only half developed or mature larvae may be found within the roots. Neonate larvae were never found in the field in spite of a search, so that it is not known how they feed in the field. PAVLAKOS claims that he reared the larvae in all the stages upon slices of melon. In the first two or three days they fed on the surface, but later bored into the tissue. According to this author the larvae moult three times before they pupate. The larval period lasts 29 days.

When ready to pupate the larva leaves the root and pupates in a cell in the ground. The pupal development lasts 8—10 days, according to LUCCHESE (1945), or 15—18 days, according to PAVLAKOS. Neither author mentions the temperature. According to MELAMED the pupal development lasted 17—18 days at 28—28.5° C.

A long preoviposition period may pass before the female lays. Of 12 females of the first generation which matured during the latter days of June, 2 females laid 13 and 19 days after maturing, 6 after 21—28 days, while 4 laid after 31—38 days. Of four females of the first generation which matured in the middle of July, one laid 17 days, two — 30 days, and one 52 days after maturing.

The number of eggs one female may lay ranges from 16—697 (PAVLAKOS). LUCCHESE claimed on the average of 300 eggs per female. In the warmer climate of Israel, MELAMED (1960) obtained a greater number of eggs as seen in Table XLVII. According to this table,

Table XLVII.

Amount of egg laying (after MELAMED, 1960)

Number of eggs per female	Number of females		
	1955	1956	1957
1— 100	10	7	12
101— 500	12	6	2
501— 800	10	8	10
801—1300	5	4	10
Total	37	25	34

of 96 females 26% laid less than 100 eggs, 26% laid 500—800, while 20% laid 800—1300 eggs.

Not all the females lay. Even hibernating females did not all lay. The number of the non-laying females was still smaller in the first generation (table XLVIII).

Table XLVIII.

The number of laying females in various years and generations (after MELAMED, 1960)

Field collected overwintering beetles			First generation beetles		
Beginning of rearing*	Number of females	% of egg layers	Date of emergence	Number of females	% of egg layers
4/4/55	57	40	21—24/6/55	21	57
24/4/56	29	86	14—23/7/56	96	9.5
17/4/57	56	57	9/8/57	8	12

* Date on which the beetles were caught in the field, after emerging from hibernation.

Number of Generations. LUCCHESE (1945) who studied the biology of this beetle in Salerno believes that the insect produces more than one generation there, while PAVLAKOS (1943) in Greece states that this insect produces but one generation. SHWEIG (1954) in Israel calculated 2—4 generations depending upon the climate of the locality. The findings of MELAMED at the Rehovot coastal plain may be summed up as follows: Egg-laying begins in April. The first generation appears in June-July. These lay in July-August and the offspring become adult in September-October. These lay only in the case of a warm November, otherwise they hibernate. There are, thus, two generations; in the inner valleys there may be three.

Limiting Factors. Both PAVLAKOS and MELAMED point out the necessity of a high percentage of moisture for the egg to develop. In fact, the egg must have contact moisture — otherwise it does not hatch. Larvae, too, need a high degree of moisture in order to thrive. However, too much moisture in the ground, as may occur immediately after excessive overhead irrigation, drives the larvae out of the ground. It becomes evident, therefore, that the quantity and type of irrigation may to a great extent regulate the infestation. This explains why, when cucumbers were grown without irrigation or watered in ditches, the beetle was less abundant, while today, with the increase of irrigation, more beetle damage is reported. Observations in infested fields have pointed out that infestation is more prevalent in the lower localities of the field.

From the point of view of agricultural practice the situation was summed up by RIVNAY (1956) as follows: In non-irrigated fields the conditions are unfavourable for both eggs and larvae. Very few eggs hatch because the upper layers of the soil are dry. In fields irrigated by canals and ditches which are, as a rule, distant from the stem of the plant, conditions are unfavourable for the eggs but may be suitable for the larvae. Where there is overhead irrigation, conditions are favourable for both eggs and larvae, and the infestation may be severe. In addition to these factors, the abundance of food should be considered. With the increase of irrigation, a continuity of hosts throughout the summer is established. Excessive overhead irrigation drives the larvae out of the ground to attack and damage the fruit.

Control. Preventive measures. In view of the above conditions it is advisable to adopt the canal irrigation system whenever possible. For these crops there are other assets in this type of irrigation.

In laboratory tests at Rehovot, DDT and dieldrin caused satisfactory kill of adult beetles. Dusting the field with these materials should efficiently control the pest. Application of insecticides is simple when the plants are young, but with older plants the foliage prevents the substance from reaching the hidden places where the beetles usually feed, and more care should be exerted in the application. No doubt other substances may also be efficient against this pest, and when there is fruit, such as cucumbers, choice should be made in favour of one material whose residues are the least toxic to human beings.

The Melon Weevil — Baris granulipennis Tour.

(Curculionidae, Col.)

In watermelon fields in Egypt and Israel young fruit may be attacked by small black weevils which gnaw small holes in the peel. Fig. 98a. This weevil is *Baris granulipennis* TOUR.

Description. The length of the body excluding the rostrum is 5 mm, the width is 2 mm (often small individuals may be found which measure 3 mm in length). The colour is entirely black. Concave dots cover the entire body. These are rough on the thorax and finer on the abdomen, which is also more glistening than the rest of the body. The head is smooth. The length of the rostrum is 2 mm, and when bent backwards it reaches the mesocoxae. The elytra have nine rows of dots which divide each elytron into 10 strips; the side strips are bent ventrad and are narrower in the centre. The last abdominal segment is exposed.

The egg is white, translucent and elliptical. Before hatching it is 0.5 mm long and 0.2 mm wide; a newly laid egg is much smaller. The oviposition hole is bent.

Fig. 98a. Young watermelons with oviposition punctures by *Baris granulipennis*. after RIVNAY – Courtesy Comm. Inst. Ent.

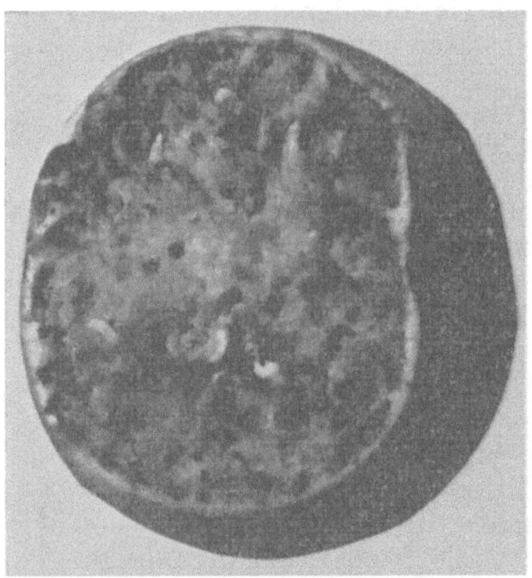

Fig. 98b. A watermelon infested with larvae of *Baris* – note the spongy contents. after RIVNAY – Courtesy Cmwlth. Inst. Ent.

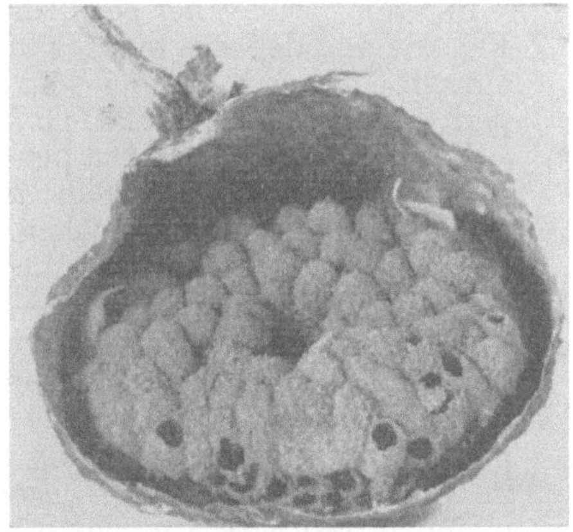

Fig. 98c. *Baris* cocoons in dried watermelon peel. after RIVNAY – Courtesy of the Cmwlth. Inst. Ent.

The larva, before pupation, is 9 mm long and 3 mm wide. It is white and a pink median line runs along the dorsum. The larva is widest at the posterior third and tapers slightly towards the head. The head is red, the labrum and mandibles brownish-red.

The pupal cell is elliptical, 6—8 mm long, 4—5 mm wide, and may be brown of various shades. They are found in the dry fruit, located one close to the other. Fig. 98c. The pupa is entirely white, 5 mm long and 3 mm wide. The rostrum reaches the third of the body the wing pads four fifths of the body. The length of antennae is one mm.

Distribution. This beetle was first described in Egypt; it is probably found in North Africa and in the neighbouring countries. In Israel it has become a pest in recent years and quite possibly it is found also in neighbouring countries such as Jordan and Syria.

Hosts. This weevil attacks only cucurbit fruit. In Israel damage is caused to watermelons, melons and cucumbers. Of the non-cultivated plants *Citrullus vulgaris* serves as a host.

Type of Injury. The female lays her eggs in the young fruit. Before doing so she girdles the stem of the fruit and thereby the growth of the fruit is stunted. The larvae feed on the pulp and seed, and gradually the fruit shrinks and dries. Fruit in which the stem was not girdled may develop to normal size but with larvae within. The taste of this fruit is tainted. Such fruit may be recognized by the scarred oviposition holes on the surface.

Life History. The adult spends the winter in diapause and awakens late in the spring. In Israel this occurs at the end of May or early June. If the amount of injury in the field is considered as a criterion then the height of activity of the hibernated generation takes place during June.

In order to lay her eggs, the female gnaws cavities in the peel Fig. 98a and deposits one egg in each. The cavity is then closed with a mash. The egg is found 1—1.5 mm deep in the peel. The incubation period lasts 3 days at 26° C, and 6 days at 22° C (Rivnay, 1960).

Upon hatching the larva penetrates deeper into the fruit, feeding upon the pulp and soft seeds. If the seeds are hard, the shell is left empty. In the summer, when the temperature outdoors reaches above 30° C, the larva may complete its development in 12—14 days. In several rearings at 26—27° C the larval development lasted 18—20 days. However, the larva may remain in this stage without pupating if humidity conditions are unfavourable. As a rule, the infested fruit dries and suitable conditions are created which allow normal pupation. When ready to pupate the larva builds a cocoon in which it pupates. This cocoon is made from mash of pulp remains, mixed with a mucus secreted by the malpighian tubes. The mucus is milked from the anus with the mandibles and is mixed with a pellet of the mash which the larva had taken into its mouth.

Three days after the completion of the cell, the larva moults and changes into a pupa. At a temperature of 27° C about 12 days pass before the adult emerges. The adult remains in the cell a few days longer and does not emerge from it or from the dried fruit until it has hardened and become black. Four days pass before the beetles mate and begin to lay.

As mentioned above, only soft fruits are chosen, as a rule from the size of a walnut to the size of a grapefruit. Many eggs may be laid in one fruit. Some samples brought from the field contained over 100 eggs each. In the laboratory, beetles which were reared on cucumbers laid a few eggs only. The highest number of eggs laid by one female was 23. It is quite possible that in nature the females are more fecund. The egg-laying period lasts a few weeks, and beetles may live several days after the reproduction period. Beetles of later generations may live many days since they enter diapause and wake up the following summer. Beetles of the first generation also have a long life. From such beetles reared in the laboratory none died before the age of 24 days while some lived three months. Table XLIX shows that nearly half of the beetles lived over 50 days.

Table XLIX.

Length of life of the weevil *Baris* of the first generation

Number of beetles	11	22	18	7	1
Lived number of days	24—30	31—50	51—70	71—90	99

Number of Generations. Both from laboratory rearings and from infested fruit collected in the field it is evident that beetles of the first generation appear in July. This is to be expected since the hibernated females lay in early June, and the life cycle lasts about a month. The second generation appears thus in August. Most of these do not reproduce the same summer; the few which may lay eggs raise a third generation which matures in September and October. These may not reproduce, but in fact fresh infestation was found in the field as late as November. However, these eggs do not mature into adults the same year; more likely they perish because of the cold and moisture.

Control. Control measures should be applied against the adult, and efforts should be made to kill them before they oviposit. From laboratory trials it was found that DDT and toxaphene have very little effect against the adults, dieldrin is better, but most effective are lindane and parathion.

Ants-Formicidae

Watermelon plants may wilt suddenly; upon examination of the plant it is revealed that ants have established themselves in the ground near the stem and fed upon it. Fig. 99. The ants which are of the habit of doing such injury belong to the genus *Tetramorium*.

On one occasion as many as 10% of the plants were thus attacked. Relief was obtained when dieldrin was dusted on the ground near the stem. As preventive measure dusting of the healthy plants was recommended.

Fig. 99. Watermelon roots injured by ants.

The Melon Fly — Myopardalis pardalina Bigot.

[Syn.: *M. caucasicus* ZAI.]

(Trypetidae, Dipt.)

The damage caused by this melon fly is not equally felt every year. For instance, in Israel between 1943—1956, damage was not noticeable; but in certain years outbreaks may occur which claim heavy tolls from the farmers, as was the case in that country in 1957 and 1958.

Description. The male fly is 4.5 mm long and the female 5.5 mm. The head is yellow, the frons near the base of the antennae is brownish; the thorax is yellow and three dark stripes run longitudinally over its notum. The pleurae are decorated with dark patches. On the scutellum there is a brown patch at its base, and two lateral spots which extend to the metathorax. In addition there is an 8-shaped spot at the tip of the scutellum. The legs are yellow, the wings are transparent and four yellow bands bordered with brown, run across the wing; the fourth band runs along the margin of the wing and is connected with the third one. The abdomen is yellow and its tip darker. The egg is white, cylindrical, and tapers at both ends. It is 1—1.2 mm long and 0.2 mm wide.

The neonate larva is 1.75 mm; before pupation it is 10 mm long and 1.75 mm wide. Like other maggots its body tapers anteriorly and the head is small, elongate and hidden in the margins of the prothorax. On its ventrum rough girdles decorate the segments. These help the maggot in its boring process. The pupal case is 5.8 mm long and 2.5 mm wide.

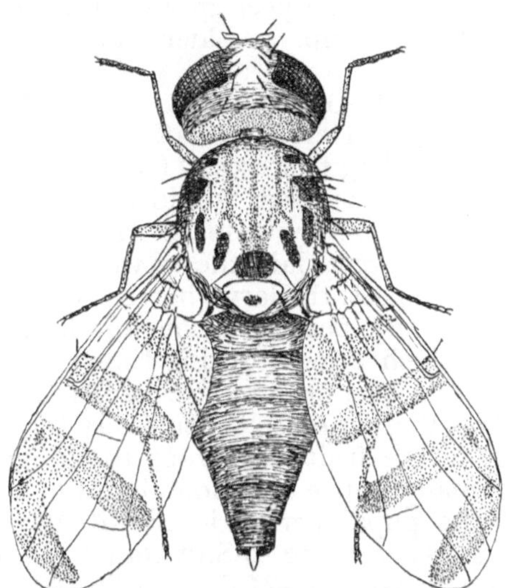

Fig. 100. The melon fly *Myopardalis pardalina.*

Distribution. This fly was first described from specimens obtained in Baluchistan in 1891. It is typical to the whole Irano-Turano Zone and damage was recorded from India, Afghanistan, Iran, Iraq and Transcaucasia. In this latter country it appeared for the first time at the end of the second decade of this century and was described as a new species — *M. caucasicus* ZAITZEV. Later the fly spread still farther west. It was recorded from Turkey, Syria, Cyprus, Israel and Egypt but in these countries it seems to be of recent arrival.

Hosts. The melon fly is restricted to plants of the cucurbit family. Melons of all varieties — watermelons, cucumbers, vegetable marrow, and even pumpkins — are attacked although not to an equal degree. Of the non-cultivated plants, *Ecballium elaterium* serves as a host.

Type of Injury. The female oviposits in the fruit, and the larvae which hatch feed upon the fruit pulp. When attacked, young fruit may dry and drop. Larger fruit may remain on the plant but growth is stunted; this is the case especially with cucumbers. Still larger fruit may mature, but it is "wormy fruit". AHMED JANJUA (1954) reports 50% damage; in Israel 90% damage was also reported.

The oviposition hole may be detected by the drops of gum which ooze from the peel. A slight scratch in this place reveals the freshly-laid egg. After a few days the peel grows thick and the egg sinks deeper and closer to the pulp. In older fruit the hole cannot be detected.

Life History. The adult appears in the spring. A few days after emerging from winter quarters, copulation takes place and soon after, oviposition. According to REKACH (1930), one female may lay about 100 eggs. As a rule one egg is laid in each oviposition hole although several such holes may be made without oviposition. Two or three days after the egg is laid the larva hatches. By then the egg is quite deep in the peel; the larva bores even further, and after a few days of feeding in the peel, it reaches and feeds upon the pulp.

In temperatures which prevail during the summer it completes its development within two or three weeks. According to JANJUA this period lasts 12—15 days only. When ready to pupate the larva bores its way out of the fruit and drops to the ground where it pupates. In Israel many pupae were found also in the fruit.

The length of the pupal period is 13—20 days according to REKACH, or 5—7 days according to SHWEIG (1954). YATHOM (1961) found that this period lasts 13—24 days at 26—28° C, and 19—23 days at 25—26° C.

Diapause. The fly overwinters in a obligatory diapause in the pupal stage. REKACH states that 25% of the pupae of the first generation and 90% of the second generation in Caucasia went into diapause and emerged as flies only the following spring. In Israel, according to YATHOM, of 4408 pupae in September 99.3% went into diapause and emerged the following spring.

Number of Generations. REKACH states that in Caucasia two and part of a third generation may develop. Flies of the overwintering generations begin to appear early in June and continue to emerge until July. There is a lapse of one month between one generation and another. There is, thus, much overlapping of generations. In Israel the flies appear by the end of May, and continue to emerge until the end of June. The flies of the first generation appear in July, those of the second in August and those of the third in September. As stated above these mostly enter a state of diapause.

Control. A cover spray with dieldrin or parathion should give satisfactory control of the pest.

In laboratory trials a bait of fresh beer yeast greatly attracted the flies, and diazinon proved to be the satisfactory insecticide used with this bait (YATHOM). Bait applications should be made over large areas and at close intervals, otherwise complete control is not obtained.

The Melon Leaf Miner — Liriomyza bryoniae Kalt.

[Syn.: *L. citrulli* RODEND.]
(Agromyzidae, Dipt.)

In Israel minute larvae mine in the leaves of all kinds of melons. Until a few years ago, before the 1950's, this pest was hardly noticeable, but today it is abundant to the extent where control measures are necessary. The larvae were reared in the laboratory and the adults obtained were identified as *L. bryoniae* KALT.

Description. The adult fly is 1.75 mm long and the wing span is 2 mm. The colour of the head, frons, face and genae is yellow. The occiput is brown, the thorax entirely yellow, and the abdomen yellow except for the tergites which are brown with yellow posterior margins. The pupal case is 2 mm long, brown and its segments are quite convex.

Distribution. Damage to cucurbits by this pest was recorded from USSR and Israel. The pest is probably also abundant in neighbouring countries.

Hosts and Injury. This pest is polyphagous, but cucurbits are its favourable hosts. The leaves of watermelons and melons may be so severely attacked that the infested area may reach 40—50% of the total area of the leaf. According to RODENDORF (1950), the plants may be attacked soon after sprouting, the cotyledons become mined and larvae also enter the young stem.

Life History. The life history of the pest has not been studied and very little is known about it.

In Israel the injury is already felt in the month of May, and from observations it is apparent that a few generation develop during the summer at three weeks intervals.

Control. Farmers who sprayed their melon fields with parathion or dimecron claim to have obtained relief from this pest.

General considerations about control practices of Cucurbitaceae.

Often irreparable damage may be avoided if measures of control are taken in due time. Cleaning the roads and margins around the field will reduce infestation of spider mites and *Agrotis* larvae.

Treating seeds against seed maggots and the soil against Tenebrionidae as preventive measures may serve the farmer the trouble of reseeding.

Ordinary applications with contact insecticides against plant lice will not always yield satisfactory control because the material will not reach the insects hiding within the folded leaves. Either a dust or a spray with systemic insecticides should be applied. The latter should be applied only when the plant is young or several days before harvest. These systemic sprays may kill other kinds of sucking insects that may be on the plant. Some insecticides such as DDT and BHC are injurious to Cucurbitaceae. Only safe materials should be chosen as well as those which may kill two or three pests together.

The farmer should watch for the first signs of fruit flies and should carry out preventive measures against them.

LITERATURE

Nazeer Ahmed Janjua, 1954. *Indian J. Ent.* **16,** *227-233.*

E. Lucchese, 1945. *Boll. R. Lab. Ent. Agr. Portici* **5,** *274-295.*

A. W. S. May, 1946. *Qd. Agr. J.* **62,** 3, *137-150.*

Venezia Melamed, 1956. *Ktavim* **7,** *83-95.*

Venezia Madjar Melamed, 1960. *Ktavim* **10,** *139-145.*

Halim Najjar, 1936. Circ. Inst. rur. Life Near East, American University, Beirut, No. **8,** 10 pp.

J. Pavlakos, 1943. *Z. ang. Ent.* **30,** *1-78.*

V. N. Rekach, 1930. Bull. Azerbaijan Centr. Agric. Plant Breeding Expt. Stat. no. **9,** 32 pp.

E. Rivnay, 1954. *Hassadeh* **34,** (1) *53-54* (in Hebrew).

E. Rivnay, 1960. *Bull. ent. Res.* **51,** 1, *115-122.*

B. B. Rodendorf, 1950. *Ent. Obozr.* **31,** *82-84.*

A. A. Shtakelberg, 1928. *Isr. Otd. pukl. Ent.* **III,** 2, *281.*

K. Shweig, 1954. Vegetable Pests. Hassadeh Publishing Co, *118-121.* (in Hebrew).

S. Yathom, 1961. *Hassadeh* **41,** *1054.* (in Hebrew).

Ph. Zaitzev, 1919. Scientific conclusions. Dept. Tiflis Bot. Gardens No. **1,** *1-4.*

CRUCIFERAE

KEY to Pests of Cruciferae

A. The roots and subterranean stem are injured.

I. The plant fails, the stem is gnawed or cut

1. Tunnels in the ground in the vicinity of the plants . . . . : *Gryllotalpa* page 44
2. Large white grubs are in the ground . . . . : Scarabaeidae page 273
3. Greenish-grey curled caterpillars are in the ground . . . . : *Agrotis* page 88

II. The plant is chlorotic, its development impaired, or wilting

1. The cortex is gnawed but is dry, ants are found around it . . . . : Ants page 282
2. The cortex is moist and rotting; many white maggots feed in it . . . . : *H. brassicae* page 309; *H. cilicrura* page 131

B. The stem is injured

1. Small caterpillars bore into the "heart" of the young plant, entrance is from the top . . . : *Hellula* page 308
2. Large legless grubs bore into the stem or stalk of both young and older plants, entrance from the side . . . . : *Lixus* page 295

C. The leaves are injured

I. The injury is by chewing insects

1. The main veins are bored, or the underside soft tissue of the leaves eaten, the net of veins being left. . . . . : *Plutella* page 304; *Pieris* page 298
2. Small numerous holes are made in the leaf . . : *Phyllotreta* page 293
3. Part or all of the leaf has been eaten by caterpillars.
   a. The caterpillar has more than 5 pairs of abdominal legs; large head . . . . : *Athalia* page 296
   b. The caterpillar is smooth, pale-green, tapering, with 3 pairs of abdominal legs . . : *Trichoplusia* page 128
   c. The caterpillar is velvety-green with 5 pairs of abdominal legs . . . . : *P. rapae* page 301
   d. The caterpillar is hairy, striped, with 5 pairs of abdominal legs . . . . : *P. brassicae* page 298

II. The injury is by sucking insects

1. Soft-bodied plant lice abound on the leaf-stalks and inflorescences . . . . : Aphids page 60; *(Brevicoryne)* page 288
2. On the leaves there are pale areas devoid of chlorophyll — long, greyish-yellow 1 mm long insects . . . . : Thrips page 74
3. The injured leaf is chlorotic devoid of chlorophyll — round red creatures are on the leaf in the corners of the veins . . . . : Acarina page 80

D. The foliage, stem and inflorescence may be attacked
1. Very numerous narrow elongate grey bugs abound on the plant . . . . : *Nysius* page 289
2. Large Pentatomid green bugs suck the plant: *Nezara* page 49
3. Large Pentatomid colourful bugs suck the plant sap. . . . . : *Eurydema* page 291

The Cabbage Aphis — Brevicoryne brassicae L.

[Syn.: *Aphis brassicae* L., *A. raphani* SCHRANK, *A. floris rapae* CURT., *Siphocoryne brassicae* GORT., *Brachycolus brassicae* L.]
(Aphididae, Hom.)

Of the various aphids which feed upon Cruciferae, *B. brassicae* alone is restricted to this family.

Description. Apterate females: The body is dark green or light green, covered with a waxy meal. The head is dark and dark maculae run across the abdomen. The cornicles are dark brown and the third segment of the antennae is twice as long as the fourth. The length of the body is 2.4 mm; the length of the antennae 1.60 mm, the length of the cornicles 0.16 mm and of the cauda 0.20 mm.

Alate: The body is entirely green, covered with a waxy meal. The head and the thorax are dark grey and dark maculae run across the abdomen. The cornicles are brown and the cauda dark green. The third segment of the antennae is more than twice the length of the fourth and within it there are 50—60 sensory pores. The cornicles are small and thick in the middle. The length of the body is 2.5 mm, the length of the antennae 2 mm.

Distribution. This is a palearctic insect which inhabits Europe from the northern regions to the Mediterranean. It spreads eastward to Japan and Formosa and reaches India and Australia. It is found also in both North and South Africa and in America. In the Near East it has been recorded from Turkey, Cyprus, Lebanon, Syria, Iraq, Israel and Egypt.

Hosts. Only cruciferous plants serve as food to this species but not all the plants of this family are equally attacked. Thus, the genus *Capsella*, for instance, is attacked less than the genus *Brassica*.

Type of Injury. In soft, young plants the feeding of the aphids brings about leaf curl. In the cabbage, aphids penetrate to the heart and disturb the formation of a normal head; in cauliflower, the presence of numerous aphids causes the head to become sticky because of honey dew, and non-attractive because of the fumagine which develops thereon. The sucking of the aphids causes injury to seed plants. Infested plants yield less or no seed.

Life History. Towards the autumn agamic alates gather close to the ground at the base of cabbage or cauliflower plants. Amphigonic females and males are born. These females bear sexual females which mate with the males and, as a result, lay 3—4 eggs each. The eggs spend the winter in diapause and hatch the following spring. The females which hatch in the spring are the agamic females which raise several asexual generations during the summer. Naturally the number of generations in a certain country depends upon its climate. According to ELZE (1944) the development of the species is 15—20 days in the winter and 8—10 in the summer. Length of life of the adult is 19—39 days in the winter and 8—25 days in the summer.

Like other aphids in warmer countries, this species may reproduce parthenogenetically throughout the year. In Israel, no sexual females are found in nature, but ELZE obtained sexual individuals in laboratory rearings, due probably to dispersed light.

The number of offspring one female produces is from two to three dozen. The first generations in the season are more prolific than the latter generations whose power of reproduction diminishes to 25% of that of the earlier generations. In Israel the species feeds upon cultivated plants in the winter, and on wild plants during the summer.

Ecological Notes. This species can withstand adverse conditions of temperature; ten degrees centigrade below the freezing point did not affect the species. As to higher ranges of temperature, the author saw healthy colonies of the species after a spell of 42° C heat which lasted at least 2 hours. These qualities cause its high range of distribution.

Still one cannot overlook the fact that in Israel, as the period of khamseen days arrives in the spring, the population of the species dwindles. The most favourable period for its existence in Israel is the winter.

Control. As stated above, the body of the insect is covered with a waxy meal which protects it against insecticides. Dust materials are less effective than sprays. In order to overcome the difficulty of killing aphids hidden deep within the cabbage folds, a systemic insecticide may be applied, provided it is at least two or three weeks before harvest.

Nysius cymoides Spin.

(Lygaeidae, Hemip.)

The seed pods of Cruciferae are often attacked by bugs. Outstanding among these are the small Lygaeidae of the genus *Nysius*. In Israel, *N. cymoides* is often very harmful.

Description. The body is about 3 mm long and dark brown in colour. The vertex, frons and scutellum are of a lighter hue; the abdomen, legs and antennae are light brown but the tips of the antennae are darker. The wings are entirely transparent. The basic half is slightly yellowish and two broken lines run over its veins. In addition, two brown veins separate the two parts of the forewing.

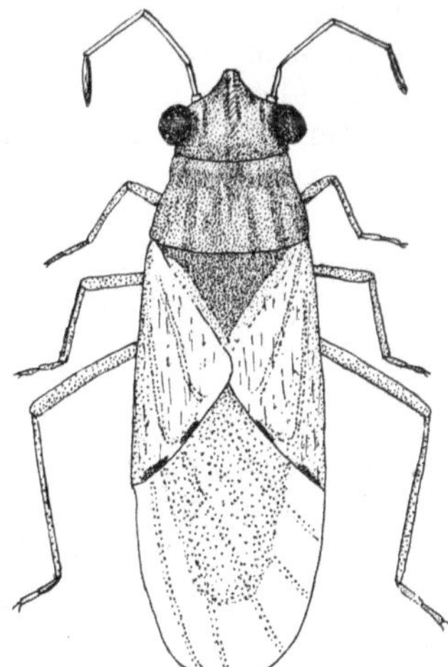

Fig. 101. *Nysius cymoides* – adult.

The eggs, according to SHWEIG (1954), are yellowish-grey, elliptical and are laid singly or in small groups on the lower side of the leaf or on the stem.

The larvae have a reddish abdomen but the thorax and wingpads are brown. The legs are yellowish-brown; the femora and tips of the tarsi are darker.

Distribution. According to LINNAVOURI (1960) this species is Holomediterranean. In Israel it is found in all parts of the country, north and Negev as well as in the mountains and in the Dead Sea Valley.

Hosts. LINNAVOURI states that it is "a common species on xerophilous vegetation". According to RIVNAY (1960) mass infestation occurred on cabbage and cauliflower sown for seeds, on wild mustard and on grape vines. During the summer it was collected in clover and alfalfa fields.

Type of Injury. The bug in all its stages attacks the inflorescences and the pods when the seeds are in the "milk" stage, and sucks them dry. The pods split prematurely and the seeds, whether injured or healthy, are spilled. Over 80% of the crop may thus be destroyed.

Life History. Very little is known about the biology of this pest. Larvae of all stages were found in the field during April and May, and a great many were brought into the laboratory for study, but the adults which were reared failed to reproduce and subsequently died. In the field adults were found throughout the summer on various crops, but no larvae were present. It is likely, therefore, that the adults go through a diapause period of the reproductive organs, and egg-laying begins only towards the winter. The number of generations which the insect may raise during the spring is unknown. The total length of development, from field observations, seems to be a few weeks only.

Ecological Factors. On certain occasions the pest appears in masses. In Maale Hakhamisha in 1947 the ground of the garden looked as if several ant hills were disturbed and the ants let loose. The plants were totally covered with the pest. In 1960 grape vines were so heavily attacked that the bark of the branches was not visible. The vines subsequently dried and the ground in the vineyard around it was crowded with millions of individuals.

In 1960, due to drought, the flora was generally very scant. All the diapausing bugs crowded to the few existing wild mustard fields. But these mustard fields were also short-lived, and the millions of larvae which had developed there began to migrate in search of more succulent hosts. The grape vines were adjacent to a large uncultivated field covered with wild mustard plants.

Control. During the attack of 1947 favourable results were obtained by dusting the seed plants with a 5% DDT dust. In 1960 DDT and dieldrin dusts checked the attack of the pest. The systemics, metaisosystox and phosphamidon, killed all the bugs on the vines. The march of oncoming bugs which continued from the neighbouring wild mustard field was checked after the strip of land separating them was dusted with DDT or dieldrin.

The Cabbage Bug — Eurydema festivum L.

(Pentatomidae, Hemipt.)

The genus *Eurydema* includes some species which are pests to cultivated cruciferous plants. While *E. oleraceum* is known as a pest in Northern Europe and *E. ornatum* in Central and Southern Europe, *E. festivum* is abundant in the Near East. *E. rugulosum* is reported as a pest in Syria.

Description. This is a bug 6—8 mm long. The colour of the body is ivory. The head is pale except for the frons and two spots on the clypeus which are black. The yellowish pronotum bears six spots, its posterior margin is reddish and the scutellum is black at its base and yellow at the margins. Each hemielytron bears a black spot while the abdomen has six pairs of spots, one pair to each segment. The egg of *Eurydema* is cylindrical, with slightly bulging sides and a lid on the top which is typical to this group. The colour of the egg is ivory which afterwards turns grey while the lid is dark in the centre surrounded by black dots.

The larva is like the adult in shape but differs in size, lack of wings and colour.

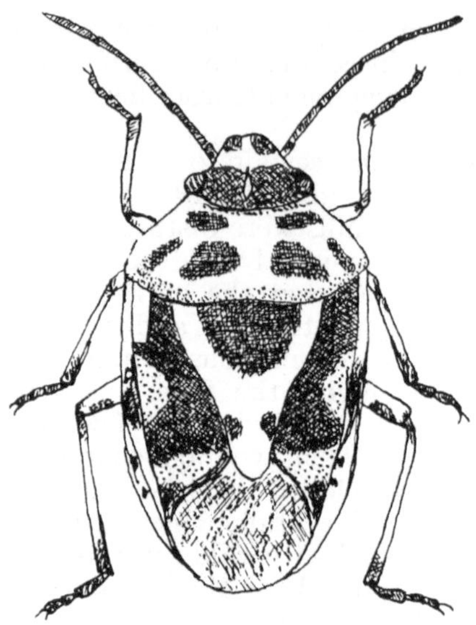

Fig. 102. *Eurydema festivum.*

Distribution. E. festivum is found in Southern Europe and South-Western Asia. Damage has been recorded from Russia, Bukhara, Cyprus and Israel.

Hosts. The species of this genus feed on Cruciferae and there are records of damage to wheat or beets. Occasionally the insect may be carnivorous, attacking the eggs and larvae of small insects.

Type of Injury. The bug feeds upon the sap of the plant. The sucking of many insects naturally weakens the plant, but large healthy plants overcome this. However, in the autumn, in the case of heavy attacks on small fresh seedlings, these may wither and dry.

Life History. The insect hibernates as an adult and awakens with the rise of the temperature in the spring. This usually occurs in April, but in Israel it may take place even in March. Mating and laying begin about a week later. The eggs are laid in groups of six; they stick to the underside of the leaf and stand one next to the other. One female may lay about 60 eggs.

The incubation period lasts about one week; at a low temperature of 15° C it lasts 15 days. The neonate larvae feed upon the plant on which the eggs were laid. Like other bugs, these nymphs moult 5 times before becoming adults. This takes place about six weeks after hatching.

Number of Generations. In Europe the adults of the first generation appear in June and a second generation may develop, the adults of which appear in August. In the southern countries of the Near East three generations may be raised. Thus, according to SHWEIG (1954) in Israel, the first generation appears in May, the second in July and the third in September-October. SHWEIG recorded a heavy attack of the species in October 1940. These were probably bugs of the second and third generations combined which invaded the irrigated vegetable gardens after the wild cruciferae had dried.

Ecological Factors. From various sources it is pointed out that these bugs are not equally distributed every year. Often they are hardly found in the fields, while on other occasions they may "alight as clouds on the plants" (SHWEIG). According to this author high temperature combined with a high relative humidity brings about outbreaks. Should high relative humidity prevail during the entire summer, all the three generations may occur in dense populations, but the population of the third is particularly outstanding.

However, the control aspect of parasites should not be overlooked. As in the case of other Pentatomidae, *Nezara* for instance, egg parasites play an important role in depressing the populations of these insects.

The Cabbage Flea Beetle — Phyllotreta cruciferae Goeze

(Chrysomelidae, Col.)

Minute elongated leaf beetles gnaw holes in the leaves of radish, cabbage and others. When disturbed they jump off the plant like fleas. These belong to the genus *Phyllotreta*.

Description. The genus *Phyllotreta* is distinguished from other genera of the Halticinae by the following characteristics: The first antennal segment is larger than the distance between the bases of the antennae, the prothorax is without a furrow, and the tibiae without an external spine.

The species *cruciferae* belongs to that group in which the colour is uniform and no pale stripes run over the elytra.

The beetle is 2—3 mm long. The frons between the antennae is slightly elevated into a crest and the prothorax and elytra are punctate. The colour is entirely metallic green except for the antennae which are pale brown.

The egg is elliptical, 0.5 mm long. The length of the larva before pupation is about 5 mm. The colour of the larva is white, its head and prothorax are dark brown, and dark spots are scattered over the body. The last abdominal segment bears a pointed lobe bent upward.

The pupa resembles the adult, but is white and the elytra not punctate.

Distribution. The species *cruciferae* is distributed throughout Eurasia. It is found from England in the west to Siberia, from Scandinavia to North Africa. Under the synonymous name of *Ph. colombiana* it is recorded also from British Columbia, Delaware and Minnesota in the U.S.A.

Hosts. The beetle feeds on cruciferous plants only. The cultivated plants which may be attacked include radish, mustard, cabbage, cauliflower, broccoli and other.

Ph. cruciferae is the most common and most injurious member of this genus. The adult feeds on the leaves into which it gnaws small holes. Often the lower epidermis is not touched. If the attack occurs when the plants are young, soon after sprouting, they may be entirely destroyed, making reseeding necessary. The larvae feed on the roots, but their injury is far less noticeable.

Life History. The adults hibernate in the ground or under rubbish. They become active in the spring when the temperature rises. In Europe this takes place in May; in more southern countries, like in Israel, this occurs in March, or even earlier if the winter is warm.

After feeding, the female lays her eggs singly in the ground near the wild mustard or wild radish. One female may lay as many as 50 eggs on the average.

The incubation period lasts about 10—12 days in Europe and about 7 days in Israel.

The neonate larva makes its way to the roots of the plant and feeds upon their outer layers. Later, it penetrates deeper in such fashion that the head is in the root while the end of the abdomen protrudes. Thus a few larvae may bore into one root.

The larval development lasts three weeks. When it reaches the length of 5 mm it leaves the root and enters the ground to pupate. The pupal period lasts about 10 days.

The first generation of these beetles appears in Israel in May. The adults of the overwintering generation are still active and thus the peak of the population is reached at this time. As the summer advances, the year old beetles die off and the new beetles continue to be active till the onset of cold weather in November when they, too, enter diapause.

Encouraging Factors. In countries with a humid climate, like England, the invasion of a crop may be so sudden, and the invading beetles so numerous, that the crop may be exterminated within a few days. MORETON (1945) studied this problem at length. He discovered that the awakening beetles do not leave their quarters immediately after emerging, but remain there as long as the weather is rainy and windy; they leave their quarters when the sky is clear, the air calm, and the soil temperature above 20° C. The beetles then migrate in masses. They are voracious, and may bring about the destruction described above.

Also SHWEIG in Israel claims that the beetles are inactive in the early morning hours and begin activity only towards nine o'clock.

Control. It has been customary to apply calcium arsenate or insecticides of plant origin, such as rotenone against these beetles. DDT, dieldrin and other synthetic insecticides are very effective against the pest but should not be applied too close to the harvest if the treated parts serve as food. Seed dressing with a solution of paradichlorobenzene 4 lbs, naphtalene 1 lb, dissolved in 1 gallon of kerosene — at the rate of 0.5—1 fluid oz. per 1 lb of seed reduced the attack very much (WALTON, 1936).

Lixus anguinus L.

(Curculionidae, Col.)

Larvae of *Lixus* spp. often bore into the stems or stalks of crucifers, especially in seed plants. In Israel *L. anguinus* is common and, in Eritrea, *L. latro* MORSH. As the biology of both is very similar that of *L. anguinus* L. will be discussed.

Description. The body is elongate and narrow, 14—18 mm long and 3—4 mm wide, tapering towards the end. The underside of the body is grey, the upper side seems bluish-grey. The following details should be noted: A grey-white band runs along each of the side margins of the elytra and extends to the sides of the prothorax. In addition, a greyish-blue band runs down the middle of each elytron which is dark brown. These bands extend to the pronotum and rostrum thus giving the beetle a striped appearance.

The egg is spherical and yellow, about 1 mm in diameter. The larva is legless, about 18 mm long and 5 mm wide, having lateral folds which obliterate the true segmentation.

The pupa is about 14—17 mm long, and is located in a cell in the tunnel made by the larva.

Distribution. This species is distributed throughout the Mediterranean region.

Hosts and Injury. The adult may feed on plants other than crucifers, thus specimens were found on barley and horsebean. Oviposition, however, takes place on Cruciferae only. The adults feed on the foliage or cortex of the stem, but this produces little injury. Feeding by larvae causes the main damage. The boring into the stem weakens the plant; the injury is particularly marked in plants grown for seed — in these the seeds either dry prematurely or they fall before the seeds mature. Fallen plants may amount to 30% in Israel.

Life History. From the short account of SHWEIG (1954) it is apparent that the life history of this species is very similar to that of *L. junci*, the beet *Lixus*, about which a larger account is given.

The adult awakens in the middle of the winter. In a mild winter it may be as early as January, otherwise in March. Oviposition begins a few days after emergence; the egg is oviposited in a cavity made by the female in the stem or stalk; the hatching larva bores into it. According to SHWEIG the incubation period is 7—10 days, while the development of the larva lasts 74—90 days and that of the pupa 14—18 days. Similar data are given by JANNONE (1946) for *Lixus latro* in Eritrea. The insect may thus raise two or three generations; the adults of the first appear in June, the second in August, those of the third overwinter. The most important, economically, is the first generation which also finds an abundance of food.

Control. BHC, lindane and parathion reduced the infestation to a great extent.

The Cabbage Sawfly — Athalia rosae L.

[Syn.: *A. colibri* CHRIST.]
(Tenthredinidae, Hym.)

Larvae similar to caterpillars but belonging to a different group may be found feeding on cabbage; these are the larvae of the sawfly *Athalia rosae* L.

Description. The adult is a sawfly 8 mm long. The head is conspicously broad and the eyes bulging. The colour of the head is black, except for the mouth-parts which are brownish. The antennae are black and the third segment is as long as the following two together. The thorax is yellowish with black spots, the legs are yellow except for the tips of the tibiae and tarsi which are brownish and the abdomen is yellowish except for the first abdominal segment.

The egg is white, elliptical, and about one mm long.

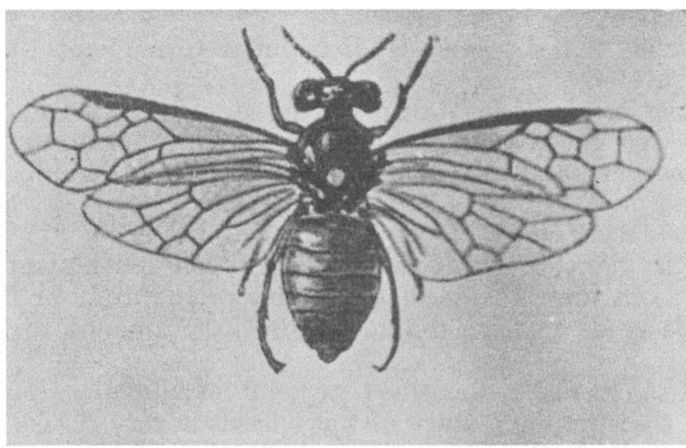

Fig. 103. *Athalia rosae* L. Courtesy Hassadeh.

The larva is 15—20 mm long. Its body is smooth and has eight pairs of abdominal legs, one on each abdominal segment except for the first. The colour is grey on the top and bright green on the sides and ventrum; the head and legs are black.

The pupa is white, the legs and wing pads are free, and it closely resembles the adult in shape. It is found in a grey silken cocoon to which plant or soil particles are attached.

Distribution. This insect is distributed throughout Eurasia: from England to Japan and from Scandinavia to the Mediterranean shores. It was also recorded from North and South Africa, from Turkey and Israel. Damage was recorded in particular from Finland, Denmark, Germany, Switzerland, Italy, USSR, Manchuria and Japan.

Hosts. The adults feed upon pollen of flowers. The flowers may be Cruciferae, Umbelliferae or Compositae. However, egg-laying takes place on Cruciferae only, wild or cultivated. Larvae were found on cabbage, cauliflower, mustard, turnip etc.

Type of Injury. The young larvae feed upon the soft inner tissue of the leaves; later they feed upon the entire leaves too and, in the case of young seedlings, may defoliate them or leave the midribs only.

Life History. The adults appear in the spring, in Northern Europe in May, in Italy in April. On sunny days they are seen on wild flowers feeding and mating and soon afterwards oviposition takes place. With the aid of the saw-like ovipositor the female deposits an egg at the edge of the leaf under the epidermis. One female may lay about 10—24 eggs per day. Thus in the two weeks of her life a female may lay about 300 eggs. Some authors mention smaller numbers of eggs. TAKIZAMA & AKIYAMA (1935), for instance, cite 50—150 eggs per female.

The incubation period lasts 4—12 days, depending upon the temperature. As mentioned above, first the larva bores into the leaf and feeds upon the soft tissue. When it becomes larger, it feeds on the entire leaf. From 10—35 larvae may be found on one plant.

The larval development lasts three or four weeks. TAKIZAMA & AKIYAMA claim that the larval development in Japan lasts only 7—12 days.

Upon completion of its feeding period the larva drops to the ground. spins a cocoon and pupates. This stage may last 9—21 days, depending upon the circumstances.

Diapause. All authors who treated the biology of this insect claim that it spend its winter in diapause in the cocoon. TAKIZAMA & AKIYAMA as well as RIGGERT (1939) claim that the diapause takes place in the larval stage. Pupation takes place in the spring and lasts 4—8 days.

Number of Generations. The number of generations differs according to the climate of the country. Thus, in northern Germany two generations develop, while in Switzerland and Italy there are three. In Israel SHWEIG mentions three generations while in Japan, TAKIZAMA & AKIYAMA claim five.

Control. The insect is easily controlled by synthetic insecticides. However, due to the poisonous residues which may remain on vegetables, it is advisable to employ non-toxic insecticides such as pyrethrum and derris which also give satisfactory control of this pest.

The Large Cabbage Butterfly — Pieris brassicae L.

(Pieridae, Lep.)

With the approach of winter, large white butterflies appear over the vegetable gardens and in the fields; their numbers increase as the winter advances. The larger of these belong to the species *Pieris brassicae* L.

Description. The body of the butterfly is 20 mm long and the wingspan 60 mm. The body colour is greenish-grey and the wings are white except for the front angle. From the third of the front margin to two thirds of the side margin the area is greyish-black. On the disc of the wing there are two spots of the same colour, and a similar spot is located on the hind wing near the front margin. In the male the spots on the fore wing are missing.

The egg is cone-like, tall and narrow. It is 15 mm high and 0.6 mm in diameter from its apex radial ridges to the base. Its colour is pale-yellow turning to yellow.

The larva, when mature, is 30—40 mm long. Its colour is greyish-green with the ventral side paler than the dorsal part. Three yellow longitudinal bands run along the body, one median and two lateral. On each body segment there are two large lateral black spots and two small spots closer to the median line. The head is black in the front and speckled grey in the rear and the prothorax is black. The entire body is covered with soft hair.

The pupa is elongate, 25 mm long, angular and tapering posteriorly. Its colour is yellowish-green mottled with black spots. Its posterior tip is fastened to the object upon which it is located, the back being held in an oblique position by a silken thread which is also fastened to the object.

Distribution. This species is a palearctic insect and, as will be discussed below, tends to migrate long distances in accord with the season. Thus, it is distributed over large areas. In Asia, it spreads from the western countries over Central Asia to the Himalaya range and the Bay of Bengal in India. In Europe its northern boundaries are Finland, South Scandinavia and Scotland, and the southern limits are the Mediterranean countries, including North Africa and excepting Egypt.

Hosts. The species feeds on plants of the mustard family only. Preference is shown, however, to cabbage and cauliflower, while radish is less attractive. Occasionally larvae may be found on *Nasturtium* also.

Type of Injury. The caterpillars may entirely defoliate the plants and thereby retard their growth and weaken the plants. When the infestation occurs in nearly mature cabbage or cauliflower, the defaecation contaminates the parts of the plant which serve as food; this is especially aggravated with overhead irrigation. Such vegetables fetch lower prices in the market.

Life History. After emergence from the pupal case, the adult is on the wing during the sunny hours of the day, feeding, courting and mating. On a rainy day or during cloudy hours the butterfly hides on the leaves. Mating does not take place at a temperature below 14° C. The preoviposition period may last 10—20 days if the temperature is 13—14° C, but only 3—4 days when the temperature is 26° C.

The eggs are laid on the leaf in clusters, one egg next to the other. Soon after the last cluster is laid the female dies. The number of eggs one female may lay is on the average of 150.

The neonate larvae remain together for some time and feed on the soft tissue of the leaf, leaving the hard veins. A few days after such feeding they moult and disperse into smaller groups of 4—5 each. At this stage they begin to feed on the entire leaf. The larva moults four times although KLEIN (1932) states that at temperatures above 20° C the larva moults three times only. The larval period lasts about two weeks in the Israeli summer, three during the spring and over four weeks in the winter.

When ready to pupate the larva fastens itself to some object; this may be a wooden post, a wall or the like. The head and thorax are then bent backwards and a silken thread is secreted and spun

Table L.

The length in days of the stages of *P. brassicae* at various degrees of temperature (after KLEIN, 1932)

Temperature in °C	Preoviposition period	Incubation period	Larval period	Pupal period	Total development	Length of life of adult
13 —14	10—21		47	48—69	48	12—17
14.5—15		10—13	31—34			
15.5—18	4— 8	6—11		11—17	44—57	
19 —20		4— 6	18—20			6— 8
25 —27	3— 4	3	12—13	7— 8	22—28	4— 6

around its back. The body then shrinks, and the skin shrivels and is shed. The pupa is exposed standing obliquely, fastened at its tip with the silken thread to the object.

As seen from table L the pupal period lasts about 7 days, but may be prolonged to over seven weeks when the temperature is low.

The butterfly may thus complete its entire life cycle in 3—4 weeks in the summer or 6—8 weeks in the winter of Israel when the temperature is 15—17° C. According to the data presented above the threshold of development is 8° C (KLEIN).

One female may lay 200—300 eggs. KLEIN finds that the oviposition depends upon the temperature. At 22—24° C the egg-laying averaged 100—120 per female, at 24°C–70, while at 25—27°C 12—14 eggs only.

Limiting Factors. The data on the ill effects of higher temperatures upon oviposition directs our attention to other physiological values at various temperatures, as follows:

1. No egg-laying takes place at a temperature above 27° C;
2. At 26—27° C eggs do not hatch, at 25° C the mortality of the eggs is about 50%;
3. No egg-laying takes place at 13° C, the optimal conditions for oviposition being 15—22° C.
4. The butterfly does not oviposit at a relative humidity below 60%, while at a relative humidity below 70% eggs do not hatch.

If we compare these data with the climatic conditions of Israel we find that the insect can raise a few generations during the Israeli winter, but finds it difficult to reproduce in the spring, and impossible to exist during July, August and September. In fact, during these months the insect is not present at all, nor is it found in any country with a climate similar to that of Israel. If there are any stages of the insect alive in June they are exterminated by the climate of the summer. Yet, in October adult butterflies begin to appear again, females lay and plants become infested anew. This is because of the migratory habits of the butterfly.

Migration. In Europe *P. brassicae* is a well known migrant and there are records of many observations on this phase. WILLIAMS in his various papers (1936—1957) gathered these records, especially from England and Germany. This author points out that the migrations of this insect are regular but the density of migrating insects varies. It may be as low as 3 each 5 minutes in a breast line of 100 metres, or several hundreds in the same time and space, or sometimes even thousands, as may be seen from the following example: In 1508, in July near Calais a swarm was "so thicke as flakes of snowe" and the town of Calais could not be seen at four o'clock from a short distance.

Certain years may be richer in migratory butterflies. In England, for instance, during 1930—'40, the years 1930, '33, '34, '39 and '40 were cabbage butterfly years. WILLIAMS shows further that there are two peaks of migratory swarms in Europe, one in May-June, and one in July-August. From the few records which are available about the spring peak it seems that the flights are mostly in a northward direction. In July-August, however, of 74 observations — 50% of the direction of flight was southwards, 25% were southwesterly and 10% in a southeast direction. This means that in the spring the swarms move from warmer countries to cooler, while towards the autumn they fly from cold countries to warmer.

The occurrence of the butterfly in Israel is in accord with this situation. In the Near East no dense swarms "as falling snow" attract the attention of the man in the street. KLEIN (1932) believes that there is no migration from Israel northwards, as the insect does not survive the spring conditions.

FLETCHER (1925) describes the migrations of this butterfly in India. According to him the butterfly spends the summer in the Himalayas near Nepal, and migrates southwards 200 km to the shores of the Bay of Bengal. It appears in Pusa in February and disappears in May regardless of the fact that cruciferous plants are plentiful.

The Small Cabbage Butterfly — Pieris rapae L.

[Syn.: *Pontia rapae* L.]
(Pieridae, Lep.)

This butterfly is smaller than *P. brassicae*, and in territories where they exist together its damage is slighter than that of the large cabbage butterfly. On the other hand, it has spread further and has

Fig. 104. Larvae of *Pieris rapae* L. after METCALF & FLINT McGraw-Hill Pub. Co., Destructive and Useful Insects, 3rd. Ed., after Conn. Agric. Exp. Sta. Bull.

penetrated into countries where the larger insect is not found.

Description. The adult is 10—15 mm long with a wingspan of 50 mm. The colour of the wings is white, except for the anterior angle of the forewing which is dark grey. On the forewing of the female there are two dark grey spots, on that of the male only one spot. The colour of the female body is grey-white, that of the male slightly darker on the top.

The egg is yellow, cone shaped, and the surface is engraved with radial ridges from the apex to the base.

The larva is 30 mm long, green and has a velvety appearance because of the dense short hair which covers the body. With a magnifying glass black dots may be seen scattered over the body. A yellowish median line runs over the dorsum and rows of yellow dots run between the spiracles.

The pupa, like that of *Pieris brassicae,* is angular and attached at its tip to the object upon which it pupated. Its colour changes with the environment from grey to green-grey or yellow-grey.

Distribution. Pieris rapae is of palearctic origin. It is found in Europe and Asia from the British Isles to China and Japan. Its southern boundaries are the Mediterranean countries; it is found in Spain, North Africa, Egypt, Cyprus and Israel. As mentioned above it has spread to other continents too. It invaded Canada in 1860, and from there it spread to the U.S.A. It reached Australia and in the nineteen-thirties established itself also in New Zealand.

Hosts and Injury. The main host plants are Cruciferae, but it may be found also on *Nasturtium* and *Reseda.* The injury is similar to that of larvae of *Pieris brassicae.*

Life History. The life history is very much like that of the large cabbage butterfly but a few differences should be noted: The female lays the eggs singly and not in groups. One female may lay on the average 350 eggs. The threshold of development is 8.4° C for the egg, 6° C for the larva and 7° C for the pupa. The thermal constants are 56 day-degrees centigrade for the egg, 217 for the larva and 150 for the pupa (MUGGERIDGE, 1943). The number of generations naturally depends upon the climate of the country. Thus, RICHARDS (1940) counts three in England though in warmer years part of a fourth may develop, and QUERCI (1932) states that the insect can exist and develop at a range of temperature from 15—32° C; at a temperature above this the larvae die; similarly the caterpillars cannot exist at a temperature below 9° C.

RICHARDS who studied the causes of the fluctuations in the population of the insect in England states that an average temperature of 15—16.5° C and 7.5—12 mm rainfall during one week cause an increase in the number of this butterfly, for these are the most favourable for its development.

Biological Balance. Upon reviewing the literature on this pest, it is striking to note that complaints about damage by this insect come from countries to which the insect has migrated. It is an established fact that in Canada, the U.S., Australia, New Zealand and Tasmania, the smaller cabbage butterfly is a more serious pest than it has ever been in Europe, its country of origin. The reason is that in its old home several parasites exist which keep the butterfly under control, while in the newly acquired territories no parasites, or only a few, are present to keep a natural balance of the pest.

The butterfly may serve as a good example of the role parasites may play in the biological balance. "Biological balance" means that the number of offspring at the end of a certain period should not exceed the number of parents. In other words, there must be agents which kill a great percentage of the eggs, the larvae and the pupae so that only one pair of offspring is left to each pair of parents.

In Europe such agents are available for the small cabbage butterfly. Thus BLUNCK (1951) who made counts of parasitism found that the parasitic wasp *Trichogramma* sp. destroyed 77% of the eggs of *Pieris rapae*, while *Apanteles rubecula* killed 90% of the larvae. If we bear in mind that in addition to these there are other parasites of other Hymenopterous genera, such as *Anilastus* and *Pteromalus*, and Dipterous genera, such as *Zenillia*, it is easy to understand how most of the offspring of the insect are exterminated. In Australia and New Zealand these parasitic insects were not available; therefore, the pest multiplied.

The governments of these countries exerted efforts towards re-establishing a balance; the parasites *Apanteles glomeratus* L., *A. rubecula* MARCH., *Pteromalus puparum* L. and others were introduced into Australia and New Zealand, and the situation improved greatly.

Due to the fact that the new synthetic insecticides are non-selective and destroy parasites together with their hosts, the attention of entomologists was directed to the exploitation of micro-organisms against *Pieris rapae*. The virus *Borellina campeoles* ST. which causes a polyhedral disease gave satisfactory results.

Bacteria which infect larvae of *P. rapae* are *Micrococcus pieridis* BURRIL, *Bacillus aerifaciens* ST. and others. Promising results were obtained with *Bacillus thuringiensis* BERL.

The Cabbage Moth — Mamestra brassicae L.

[Syn.: *Barathra brassicae* L.]
(Noctuidae, Lep.)

This species is European. In the Near East it is reported as a pest from Lebanon and Syria only. In view of this a short account will be given here.

Description. The wing span is 40—50 mm; the forewings are greyish-brown, with a few dark brown undulating transverse lines. The light reniform and orbicular spots have a dark border and the hind wings are grey. The larva is 40—50 mm long and its colour varies from grey-brown to light green depending upon the food. The head and prothoracic plate are shiny black and on each side of the body there is a yellowish longitudinal line. Above this line, on each abdominal segment, there is a yellowish line adjacent to a short black band. A light band transverses the last abdominal segment and four setigerous dots are found on the disc of the body segments.

Distribution. According to the literature this species is European and extends to the Mediterranean basin. It is found on the Canary Islands, "Asia and Japan". Apparently the pest is limited only to the northern countries of the Mediterranean basin. As mentioned above, it was reported from Syria and Lebanon (TALHOUK, 1950, 1954) but it was not found in Israel or Iraq nor was it reported as a pest from Greece or Cyprus.

Hosts and Injury. In addition to cruciferous plants which are the preferred hosts, larvae may be found on beets, peas, beans, lettuce, etc. The damage consists of defoliation of the host plant.

Life History. The adult appears late in May; towards July, after feeding and mating, egg-laying begins. The eggs are laid in groups of from 20 to 100 on the underside of the leaf. Ten to twelve days later the larvae hatch. Under European summer conditions the development lasts about two months. Pupation takes place in the ground, and the pupa overwinters and does not mature until the spring of the following year. In other words in Europe this moth raises but one generation.

The Diamond Back Moth — Plutella maculipennis Curt.

[Syn.: *P. cruciferarum* ZETT.]
(Plutellinae, Tineidae, Lep.)

Often small green caterpillars bore and mine into the main fleshy veins of cabbage leaves. These are the larvae of the diamond-back moth.

Description. The body of the adult is 12—13 mm long and the wingspan 30 mm. The margins of the wings are rounded and bear long fringes of hair. The colour of the body is dark grey; the colour of the wings is yellowish-brown mottled with black dots. Along the hind margin of the forewing there is a white band with undulating borders, so that when the wings are close together widened rhomboid areas are formed (hence the name). The hind wings are pointed, their colour is dark grey, and their fringes are quite long.

The egg is oblong with dimensions of 0.5 × 0.25 mm. Its colour is yellowish-green and it is laid in groups of 3—6 on the underside of the leaves near the veins.

The caterpillar is 8—9 mm long and wider in the middle. Its

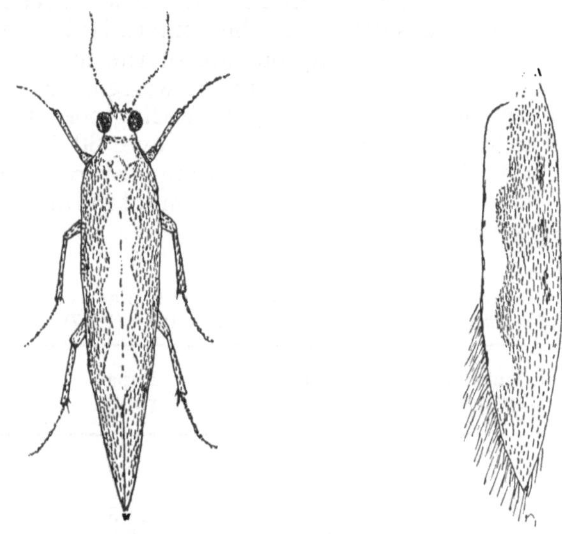

Fig. 105. *Plutella maculipennis*.
(a) adult, seen from above, (b) single forewing

colour is pale green mottled with black setigerous dots. In younger larvae there is a darker green band along the back and two lateral bands of paler colour.

The pupa is found in a loosely-woven cocoon which is fastened, as a rule, to the underside of the leaf. The length of the pupa is about 6 mm.

Distribution. According to HARDY (1938), this moth occurs throughout Europe, North and South Africa, Asia, India, Malay and Western Asia, Australia, New Zealand, Canada, the U.S.A. and South America. It is found in the Near East, too. Some entomologists claim that the moth is found wherever cabbage grows.

Hosts. Many cruciferous plants may be attacked, but this pest seems to prefer cabbage above all others.

Type of Injury. The larvae bore into the main veins of the leaf or feed on the underside tissue. Should the attack be severe, when the plants are young and small they may be destroyed. In the Near East this may happen particularly at the end of the summer when the population of the pest has risen due to favourable climatic conditions and when the seedlings are still soft and young.

Life History. The eggs are laid in small groups on the leaf, usually on the underside. Under favourable conditions of temperature the incubation period lasts 3—4 days or, according to HARDY, 13 days at 10° C and 1.5 day at 35° C (see table LI). When the larva hatches it is pale green and 2 mm long. At first it wanders over the surface of the leaf and then begins to bore into its tissue. It feeds upon the internal tissue which is softer than the epidermis; however, two or three days later it feeds on the outside of the leaf. The larva is "nervous"; at the slightest disturbance it wriggles and drops from the leaf by means of a silken thread. Its feeding period is a week to 10 days. At the low temperature of 10° C, over 60 days may lapse before it pupates; however, at higher temperatures this period is much shorter (see table LI). The development of the pupa lasts 4 days at 28° C or 8 days at 20° C.

Table LI.

Effects of temperature on the rate of development in days of various stages of *Plutella maculipennis*. After HARDY (1938) and MINER (1947) (in Parenthesis)

Temperature in °C	Incubation period	Larval period	Pupal period	Number of laying days
10	13	60+		
13.5		(22)		
15	6	17	11.5	35
20	4.5	10.5(10.5)	8(8)	13
24.5—25.6	3	7.5(8)	3—6	
28	2.6	(5)	4	8
29			(3)	
30	2.5	6.5		
35	1.5	5.25	3	3

The adults are hidden and in their flight are also inconspicuous. Mating takes place a day or two after emergence (at a temperature of 23—27° C). The number of eggs, according to GUNN (1917) is 35—50. A more detailed account of the fertility of this moth is found in the work of HARDY (1938) and of MINER (1947), who raised the moth under various conditions of temperature (see table LII).

Table LII.

Number of eggs per female at various degrees of temperature. After HARDY (1938) and after MINER (1947) (in Parenthesis)

Temperature °C	7	15	20	23—25	28	35
Max. & min. number of eggs	0—0	18—57	69—248	(113—140)	157—217	3—56
Average number of eggs	0	40	172	—	95(85)	26

Number of Generations and Diapause. According to HARDY's account of the life history of the adult, the moth completes its development in 12 days at 30° C, in 16 days at 25° C and in 35 days at 15° C. Therefore, the moth could raise a great number of generations during the year. Actually this is not so since the insect goes into diapause for some time. Thus in England there are only two generations a year, in the U.S. from 2—6 (depending upon the locality) and in arctic USSR only one.

It was first believed that this insect hibernates in the pupal stage, but HARDY found that the diapause occurs in the adult stage, and it awakens in May, or even in June in more northern places.

Limiting Factors. Mention is made in the literature that rain brings about a decrease in the population of this pest. HARDY believes that this is due to mechanical injury or because caterpillars are washed away. *Plutella maculipennis* occurs in many regions where climatic conditions are quite different. It is found in India, in Israel and also in the arctic region where, according to SEMENOV (1950), the pest is active one month only. This is due to the fact that the adult is especially adapted to countries with very warm climates. According to HARDY the moth may stay alive several days at a temperature of 38° C and lay eggs (from 3—56 per female). MINER finds that moths which stay two days only at a temperature of 36—37° C produce non-viable eggs. Of 667 eggs laid by 7 females (which is a normal rate of reproduction) hardly any hatched. Normal hatching takes place from 15—30° C, and any deviation from this range of temperature, in either direction, affects the percentage of hatching. A mortality of 80—90% of the eggs occurs at 35° C or 10° C. This latter degree of temperature is considered the threshold of reproduction. At this temperature a female may exist for several months without food, and awakes and resumes activity with the onset of favourable temperature.

Control. Some parasites of the genus *Angitia* keep the species in balance to a certain extent; one of them *Angitia cerophaga* GRAV. was introduced into Indonesia against this pest (VOS, 1953).

Many writers point out that the insecticides which kill the caterpillars of the white cabbage butterfly kill the larvae of this moth too. HENDERSON (1957), however, points out that DDT, dieldrin and BHC became ineffective against *Plutella.* When these insecticides were compared with diazinon and malathion, diazinon proved to be the best. A weekly application of 0.04% a.i. of this insecticide reduced the infestation to 13 larvae as compared with 758 in the non-treated plants, and the yield was 4 times greater than in the non-treated parcels.

The Oriental Cabbage Web-Worm — Hellula undalis F.
(Pyralidae, Lep.)

In the autumn young seedlings of Cruciferae, in the nursery or in the field, may wither and dry. Examination of such plants may reveal a caterpillar boring into the centre of the stem. This is the larva of *H. undalis*.

Description. The insect is 6—9 mm long, and the wingspan 18 mm. The colour of the forewings is reddish-brown, with undulating bright and dark lines running across them. On the disc close to the fore margin there is an oblong brown spot. The hind wings are light brown; fringes of short hair decorate the margins of the forewings and fringes of longer hair are found on the margins of the hind wings.

The egg is white, flat and 0.3 mm in diameter. The larva is yellowish with longitudinal brown lines running along the body. Its head is brownish-red. When mature it reaches the length of 15 mm.

The pupa is found in a loose cocoon of silk threads and is 10 mm long.

Distribution. It would be incorrect to consider the literature records as a criterion for the distribution of *H. undalis* because until 1953 the common name "cabbage web-worm" referred to this species. Recently, however, a paper was published (CAPPS, 1953) in which it is pointed out that this species does not exist in the Western Hemisphere, and that the common name refers to another related species, *H. rogotalis* HULS, distributed in America and the Pacific Islands, *H. undalis*, therefore, is distributed in the Old World: In the western countries of Asia, the Mediterranean countries, Sudan, Malay, Java, Australia and New Zealand.

Hosts and Injury. Cabbage, cauliflower, mustard and radish are the preferred hosts. Boring of the larva into the centre of the stem harms the plant and if the seedlings are young and small they may be killed. When neglected, all of the seedlings in a nursery may be exterminated. In mustard and radish the larva may reach the roots and fill the galleries with faeces which stimulate rot, and in irrigated heavy soils the loss is quite marked.

Life History. This pest is quite common, yet its life history has been studied very little, and information on this subject is scant. The present discussion is based on the work of BOUHELIER & HUDAULT (1935) in Morocco, and SHWEIG (1954) in Israel.

The adult female lays her eggs at night, singly or in small groups, between the young leaves. The egg hatches after a period of 3—5 days. The neonate larva bores into the stem from the top. In the course of 20—30 days (in Morocco — BOUHELIER & HUDAULT) or 17—20 days in the Israeli summer (SHWEIG), the larva completes its development. Late in the autumn this period lasts 50 days. The larva leaves the tunnel, enters the ground or remains under rubbish near the plant to pupate. This takes place in a loose silken cocoon.

The pupal stage lasts 5—7 days while the preoviposition period is 3—4 days. On the average one female may lay about 100 eggs. In Israel a generation lasts about 25—30 days and in Morocco 35—40 days in the summer and 10—16 weeks in the colder season.

Phenology. In Morocco the moth appears in May, and until January it is capable of raising four generations. During the winter the moth is not seen.

In Israel, according to SHWEIG, the moth appears in June and raises about seven generations.

The Cabbage Fly — Erioischia brassicae Bouché

[Syn.: *Anthomyia brassicae* BCH., *Hylemyia brassicae* BCHÉ]
(Anthomyiidae, Dipt.)

Since in several countries this fly is rated as one of the most important pests of cabbage and related vegetables, it has been studied more intensively than other flies. In the Near East, however, this fly is of lesser importance, and in the southern countries of this region it is questionable whether it is found at all. A survey carried out by YATHOM (1962) in Israel revealed that the damage occasionally caused to cabbage or turnip in Israel is caused by maggots of flies other than *H. brassicae.* In the literature reports are seldom published about damage in warmer countries.

Description. The adult looks very much like a housefly except that it is smaller; the length of the body is 5—6 mm. Its colour is light-grey and on the thorax there are three darker median longitudinal lines and two lateral shorter ones. The eyes are smooth; in the male they touch each other, in the female they are separated by a narrow reddish strip.

The egg is small, white, elongated, cylindrical and tapers at both ends.

The mature larva is 7–8 mm long, cylindrical, with a pointed anterior end and a truncate posterior end. On the last abdominal segment there are 12 minute papillae. The central ones are the largest and each is divided into two. The pupal case is 6 mm long and reddishbrown.

Distribution. Most complaints about damage by this fly come from the U.S., Canada, England and countries of Northern and Central Europe. During the years 1945—1960, one complaint only was recorded from Spain and one only from Morocco.

According to CHITTENDEN (1916), the fly is not found south of New Jersey. It is true, FULTON (1942) claims that the fly is present also in North Carolina, but this is only in altitudes above 1000 metres. It may thus be summarized that the fly is distributed in the north of the U.S., Canada, and North and Central Europe, and

Northern Siberia. Since in more southern countries it may be found in higher altitudes, it may then be present in the northern countries of the Near East.

Hosts. The plants which serve as hosts to this fly are usually of the mustard family. Cabbage, cauliflower, radish, turnip, etc. may all be infested, but occasionally also celery and beets were found to harbour maggots of this fly.

Type of Injury. The maggots bore into the stem near the crown or even into the subterranean parts of the plant. In case of heavy infestation the plant may be entirely destroyed. In cases when the attack is less severe the growth of the plant is hampered, the leaves become yellow and wilted, and the plant as a whole does not form a head.

As many as 125 larvae were found on one plant, and as many as 75% of the plants in a field were found infested.

Life History. The insect spends the winter in diapause in its pupal stage. The adults emerge in the spring. Depending upon the climate of the country the emergence may be early or late. Thus, in England and in North America it appears in April while in Siberia in June. According to BREMOND (1937), in Morocco the pupa may enter into a state of diapause from April till October.

After feeding and mating the females begin egg-laying; this happens only during the sunny and warm hours of the day.

The eggs are laid in the ground around the host plant. The incubation period in May lasts from 3—7 days. The larva makes its way between the soil particles until it reaches the plant and penetrates into the tissue of its subterranean parts. The larvae feed for three weeks (in England and Massachusetts) before they are ready to pupate. In colder climates this period may last 4—5 weeks. Pupation takes place in the ground.

The pupae of the first generation all develop during the same summer, while pupae of the second generation may either continue their development and mature during the same summer, or may go into diapause and mature the following spring. The later the generation, the higher the percentage is of pupae which remain in diapause.

The pupal development lasts two or three weeks under the conditions of the English summer. If we add the preoviposition period we find that one generation of this fly lasts 40—50 days in England. In this country it can raise three generations per year. In warmer climates, in Washington, for instance, 4—5 generations may develop. In colder regions, however, the development is far slower. In Siberia (according to SEMENOV, 1950) the larva pupates in August, remains in diapause until the following June, in other words, the fly raises only one generation per year.

Observations on the phenology of the fly were carried out in

England by Mrs. MILES. According to this author (1951) the fly is active only in the warm hours of the day; during cloudiness, rain and cold, and during the early morning hours the fly hides. The threshold of oviposition is at 15° C. For these reasons, in a year when the spring is cool, the fly may begin its activity later in the season, and thereby may raise fewer generations. In England the number of sunny days is the decisive factor for outbreaks. Thus, when the temperature in England rises suddenly from 9 to 18° C, a sudden increase of egg-laying occurs in which about 40 eggs per plant were counted. The highest percentage of egg-laying takes place when the temperature is between 18—21° C. When the temperature droped to 14° C only 6 eggs per plant were counted (MILES 1955).

Control. Before the existence of the synthetic insecticides it was customary to control the maggots by applying mercurous chloride (Hg_2Cl_2), calomel mixed with gypsum at the rate of 1:1 or 3:2. This powder was dusted around the plant at the rate of 10 kg/dunam. It was also customary to spray around the plants with a mercuric chloride ($HgCl_2$) solution, one litre of a 0.1% solution to each 3 metres.

Of the synthetic insecticides BHC, aldrin, heptachlor and parathion yielded satisfactory control of the fly. According to GOULD (1955) dieldrin and endrin were most effective. These materials should be used before setting, and water thoroughly. It should not be forgotten that these substances leave toxic residues. Lindane and methoxichlor cause retardation of growth and should not be used.

General Considerations on Control Measures

Measures of control against *Hellula undalis* should be made prophylactically. Dusting the young seedlings in the beds or in the fields with DDT, dieldrin or endrin may reduce the infestation to a great extent. The aphids which feed on cabbage or cauliflower may be hidden in folds and cracks and, therefore, are hard to reach. A systemic insecticide is more advisible in such cases, but not too close to harvest time.

From the literature one learns that the various caterpillars which attack plants of this family are more abundant in the northern countries than in the Near East. In this region control measures are often superfluous. One learns also that most of these caterpillars may be killed by the same insecticide. Thus, in many earlier references DDT, BHC, dieldrin and endrin are mentioned as efficient substances against all cabbage caterpillars. Lately, however, there are many reports from which it is evident that some of the insecticides have become less effective. In view of this no general recommendations are given here, as conditions may be different in various localities, and where needed, the insecticide should be chosen by trial. Fortunately, it was found that the *Bacillus thuringiensis* is effective against all these caterpillars and should be exploited whenever possible. It should be remembered that on these vegetables no insecticide, except derris, pyrethrum and nicotine should be applied too close to the time of harvest.

LITERATURE

A. Balachowsky & L. Mesnil, 1936. Les Insectes Nuisibles aux Plantes Cultivées. Paris, *1182-1222.*

H. Blunck, 1951. *Z. Pflanzenkrankh.* **58,** *25-26.*

F. S. Bodenheimer & E. Swirski, 1957. The Aphidoidea of the Middle East. The Weizmann Science Press of Israel, Jerusalem. 378 pp.

R. Bouhelier & E. Hudault, 1935. *Rev. Path. vég.* **22,** *123-130.*

P. Bremond, 1937. *Rev. Path. ent. Agr.* **24,** *172-174.*

H. W. Capps, 1953. *Bull. S. Calif. Acad. Sci.* **52** (2), *46-47.*

F. H. Chittenden, 1916. *J. econ. Ent.* IX, **6,** *571.*

D. L. Elze, 1944. *Bull. Soc. Fouad Ent.* **28,** *33-43.*

T. B. Fletcher, 1925. *Bull. ent. Res.* **16,** *177-181.*

B. B. Fulton, 1942. *Bull. N. C. Agr. Exp. Sta.* **335,** 24 pp.

G. E. Gould, 1955. *Purdue Univ. Exp. Sta. Bull.* **616,** *1-8.*

D. Gunn, 1917. *Bull. Univ. S. Afr. Dept. Agric.* **8,** *1-10.*

J. Eliot Hardy, 1938. *Bull. ent. Res.* **29,** *343-372.*

M. H. Hassanein, 1958. *Bull. Soc. Ent. d'Egypte,* **42,** *325–337.*

M. Henderson, 1957. *Malay agric. J.* **40,** 4 pp.

G. Jannone, 1946. *Int. Bull. Plant Prot.* **20,** 11-12, *115M-121M.*

H. Z. Klein, 1932. *Z. ang. Ent.* **XIX,** *395-448.*

R. Linnavouri, 1960. *Ann. Zool. Soc. Zool. Bot. Fenn. 'Vanamo',* **22,** *1-71.*

M. Miles, 1951. *Ann. appl. Biol.* **38,** 2, *425-432.*

M. Miles, 1953. *Ann. appl. Biol.* **40,** 4, *717-725.*

M. Miles, 1955. *Ann. appl. Biol.* **41,** 4, *586-590.*

H. E. Milliron 1953. *J. econ. Ent.* **46,** *179.*

F. D. Miner, 1947. *J. econ. Ent.* **40,** 4, *581-583.*

B. D. Moreton, 1945. *Ent. mon. Mag.* **81,** 970, *59-60.*

J. Muggeridge, 1943. *N. Z. J. Sci. Tech.* **24,** *107-129.*

O. Querci, 1932. *Ent. Rec.* **44,** *168-176.*

O. W. Richards, 1940. *J. Anim. Ecol.* **9,** *243-288.*

E. Riggert, 1939. *Z. angew. Ent.* **26,** 3, *462-516.*

E. Rivnay, 1960. *Hassadeh* **40,** *1253.*

A. E. Semenov, 1950. *Dokl. vsesoyuz. Akad. sel. khoz. Nauk Lenina* **15,** 2 *39-42.*

K. Shweig, 1954. Vegetable pests Hassadeh Library, *110-111.*

F. Silvestri, 1939. Compendio di Entomologia applicata Portici *479-482.*

M. Takizama & T. Akiyama, 1935. *Kontyu* **9,** 5, *207-220.*

A. S. Talhouk, 1950 and 1954. *Bull. Soc. Fouad 1er Ent.* **34,** *133-141*; **38,** *305-309.*

H. C. C. A. A. Vos, 1953. *Contr. gen. Agr. Res. Sta. Bogor.* **134,** 32 pp.

C. L. Walton, 1936. Progress Report *Rep. agric. hort. Res. St. Bristol 80-86.*

W. D. Whitcomb, 1944. *Mass. Agr. Exp. Sta. Bull.* **412,** *1-28.*

C. B. Williams, 1958. Insect Migrations I-XIII, 235 pp. The New Naturalist, Collins, London.

E. P. Wiltshire, 1957. Ministry of Agriculture - Govt. of Iraq. *1-162,* plts I-XVII.

Sh. Yathom, 1962. *Hassadeh,* **42,** *582-584.*

Pests of Peas (*Pisum sativum*) and Beans (*Phaseolus vulgaris*)

A key to the pests of these vegetables is found in the chapter on leguminous fodder crops — page 205.

General Considerations on the Measures of Control.

In small vegetable gardens it may be more advisable to collect *Agrotis* larvae by hand or use bait rather than to apply insecticides to the soil.

Systemics against aphids should not be used within three weeks before the harvest. A short time before pod picking, nicotine or malathion should be applied if necessary. Other insecticides, such as endrin, may be employed if the pods themselves are not destined for consumption.

Contact insecticides, such as DDT, endrin, sevin, etc., when applied during the blossom or early pod period, may prevent infestation of the pods by caterpillars of moths and butterflies and by beetles but they may also interfere with the pollination and should not be employed unless absolutely necessary.

Pests of Okra *(Hibiscus esculentus)*

Being a malvaceous plant okra is subject to attack by most of the pests which infest cotton, a key to which is given on page 398.

As a rule the nature of infestation of okra necessitates measures of control very seldom.

Pests of Lettuce *(Lactuca sativa)* and of Artichoke *(Cynara scolymus)*

In the Near East lettuce is grown in the winter when pests are less abundant. Consequently, it escapes the injury by some insects which might have attacked it had it been grown later. Still, various caterpillars may injure lettuce. *Agrotis segetum* and *Triphaena pronuba* injure the root crown, while *Plusia chalcites*, *Trichoplusia ni* and *Autographa gamma* feed on the foliage. Occasionally some plant lice may be found on the soft leaves, among them *Dactynotus sonchi* GEOFF. Later in the season, plant bugs, such as *Nezara*, attack the plants, and in the inflorescences *Trypanea stellata* may be found. Discussions of these pests and their control are found in their respective chapters.

Artichoke plants may be infested with plant lice, among them *Brachycaudus cardui* L. The leaves may be injured by the larvae of

Pyrameis cardui, Cassida palestinae REICH and *Agapanthia*. Some Trypetidae may be found in the inflorescence. An account of some of these will be in the chapter on pests of Compositae Industrial Crops, page 384.

The Artichoke Beetle — Sphaeroderma testaceum F.

[Syn.: *Sph. cardui* GYLL.]
(Chrysomelidae, Col.)

Small larvae which mine into and feed between the two layers of the epidermis may be found in the leaves of artichoke or other plants of this family. These are the larvae of *Sphaeroderma testaceum*.

Description of adult. The body is elliptical, 4 mm long and 3 mm wide. The elytra are convex and the colour is dark red. The head is wide and the eyes are large and kidney-shaped. The antennae are filiform, slightly broad at the tip and inserted between the eyes. A curved groove connects the eyes and the frons is elevated. The prothorax is twice as wide as long; its side margins are curved inward.

The egg is elliptical, yellow, one mm long. The larva is 7—8 mm long, white, with a brown head and pronotum. Small plates are located on the meso- and metanotum.

Distribution. The beetle occurs in Europe, North Africa and Israel.

Hosts and Injury. The beetle and larvae feed upon the leaves of artichoke and wild composites, such as *Cirsium*. The injury is caused by the larvae mining between the layers of the epidermis. When heavy infestation occurs most of the leaves become brown and wither; but heavy infestation is rare.

Life History. There is little information on the life history of this species; but from observations in Israel plus our knowledge of the closely allied *Sph. rubidum* it may be surmised that the beetle spends the summer in diapause in the adult stage, and awakens early in the spring — in March in Israel.

A week after copulation egg laying begins. The eggs are attached to the leaf and the incubation period lasts 3—4 days. The larva bores into the leaf directly from the egg and feeds upon the inner soft tissue. No data are available on the length of development of the larva and pupa.

Larvae were observed in Israel only in the spring (in April) and adults were found only in April-May. It may be assumed that only one generation is raised a year.

Control. No recent information is available as to control. Synthetic contact insecticides probably will kill the adult, and perhaps BHC and parathion may kill the mining larvae.

The Painted Lady — Vanessa cardui L.

[Syn.: *Pyrameis cardui* (L.)]
(Nymphalidae, Lep.)

This is one of our decorative butterflies; in the spring it displays its attactive beautiful colours, but its numbers may rise to undesired proportions and its larvae may destroy various crops.

Description. The butterfly is 15—18 mm long. The forewing is coloured as follows: The basic half is red with a row of 3 or 4 black spots, one large and four smaller, arranged in an arc opposite the larger spot. The hind wing is red and its base brownish. Three rows of black spots run parallel to the hind margin; the spots in the middle row are narrow, elongate and less conspicuous, the dots of the last row are located on each indentation of the wing.

Fig. 106. *Vanessa cardui.*

The egg is bright green. It is cone-shaped with 12—18 ridges engraved in its surface running from the top to the base.

The larva is 35—40 mm long. Its colour is yellow below the line of spiracles and reddish-brown above it. This colour is not uniform but is interrupted by pale short bands on each segment and by one median longitudinal band along the entire body in the middle of which the heart tube is visible. The head is small and black. In light coloured specimens the brown area diminishes so that only single brown spots remain. In such caterpillars the head is yellow, mottled with brown dots.

The body is covered with hair and ramified spines. The latter are arranged in rows, six on each thoracic segment and seven on each abdominal segment. Setigerous papillae decorate the thoracic and abdominal legs.

The pupa is attached with the tip of its abdomen to the object upon which it is found, head downward. Two yellow processes adorn the head and each abdominal segment.

Distribution. V. cardui is a migratory insect and is therefore distributed over wide areas. It is found in Europe and reaches as far north as Iceland. In Asia it occurs in India, Pakistan, Iraq and Israel. In North America it reaches as far north as Canada, and it inhabits several Pacific Islands. The Mediterranean area is one of its breeding places.

Hosts. Compositae seem to be the preferred food plants. The larvae feed upon *Helianthus*, *Cirsium*, *Xanthum* and *Cynara*. Second in preference are the leguminous crops. *Lathyrus* beans, and alfalfa crops are often attacked as well as malvaceous plants.

Type of Injury. The injury consists of the defoliation of the plant by the larva. Since the attack is often made by a dense population, entire fields may be defoliated within a short period, as described by JANNONE (1945) on artichoke in Eritrea.

Life History. The eggs are laid on the leaves upon which the hatching larvae feed. The development of the larvae lasts about one month under conditions of the Israeli early spring, and 6 weeks in the European summer. At the end of this period the larva pupates while hanging head downward.

The development of the pupa is 10 days in Israel or Eritrea (JANNONE, 1946) and 15 days in Europe.

Phenology. According to JANNONE the insect is found in the high plateaus of Eritrea from July to September during which time it raises two generations. From November it disappears and is not found throughout the winter months.

In Israel the butterflies appear in March and continue during the summer.

Migration. There is no need for further evidence about the migratory habits of this species. WILLIAMS (1958) mentions the fact that the larvae cannot survive the winter in England; yet the butterfly frequents that country every year. Furthermore, many naturalists reported observations of its flights, in some cases far at sea (in one case as far as 800 km from the nearest coast).

Although the insect breeds in certain countries in Europe, its main breeding places, according to WILLIAMS, are the margins of deserts in Africa or in similar places in America.

From Africa the insects migrate northward to Europe, through Spain to England, and as far north as Iceland; through Sicily and Italy to Central Europe and through Sinai and Israel to Syria, Asia Minor and the Caucasus. The migrations of the insect are not necessarily northward; they may migrate eastward. According to JANNONE (1945), butterflies from the valleys of the Sudan, Ethiopia and Eritrea migrate eastwards to the plateaus of Eritrea, raise two generations there and then proceed in the direction of the Red Sea. In October-November numerous flocks move from the Sudan and Ethiopia eastwards to the Red Sea.

There is one observation recorded about the beginning of a flight. WILLIAMS quotes SKERTCHLEY who, while riding on a camel on the margins of the desert near Suakin on the Red Sea, "noticed that the whole mass of grass was in a state of violent agitation although there was no wind. When he dismounted he found that the cause was the emergence from the pupal cases of myriads of Painted Lady butterflies, which dried their wings and about half an hour later flew off together eastward towards the sea".

There is probably also a migration southward from northern countries back to Africa (WILLIAMS). The migration of *V. cardui* does not occur in equal numbers every year. On certain years the invasions are denser than on other occasions. In Europe, 1879, 1880, 1906, 1913 and 1952 were *V. cardui* years.

In Israel conspicuous flights were recorded in 1917, 1922, 1927 and 1935. In 1951 many alfalfa fields were damaged by the caterpillars of this butterfly.

LITERATURE

A. BALACHOWSKY & L. MESNIL, 1936. Les Insectes Nuisibles aux Plantes Cultivées, *1376*, Paris.

G. JANNONE, 1945. *Boll. Soc. ital. Med.* (Sez. Eritrea) **IV**, 3, *451-61*, Asmara.

G. JANNONE, 1946. *Int. Bull. Plant Prot.* **20**, 11-12, *115M-121M*.

C. B. WILLIAMS, 1958. Insect Migrations, 235 pp., *The New Naturalist*, **13**, Collins, London.

SOLANACEAE

KEY to the Pests of Solanaceae

A. Subterranean parts of the plant are injured

I. In potatoes there may be external cavities in the tuber; in other plants the subterranean part of the stem is gnawed or cut

1. There are tunnels in the vicinity of the plants: *Gryllotalpa* page 44
2. Large white grubs are found in the ground . . : Scarabaeidae page 273
3. Curled grey-green caterpillars are found in the ground . . . . : *Agrotis* page 88

II. In potatoes the tuber is tunnelled

1. Purplish-white, black-headed larvae bore in it . . . . : *Gnorimoschena* page 327
2. Cream white, reddish-brown headed larvae bore in it . . . . : *Euzophera* page 331

B. The stem is injured

1. The stem near the ground is gnawed . . . . : Tenebrionidae page 138; *Agrotis* page 88
2. A larva bores into the stem, or into the shoots of potato or tomato . . . . : *Euzophera* page 331

C. The foliage and blossoms are injured

I. The injury is caused by biting insects

1. Large caterpillars with a horn on the end of the body are on the plant . . . . : *Acherontia* page 333
2. Striped dark green caterpillars with dark spots on the abdomen are present . . . . : *Prodenia* page 102
3. Pale green caterpillars, tapering anteriorly, with 3 pairs of abdominal legs are present . . : *Plusia* group page 125
4. Small, purplish-green caterpillars bore into the main veins and petioles or gnaw the leaf leaving the epidermis . . . . : *Gnorimoschena* page 327
5. Brick-red, grub-like larvae or striped hemispherical beetles feed upon the leaves (potatoes): *Leptinotarsa* page 322

II. The injury is caused by sucking insects

1. Small, soft-bodied, globular plant lice abound on the stem or leaves . . . . : Aphididae page 60
2. Minute, elongate, greyish-yellow insects, less than 1 mm long, are on the leaves; their feeding leaves pale or silvery areas . . . . . . : *Thrips* page 74
3. Green, elongate Homoptera move quickly over the leaves . . . . : *Empoasca* page 53
4. The leaves become chlorotic, devoid of chlorophyll, a delicate spider web and minute red creatures may be found on the leaves . . . . : Acarina page 80
5. The plant (tomatoes) or part of the plant wilts, becomes brown or rusty; with the magnifying glass minute, yellowish, spindle-shaped creatures are seen . . . . : *Vasates* page 319

6. Silvery patches with overgrown pubescence is found on the leaves (tomatoes) . . . . . . : *Aceria* page 320
7. Large Pentatomid bugs, 5—10 mm long, feed upon the stem, foliage or blossom . . . . . : *Nezara* page 49; *Dolycoris* page 51; *Eurydema* page 291
8. Blossom drops, small, delicate, pale-green bugs are found on the plant . . . . : *Engytatus* page 320

D. The fruit is injured
1. Puncture spots or brown scars are seen on the fruit (tomatoes) . . . . : *Nezara* and others page 49
2. Caterpillars gnaw or bore into the fruit
   a. green striped caterpillars . . . . : *Heliothis* page 119
   b. striped olive green caterpillars with black spots over the body . . . . : *Prodenia* page 102

The Tomato Russet Mite — Vasates leucopersici Massee

[Syn.: *Vasates destructor* KEIFER]
(Eriophyidae, Acarina)

In recent years a type of injury was discovered on tomatoes which was not formerly known in Israel nor in the Near East. The leaves become bronze-rusty, wither and die. Examinations with a magnifying glass revealed numerous mites which were responsible for this injury.

Description. The adult is a minute creature, 170 μ long, hardly visible with the naked eye. Its body is wide anteriorly and tapers posteriorly, assuming the shape of a carrot. Its colour is ivory-white. A higher magnification shows a few setae on the body and two pairs of legs at the anterior end. The body is divided into numerous ring-like narrow segments, 27 of which may be counted from above.

The egg is spherical and smooth. The newly-hatched larvae are elongate, 80 μ long, while the second instar larva is 130 μ long. Both are of paler colour than the adult.

Distribution. The pest is recorded from various parts of the world. First it was discovered in Australia, then it was found in the U.S.A., where it is known as *V. destructor;* it has been found also in North Africa, Egypt and Israel.

Hosts. Of the cultivated plants, tomato seems to be the host most prefered, although it may feed on various types of potato. It has been found on the nightshade *Solanum nigrum* and on other related plants.

Type of Injury. The mite feeds on the outer cells of the plant. As a result the injured parts become silvery, then bronze-coloured or rusty and finally wither and die. The attacked stem becomes brown, bare, and cracks. At first the lower, older leaves are attacked, then the rest of the plant becomes infested and the leaves drop, except for the most recent ones. Injured plants yield little or no fruit at all.

Life History. Very little is known about the biology of this species except that it lays eggs and may raise several generations during the year. From the knowledge of related species, the citrus rust mite, for instance, it may be surmised that its development is very speedy and its rate of reproduction high, which explains the fast spread of the species during the summer.

Control. Dusting the plants with sulphur, or spraying with wettable colloidal sulphur 0.5—1% in water once gave relief from this pest (SMITH, 1955). Lately sulphur has become less effective. Substances like parathion, endrin and toxaphene have given more satisfactory control (TUFT & ANDERSON, 1954).

The Tomato Gall Mite — Aceria leucopersici Wallenstein

In North Africa, including Egypt, another mite attacks tomatoes. Earlier it was described as *Phytoptus calacladophora* NAL and also as *Ph. cladophthirus* NAL. Later, comparisons revealed it as synonymous to *Aceria leucopersici* WALL. (LAMB).

The feeding of this pest causes hypertrophic growth of the hair, which gives the plant a silvery appearance. A severely attacked plant withers and dies. The pest may be controlled with the same methods applied against the russet mite.

The Tomato Bug — Engytatus tenuis Reut.

[Syn.: *Cyrtopeltis tenuis* REUT.]
(Miridae, Hemipt.)

A healthy stand of tomatoes fails to yield fruit because of a premature drop of the blossoms. The cause of such injury may be a little bug, *Engytatus tenuis.*

Description. Like other members of this family, the body is elongate and wider in the middle, 3 mm long, pale green except for the tips of the antennae and the tip of the rostrum which are brown, and the legs which are greenish-wite. The wings are delicate, translucent and punctate at the tips.

The eggs are cylindrical, slightly bent, and are deposited in groups in the plant tissue.

The larva is pale green and resembles the parents except that it is smaller and wingless.

Distribution. The pest is found in South and Western Asia, in Australia, the West Indies, North Africa and the Near East.

Damage to tobacco has been reported from Indonesia and the

Solomon Islands, and to tomatoes from Queensland, Egypt, Libya, Cyprus and Israel.

Hosts and Injury. Various solanaceous plants may serve as food;

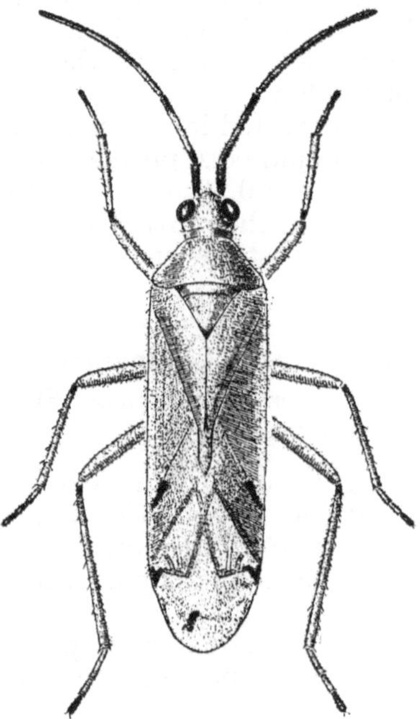

Fig. 107. *Engytatus tenuis.* adult. After SLOAN, Qd. agric. J.

of the cultivated plants, eggplants, tomatoes, potatoes and tobacco are preferred.

Injury is caused by the sucking of the plant sap (but this bug prefers, for the site of feeding, the delicate parts of the blossom). Apparently the enzymatic secretion into the plant in the course of feeding causes this premature blossom drop.

The bug is also suspected of transmitting leaf curl of potatoes in India and certain diseases of tomatoes in Libya.

Life History. The adult and nymphs spend the winter in the stage of a facultative diapause. On warm days they may be seen occasionally on wild plants. However, the mass awakening takes place in the spring when they begin feeding, mating and laying. The incubation period of the egg lasts 6 days in the spring and 4 days in the summer. The larva moults five times before it matures and this development lasts 3—4 weeks. Thus, the generation lasts more than one month in the spring and less than one month in the summer.

Number of Generations. In accordance with the speed of development the bug may raise five generations in the coastal plain of Israel from May to September-October. In the inner valleys a sixth may arise. The winter generation is the least numerous and the damage not noticeable. With the advance of the generations the damage increases. In June-July when the second and third generations are active the injury is very great, especially when the weather is warm and the humidity not too low.

Control. The eggs are protected in the plant tissue and cannot be eradicated. Insecticides should be applied against larvae and adults. The larvae may be killed by a 0.2% solution of nicotine; the adults are killed by DDT. As a rule, the preventive applications which are carried out against other pests, such as *Prodenia* and *Heliothis*, control this pest too.

The Colorado Potato Beetle — Leptinotarsa decemlineata Say.

[Syn. *Doryphora decemlineata* SUFFR.]
(Chrysomelidae, Col.)

This beetle, whose origin is in the Americas, does not occur in the Near East. However, it has spread widely and is established in Europe, and recently has reached the outskirts of our zone. In view of this fact as well as its economic importance it will be discussed here.

Description. The adult has an oval, hemispherically shaped body, 10—12 mm long. The eleven-segmented antennae reach the posterior

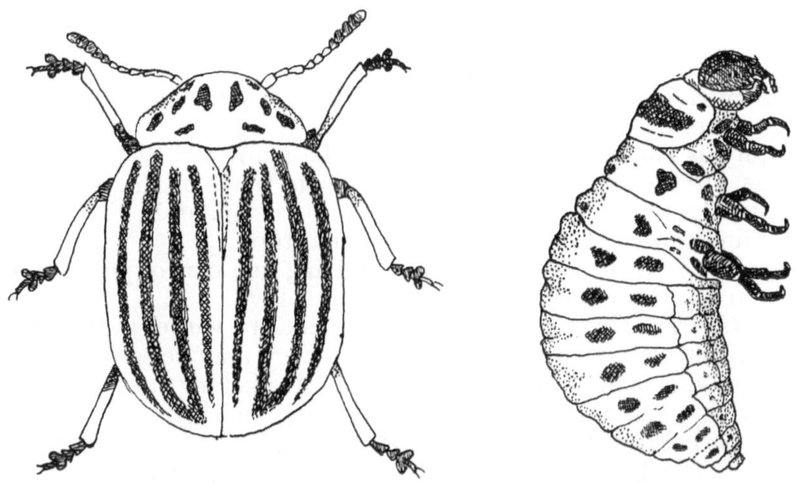

Fig. 108. *Leptinotarsa decemlineata*; (a) adult; (b) larvae. After BALACHOWSKY & MESNIL.

margin of the pronotum; the first antennal segment is large, the second smaller than the third and 7—11 segments are large and wider than long. These are black-brown, while the others are reddish-brown. The head is yellow and on its frons there is a triangular spot. On the yellow thorax there are several spots whose shape and number vary. On each elytron five longitudinal black bands alternating with yellow converge at the apex. The underside of the body is reddish and black spots decorate the abdominal segments.

The egg is elliptical, 1.2 mm long. It is glued to the leaf and laid in groups standing one near the other.

The larva is hemispherical, when mature 12—14 mm long and brick-red. The segments bear lateral setigerous papillae. A large black spot decorates the pronotum, while lateral smaller spots, two on each side, adorn the meso- and metathorax. Two lateral black spots are located on each side of the abdominal segments.

The pupa is 10 mm long, yellow, with legs folded at right angles to the body. It is found in a cell in the ground.

Distribution. Until 1850 this insect was found only on the eastern slopes of the Rocky Mountains in the U.S., in the neighbourhood of Colorado and adjacent states. The climate of that country and the scarcity of host plants prevented the build-up of dense populations of the pest. The common host plant then was *Solanum nigrum*. With the advance of settlers and the conversion of large areas into potato fields conditions changed. The beetle quickly adopted the potato as a host. A sudden increase of food supply brought about an increase in the population. The insect spread eastward and occupied territories which it did not inhabit before. By the seventh decade of the century alarming reports appeared in the local papers about destruction and damage in Nebraska, Illinois, Wisconsin and Ohio. In 1875 the beetle reached the Atlantic states and continued to spread both southwards and northwards to New England and Canada. At the end of the century it appeared in Germany. Strong measures taken at that time exterminated it.

In 1922 it was discovered again in Bordeaux, France. In spite of strong measures to eradicate it, the beetle spread further into Europe. By 1934 half of the area of France was invaded and it continued to spread to Spain, Belgium, Holland, Germany, Czechoslovakia and Poland. By the middle of the twentieth century its distribution in Europe was as follows: Portugal, Spain, France, Belgium, Holland, Luxemburg, Germany, Denmark, Switzerland, Austria, Italy and Yugoslavia. The following countries are still free of it: Great Britain, Ireland, Norway, Sweden, Finland and U.S.S.R. Up to the present time no records of its occurrence have come from Asia and Africa.

Hosts. The plants which serve as hosts to this beetle are of the family Solanaceae. Not all members of this family are equally attacked. The most favoured is the potato plant, the least favoured of cultivated plants is the pepper. After the harvest of potatoes, damage may be caused to eggplants, too. Of noncultivated plants *S. nigrum*, *Datura* and *Nicotiana* may be attacked. Occasionally cabbage and petunia may be injured, too.

Type of Injury. Both adults and larvae feed on the foliage. In the case of newly sprouting potatoes the plants may be destroyed in a short time. But older plants also may be entirely defoliated leaving only bare stalks.

Life History. The beetle overwinters as an adult and emerges from the ground in the spring. In Bordeaux, France, and Turin, Italy, this takes place in Apiil-May; in northern France, Germany, Denmark, Switzerland, Hungary, and Yugoslavia it occurs in May-June. The beetle feeds, mates and begins egg laying.

The eggs are laid in clusters of 4—80 eggs each. One female may lay several such clusters so that, on the average, about 500 eggs are laid. The maximum number of eggs laid by one female appears to be 2400—3000. According to GRISON (1944), overwintering females laid more than first generation beetles.

The incubation period lasts about 6—8 days at a temperature of 20—24° C. Upon hatching the larva feeds upon the foliage, and after moulting four times it reaches maturity and full size. It then enters the ground for pupation. The duration of the larval period is 16 days at 20—24° C. The larva remains in the ground two or three days after which it moults into a pupa. The duration of this stage is 10 days at the temperatures mentioned above. If the eleven days of the preoviposition period are added, then the complete life cycle totals about 48 days, at 20—24° C. ALFARO (1943) calculated the threshold of development as 11.5° C, while the thermal constant is 335 day-degrees centigrade.

Number of Generations. The number of generations depends upon the time when the beetles emerge from their hibernation quarters and upon the length of development. Both of these depend upon the climate of the country. Thus, in Central Europe the first generation appears in July-August while in more southern countries in July. The second generation appears in August-September. In Bordeaux, Austria, Hungary and Italy there are two generations while in Northern France, Northern Germany and Denmark there is one generation only. The beetles of the second generation enter into a state of diapause shortly after emergence.

Diapause. Diapause in this beetle is characterized by a few criteria: Positive geotaxis, diminished positive phototaxis, strong stereotaxis, no reproduction and burrowing responses. The passage from activity to diapause does not depend upon the season, generation or inheritance. Thus GRISON found that beetles of the first generation entered the stage of diapause while hibernated beetles continued to lay. This author reared three generations in the laboratory and found that 10% of the beetles of the first generation laid that summer while the rest entered the state of diapause, 4% of the second generation and none of the third generation laid that summer. As to the state of the beetles in the field, he found that, of 1000 beetles collected in the field, only 18 laid that summer. FABER (1949) of Austria, who also studied this subject, gives a different picture. Of the offspring of the first generation 90% laid the same summer while the rest entered the stage of diapause. 75% of the offspring of the second generation, 50% of the third and none of the fourth oviposited the same summer. This author points out food and temperature as factors inducing diapause. Interruption of food and exposure at 10° C for a few days, exposure of pupae for 10—14 days at 5° C, exposing pupae for 5—8 days at 35° C and also exposing larvae at low temperatures induce diapause.

More recent studies by DE WILDE (1956), GORYSHIN (1956) and others, point out the photoperiodic effects upon the diapause of the beetle and also explain certain aspects of the phenology of the beetle. GORYSHIN states that, when the beetle was exposed to a day shorter than 16 hours, the percentage of diapausing beetles was high, and with longer than 16 hours the percentage decreased. Thus, at 16 hours 34% entered the soil while at 17—18 hours very few did so. On the other hand, at a temperature of 28° C at 16 hours 9% entered

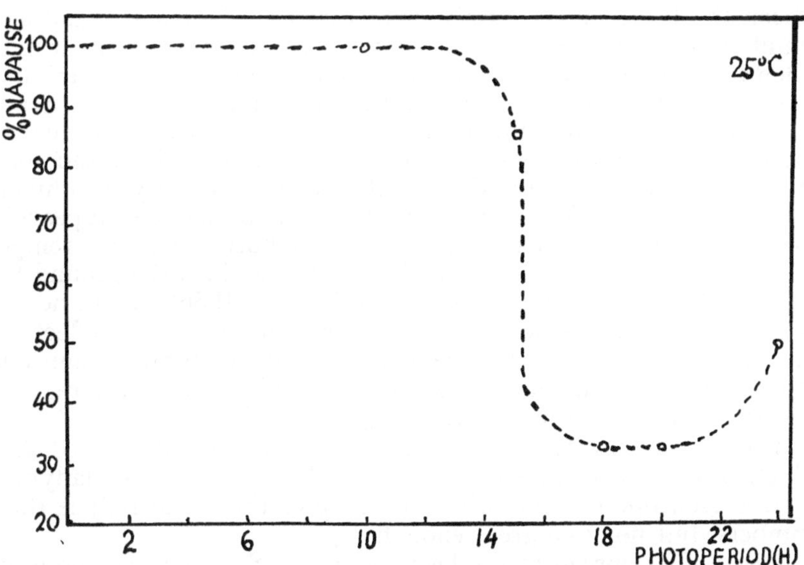

Fig. 109. Effects of Photoperiod on the percentage of diapausing beetles of *L. 10-lineata*. After DE WILDE.

the soil, at 15 hours — 85% and at 14 hours all went into a stage of diapause.

Similar results were obtained also by DE WILDE as seen from fig. 109. This author confirms the findings of previous writers (WIGGLESWORTH and others) about the role which the corpora allata play in the diapause and states that "implantation of 4—6 'active' glands into pre-diapause adults grown at 10 hour period, resulted in a prompt response, oviposition may start within 48 hours". It is evident, thus, that a long day activates, while a short day inhibits, the activity of the ovaria. From DE WILDE'S work we learn further:

Experiments in which the eyes of the beetle were either cauterized or covered with black paint and exposed to photoperiod indicated that the eyes are not the main perceptors of the photoperiod.

Limiting Factors. Cold and rainy weather retard the emergence of the beetles from the ground. This may reduce their population, interfere with productivity of the first generation, and reduce the chances of the development of a second generation.

The beetle is sensitive to the effects of high temperature coupled with low relative humidity. For this reason its population is not dense in certain Swiss areas or in southern France and in Spain. The pupae, too, are sensitive to low humidity, and when rains are scant their mortality is high. The beetles also need rainy weather for their well-being, but not too much of it. Excessive moisture and low temperatures reduce its activity and reproduction. For this reason the beetle is less of a pest in Brittany, North Holland and Hamburg. ALFARO (1943) states that the beetle takes to flight only when the temperature is above 25° C and at low relative humidity; this factor probably accounts for the fact that the beetle spreads but little in the above mentioned places. Snow, when it occurs in the active season, as may happen in some places in Switzerland, may kill the eggs and larvae and drive the beetles too early into hibernation. A similar situation may occur when a hot dry spell appears in a certain territory since eggs and larvae are both killed by a temperature above 35° C, and the beetles are driven into the ground for protection. Finally, the opinion of DE WILDE (1956) should not be omitted in this discussion. From the point of view of the effects of the length of the day, the beetle will not develop more than one generation a year in the zone south of 35° NL, and, therefore, is of less economic importance in those regions.

In the literature there are many references regarding predators which prey upon the eggs and larvae of the beetles. Bugs, ladybird beetles, Neuroptera and others feed upon them and reduce their numbers, but not to a great enough degree.

Control. The spread of the beetle in the U.S. a century ago gave the first stimulus to the use of chemicals as insecticides when it was discovered that Paris green killed the beetle when dusted upon the food plant. All substances containing arsenates or fluor compounds killed this pest. The common applications against the potato beetle were spray of 0.5% lead arsenate, or dusting with 3—5 kg per dunam with calcium arsenate, barium fluosilicate or cryolite. Also, the synthetic insecticides proved to be efficient. DDT, BHC, dieldrin as sprays or dusts were successfully applied against this pest. Lately there are reports of some materials becoming less active against this pest. The farmer should select the more effective insecticide by trial on beetles in his own territory.

The Potato Tuber Moth — Gnorimoschema operculella Zell.

[Syn.: *Phthorimaea operculella* ZELL]
(Gelechiidae, Lep.)

Potatoes in storage may be infested by worms to the extent that the weight of the stored tubers is greatly reduced. These worms are the larvae of the potato tuber moth.

Description. The length of the body is 10—12 mm, the antennae

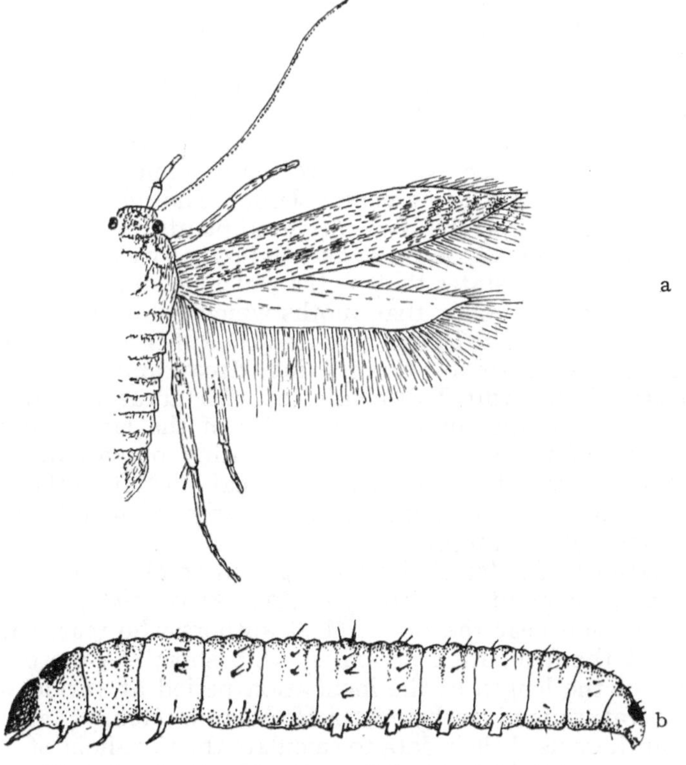

Fig. 110. *Gnorimoschema operculella.* (a) adult; (b) larva.

reach the tip of the abdomen, and the forewings are narrow, pointed at the tip and longer than the hind wings. The colour of the wings is light grey mottled with black dots and darkened at the tips. Three black spots near the hind margin of the forewing are typical of the species. The hind wings are narrow and have a fringe of long hair.

The egg is elliptical, 0.5 mm long, 0.37 mm wide and cream-coloured with an iridescent lustre. The larva is 10—15 mm long and at the prothorax 15 mm wide. Its colour is purplish-white but the head, prothoracic plate, legs and last abdominal tergite are black. Each abdominal segment bears a row of black setigerous spots; on the last segment the spots coalesce.

Distribution. The species apparently originates from tropical America. Today it is found in North, Central and South America, Cuba and Bermuda; in the Pacific it is found in Hawaii and the Fiji Islands; in Australia, New Zealand and New Caledonia; in Sumatra, India and Ceylon, and throughout Africa. In Europe it is found in Spain, Southern France, Italy, to a less extent in Germany and in Southern Russia.

In the Near East damage has been recorded from Malta, Egypt, Cyprus and Israel.

Hosts. The most important host is the potato plant, *Solanum tuberosum.* The moth lays its eggs on the tuber both in the storeroom and in the field. But should these be unavailable, oviposition may take place on the leaves where the hatching larvae bore into the main veins, petioles or young shoots. The moth may lay also on tobacco and eggplants. Noncultivated plants mentioned as hosts are *Lycium, Nicotiana, Solanum* and *Datura,* as well as *Hyoscyamus.*

BRANNON (1943) states that moths which as larvae fed on potato tubers were more prolific than moths which fed on tobacco leaves.

Type of Injury. The larvae bore into the tuber and in addition to the direct loss caused by this feeding, decay and rot take heavy tolls from infested fruit. Tubers infested with larvae are refused even by cattle. As a rule the entrance holes of the larvae are in the depressions of the eyes; thus the sprouting potentialities of the potato seed may be impaired. In case of leaf infestation the attacked leaves dry because of petiole injury. Similarly, young tobacco plants wilt because of stem injury.

Life History. The female lays her eggs in cracks, corners or folds in the tuber or plant. According to ATHERTON (1936) she may lay also in the soil near the tuber. Most authors who reared the pest state that the incubation period is "very short". A more detailed account of the length of the incubation period at various degrees of temperature is given by ATTIA & MATTAR (1939) (Table LIII). These authors used their data to calculate the threshold of development and found it to be below 13° C. HOVEY (1943) claims a threshold below 10° C while ATHERTON points out that the threshold of development of all stages is 11° C. HOVEY further points out that humidity does not affect the length of development of the eggs. No difference was obtained when eggs were reared at humidities from 0 to 100% at a temperature of 20° C.

Immediately after hatching the larva bores into the tissue upon which the egg was laid. It develops quickly and under favourable conditions it may complete its development in 7 days. When ready to pupate the larva leaves its food and seeks a suitable place to pupate. In the field this may take place under rubbish or between leaves; in the storeroom, in cracks or folds of bags and the like. It then spins a cocoon and pupates. ATTIA & MATTAR, who studied the pest in the Near East, record the data for the length of development of both larva and pupa together. Their data at various degrees of temperature are also given in Table LIII. According to these authors the ratio between the length of periods of larva and pupa varied with the temperature. Thus, at 19° C it was 3:1, at 21° C — 2.5:1, and at 29° C — 2:1.

Table LIII.

The development of the potato tuber moth at various degrees of temperature. After ATTIA & MATTAR and others (in brackets).

Temp. in C°	Incubation period in days	Larval and pupal development in days	Development from egg to adult	Preoviposition period in days
15—16	18 (HOVEY, 1943)			
18	12	80	54 (LLOYD, 1934-44)	12
20	9			
25—26	5	25—27	27 (LLOYD,	2—3
27—30	4	18—23	21 (MORRIS, 1933)	3—4
31—35	3	15—17	19—20 (LLOYD)	3—4

MORRIS (1933) finds that the pupal stage alone may last 50 days at 13—16° C and only 4—5 days at the most favourable high temperature. The adult begins to lay 3—4 days after emergence from the cocoon, and the eggs are laid in small clusters. The number of eggs laid by one female is 50 and 65 according to LLOYD and ATHERTON respectively, while MORRIS in Cyprus claims to have obtained a maximum of 150 eggs per female. The number of eggs laid at various degrees of temperature as obtained by ATTIA & MATTAR is given in Table LIV.

Table LIV.

Egg laying of the potato tuber moth at various degrees of temperature after ATTA & MATTAR.

Temperature range (°C)	11—16	15—22	19—28	22—29	21—34	30—32	33—35	38—42
Average temperature (°C)	13	18	22.5	26	27.5	29.5	34	40
Number of females	2	3	15	8	7	8	7	5
Average number of eggs	22	35	45.3	65	87.5	29	18	0

These writers state that, of all the females reared by them, only two laid more than 100 eggs each, one laid 121 and the other 134 eggs. These were reared at 21—34° C; they also claim that 38° C is apparently the temperature at which the moths cease to reproduce.

The life-span of adults is about 4—5 days (LLOYD) or 38—40 days (ATHERTON, ATTIA & MATTAR). The latter occurs at 13° C.

The moths are nocturnal in habit, but ATTIA & MATTAR claim to have seen them fly in daylight.

Number of Generations. To sum up the development records at various degrees of temperature it is noted that at 15—20° C the life cycle of the moth may last 2—3 months, but only 3 weeks at 27° C. In view of the fact that the pest does not go through an obligatory diapause (EDWARDS, 1929), it may thus raise many generations during the year. For instance there are 6 generations in France, the adults appearing in January, April, June, July, August and October.In colder climates the number of generations is smaller than in warmer areas; thus 9 generations occur in Egypt (ATTIA & MATTAR).

Limiting Factors. The moth ceases to develop at 10° C (ATHERTON). When HOVEY took larvae ready to pupate, which were reared at 25° C, and placed them at 5° C, only 1% pupated. Similarly when pupae, reared at 25° C, were placed at 5° C only 2% matured into adults. In view of this, potatoes may be safely stored at a temperature below 10° C, and the pest in the tubers will not develop beyond the stage in which it was when placed in storage.

In northern countries warm and dry weather with less than average rains cause outbreaks of the pest (LANGFORD 1932). In countries of the Near East the spring and autumn are the most favourable for this moth. At a temperature above 30° C egg-laying is a third, at 33° C it is a fifth of the highest rate of oviposition, while at 38° C it ceases (ATTIA & MATTAR).

The high temperatures surely affect other physiological features. Rain, as a factor which prevents outbreaks is due to the fact that soil becomes compact and hinders the penetration of larvae into the subterranean tubers.

Control

1. In the field. To prevent egg-laying on the plants they may be sprayed with DDT suspension 150—200 g a.i. or endrin emulsion 60 g a.i. per dunam. In order to prevent egg-laying on the tubers, these should be well covered with the earth.

After harvest, infested plants should be removed from the field and all tubers gathered.

Harvest should be made as quickly as possible, and no potatoes should be left exposed in the field at night.

2. In the store-room. The storage area should be clean from remains. Bags and containers may be sprayed with 1% suspension or dusted with 10% DDT. Tubers may be disinfested with 70 g of carbon bisulphide per cubic metre, at a temperature of 20° C for 24 hours.

The Eggplant Stem Borer — Euzophera osseatella Tr.

(Pyralidae, Lep.)

The base of the stem of the tomato plant may be infested with a boring larva; the same kind of worms may also be found in tubers of potatoes or in the tobacco plants. These larvae are of *Euzophera osseatella*, a pyralid moth.

Description. The moth is 6—8 mm long, and the wingspan is 14—22 mm. The colour of the body is yellowish-brown; an undulating line runs across the middle of the forewing. Two other lines parallel to this run across close to the side margin. On the disc, close to the hind margin, there is a large dark-brown spot. The hind wings are translucent except for the veins which are yellowish. Along the margins of both fore and hind wings there is a reddish line and fringes of hair.

The egg is slightly oval, 0.5 mm long and about 0.4 mm wide. It is yellowish-brown or rust coloured and its surface is rough.

The larva is ivory coloured, 13—15 mm long, with a brown-red head. The hooks on the abdominal legs are in concentric circles.

The pupa is narrow, 8—10 mm long and is encased in a loose spindle-shaped silken cocoon 15 mm long.

Distribution. So far this insect has been recorded as a pest from Egypt (Willcocks, 1925) and Israel only (Klein, 1941). It occurs also in Italy, Dalmatia, Cyprus, Tunis, Algeria and Morocco.

Hosts. The insect is restricted to Solanaceae only. The following cultivated plants are subject to attack: tomato, potato, pepper and tobacco. It was found on the night-shade, *Solanum nigrum* as well.

Type of Injury. The larvae bore into the stem of the plants. In young tobacco plants this boring may start from the base and the plant which is attacked withers. In tomato too the attack is usually at the base of the stem, and the entire plants is thus weakened. In potatoes they bore into the main stems, but the feeding of young larvae may kill the shoots before they develop into stems. The yield of infested plants is greatly reduced; boring larvae may penetrate even into the soil and into new tubers.

Fig. 111. *Euzophera* (a) adult; (b) larva in tomato stem – after Avidov; (c) pupal cocoons, attached to eggplant stem.

Efforts to plant potatoes in the Negev very early in August, in order to obtain an early crop, failed because of *Euzophera,* whose attack was uncontrollable. Only potatoes, planted after the 15th of that month could develop and overcome the injury of this pest.

Life History. The biology of this species was studied by RIVNAY and this discussion is based on that study.

The insect spends the winter in the larval stage. Pupation takes place, in a cocoon in the ground close to the stem, toward the end of the winter and adults appear in late March or early April. This rest is not an obligatory diapause because, when placed in a higher temperature, development was resumed and the adults appeared at an earlier date than those which remained in the colder temperature.

After a period of 3—5 days egg-laying takes place. The eggs are laid on various parts of the plant and the incubation period lasts about 13 days at 18° C, 9 days at 22° C and 6—7 days at 26—30° C.

Immediately upon hatching the larva crawls about in search of a suitable place at which to bore into the plant, either in an angle, a corner, a furrow or pit in the plant. The development of the larva lasts 30—40 days at a temperature of 22—26° C. In a cold temperature the larva spins a lining over the walls of the tunnel and remains there in a quiescent stage. In the case of too much moisture in the tunnel the spinning may extend even to the outside, thus securing favourable environmental conditions. Pupation, under the conditions of temperature mentioned above, lasts 10—20 days.

Number of Generations. According to RIVNAY six generations develop annually in Israel. One generation develops during the winter, one generation during the spring and one during the autumn; three generations develop during the summer.

The Death's Head Hawk — Acherontia atropos L.

(Sphingidae, Lep.)

Quite often potato plants in the field are entirely defoliated; large pellets of faeces may be found on the ground or a large caterpillar clings to a branch. This is the larva of the death's head hawk.

Description. The adult is a large moth, 50—60 mm long, with a wingspan of 100—130 mm. The thorax is large, covered with brown hair, and on the notum there is a large spot shaped like a human skull. The forewings are brown. A few undulating rusty coloured lines cross the wings, two pairs of which are yellow on the anterior half. The hind wings are yellow with two bent bands along the margin, the posterior of which is broader. The abdomen is broad, brown with a wide yellow band on each of the abdominal segments.

The egg is greyish-green or greyish-blue. The larva, when mature, is 150 mm long. The body colour is yellow, but on the abdominal segments (except for the last) there are, on each side, oblique bands of blue bordered with brown or green lines. Groups of 4—11 dots decorate the body segments and the spiracular margins are black.

The pupa is 50—60 mm long, red brown, and is found in a cell 15—20 mm deep in the ground.

Distribution. This insect is a tropical and subtropical species. In Eastern Asia it is found from the south-eastern part of the Continent to Japan; in Western Asia it is found as far north as Asia Minor — but it does not occur in Iraq. It is found in North Africa, Southern Europe and, in the summer, also in Northern Europe.

Hosts. The insect is a species of the Old World, but its most common host among the cultivated crops is the potato plant which originates from the New World. It seems, therefore, that this is an adopted host. Other plants which serve as food for the larvae, and which seem to be the original host plants, belong to the genera *Datura*, *Lycium* and *Jasmine*.

Type of Injury. The insect is a voraceous feeder: one larva may consume the foliage of an entire medium-sized potato plant. But they do not occur in large numbers. As a rule one or a few are present in one spot, hence their damage is not conspicuous.

Life History. The adult appears in the spring and begins its egg-laying. In Israel this may occur in April-May, while in Europe in May-June. Under climatic conditions of Central Europe two or three months are necessary for the development of the species in all its stages. Thus, it is capable of raising but one generation there. In Israel larvae are found both in the summer and in the autumn.

The adults feed not on flower nectar, like other moths of this family, but on sweet exudates of various kinds. Their liking for honey causes the moths to invade bee-hives.

Migration. The moth is a well-known migrant. Every spring it leaves the winter breeding territories, probably North Africa and Israel too, and migrates northwards. The peak of its occurrence in England is in May, June and in September. The autumn moths are probably specimens bound southward. WILLIAMS (1958) states, that the insect does not overwinter in any stage in Europe, except perhaps in Spain and Italy.

Limiting Factors. A great deal of the caterpillars collected are parasitized by various parasites. Outstanding is the Tachinid, *Sturmia atropivora* R.D., which probably reduces its population considerably.

General Considerations on the Control of Solanaceous Pests.

Tobacco seedlings are liable to be attacked by adult Tenebrionidae. Preventive measures are advisable where these beetles abound.

In rich humus fertile soil, tomatoes are attacked in their early stages by *Gryllotalpa*. Treatment of the soil with BHC or aldrin on both sides of the plant row may prevent serious damage.

Tomatoes are usually treated with fungicides prophylactically. It is advisable to mix insecticide with the fungicide against *Engytatus*, *Heliothis*, *Prodenia*, etc. The injury by these latter is usually caused by larger caterpillars which are harder to kill with insecticides. Weeds should be eradicated in order to avoid invasions of larger caterpillars from non-treated plants.

LITERATURE

A. Alfaro, 1943. *Bol. Pat. veg. Ent. agric.* **12,** *9-30.*

D. O. Atherton, 1936. *Qd. agric. J.* **45,** 1-4, *12-31, 131-145, 239-248, 331-344.*

R. Attia & Bishara Mattar, 1939. Ministry of Agr. Egypt, Techn. and Sci. Bull. **216,** *1-136.*

A. Balachovsky & L. Mesnil, 1936. Les Insectes nuisibles aux Plantes cultivées, Paris.

L. W. Brannon, 1943. *J. econ. Ent.* **36,** 3, *469-470.*

W. H. Edwards, 1929. *Bull. Dept. Agr. Ile Maurico Sir. sci.* **13,** *1-8.*

W. Faber, 1949. *Pflanzenschutzberichte* **3,** 5-6, *65-94.*

N. I. Goryshin, 1956. *Dokh. Akad. Nauk SSSR* **109,** *205-208.*

P. Grison, 1944. *C.R. Acad. Sci.* **218,** *342-344.*

P. Grison & A. Couturier, 1942. *Cah. Path. veg. Ent. agric., 38-41.*

A. S. Hassan, 1934. *Bull. Soc. Fuad 1er Ent. Egypt,* **18,** *443-444.*

C. L. Hovey, 1943. *J. econ. Ent.* **36,** 4, *627-628.*

H. Klein, 1941. *Hassadeh* **21,** *143-144.*

K. P. Lamb, 1953. *Bull. ent. Res.* **44,** *343, 401.*

G. S. Langford, 1932. *J. econ. Ent.* **25,** 3, *625-634.*

G. S. Langford, 1934. *J. econ. Ent.* **27,** 1, *210-213.*

N. C. Lloyd, 1934-44. *Agric. Gaz. N.S.W.* **54,** 7 & 9, *323-327, 337, 417-421;* **55,** 3 & 5, *107-110, 126, 193-196.*

H. M. Morris, 1933. *Cyprus agric.J.* **28,** 4, *111-115.*

E. Rivnay, 1962. Unpublished MS. on *Euzophera.*

W. J. S. Sloan, 1945. *Qd. Agric. J.,* **61,** *38-39.*

W. A. Smith, 1955. *Queensland agric. J.* — August, or *Queensland Dept. Agr. and Stock Div. of Plant. Ind leaflet* **389,** *1-2.*

T. O. Tuft & L. D. Anderson, 1954. *J. econ. Ent.* **46** (3), *502-504.*

O. Watzl, 1947. *Pflanzenschutzberichte* **1,** 3-4, *33-48.*

F. C. Willcocks, 1925. Important Economic Insects and Mites of Egypt, *150-152.*

J. de Wilde, 1956. *Tenth Int. Ent. Congr. Ontario, 213-218.*

C. B. Williams, 1958. Insect migrations I-XIII, 235 pp. *The New Naturalist,* Collins, London.

Pests of Beets *(Beta vulgaris)* and of Spinach *(Spinacia oleracea)*

Because beets and related plants are attractive to many insects the list of their pests is quite long. However, not all the pests are equally abundant. Some of them are only occasional visitors while others are restricted to a certain season only.

In the chapter dealing with pests of Chenopodiaceae of industrial value the reader will find a key to the pests of these plants. Of these only a few are of primary importance. In Israel, for instance, Acarina, *Prodenia* and *Lixus* are the major pests. All of these usually occur in summer, and attack fodder and sugar beets. Table beets and spinach which grow in winter and spring are free of these major pests; autumn crops are attacked by *Prodenia, Hymenia* etc. Aphids may infest the winter crops; certain winter caterpillars may feed upon the foliage, *Lixus* may cause some scars in the petiole, and *Cassida* may feed upon the leaves.

General Consideration on Control Measures.

Systemic insecticides should not be used against aphids during the three weeks before harvest. On beets, where the root is eaten and the leaves discarded, several insecticides such as endrin, sevin, toxaphene, dieldrin, lindane etc., may be used against caterpillars; on rhubarb and spinach, sevin or dipterex should be applied, but not during the last week before harvest.

Scars of *Lixus* on the petioles may be ignored as the infestation in table beets is negligible.

Sweet Potato Pests *(Ipomoea batatas)*

The leaves of the sweet potato may be infested with various aphids, among them *M. persicae* (see page 66). *Prodenia litura* may feed upon them (see page 102), as well as *H. convolvuli*. WILLCOCKS (1925) mentions the leaf mining caterpillar of *Bedellia somnulentella* Z. which occasionally infests sweet potato without much injury (the common host being (*Convolvulus arvensis* L.). The stem and tuber may be infested with larvae of *Cylas formicarius* (not in the Near East).

The Sweet Potato Weevil — Cylas formicarius L.

(Curculionidae, Col.)

This weevil is not found in the Near East, but it was intercepted in Israel when live specimens of larvae and adults (which were duly exterminated) were found in a shipment of sweet potato seeds. In view of the favourable climate of this region this beetle may establish itself in the Near East. Thus, a short note will not be superfluous here.

Description. The adult beetle is 6—8 mm long. The head and thorax are quite narrow; the legs are long and the rostrum is slightly bent, broad and twice as long as the head. The antennae

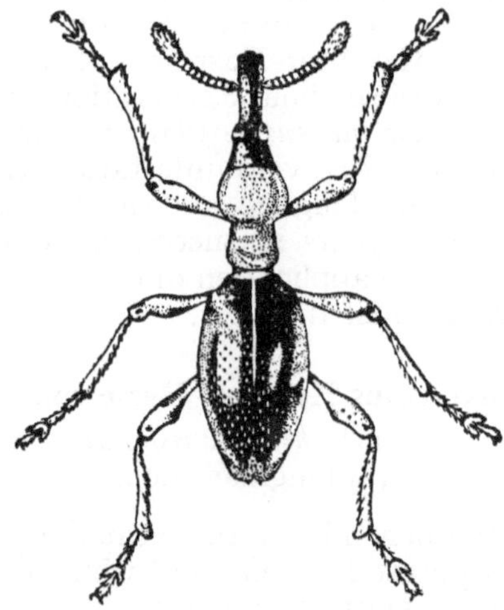

Fig. 112. *Cylas formicarius*; after VAN DER MERWE, *U.S. Africa.* D.Y. Bull. 14.

are reddish-brown, the last segment pubescent. The head is black, the prothorax red, the elytra bluish and the legs red except for the tips of the tibiae and tarsi which are dark.

The egg is 0.65 mm long, 0.45 mm broad, and its colour is white.

The neonate larva is 1 mm long; when mature it is 5 mm long, subcylindrical, legless, tapering posteriorly and wrinkled. Its setigerous papillae aid its tunnelling actions.

The pupa is white at first but turns dark. Anteriorly it is rounded, and posteriorly pointed. The legs are stretched over the ventrum. It is 5 mm long and 1.5 mm broad.

Distribution. The beetle is found in India, the Philippine Islands, Australia, Hawaii, South America, the West Indies, Southern U.S., West Africa, Uganda and Madagascar.

Hosts and Injury. The species attacks plants of *Ipomoea* and *Convolvulus*. The damage is caused by the larva which bores into the stem and tubers. In addition to direct injury infested tubers are more subject to decay. As much as 25—50% damage has been recorded.

Life History. The female gnaws a cavity in the stem or the tuber in which she lays an egg. The incubation period lasts 21 days at 12° C and 5—6 days at 26—30° C. The hatching larva tunnels into the plant and completes its development in 25—28 days at 25° C and 14—18 days at 26—36° C.

The pupal stage lasts 18—20 days at 25° C. The preoviposition period lasts 2—3 days during the summer, or one to four weeks during the winter.

Ecological Notes. This beetle does not have an obligatory diapause and is active as long as the temperature allows. The threshold of reproduction is 15—16° C. In other words, it will not reproduce during the winter of the Near East should it become established here. But, if so, it may raise about 6 generations during the summer.

Control as prophylactic measures. Plant remains should be exterminated from the field to avoid infestation the following year.

The soil should be well ploughed, and the planting of sweet potato should be avoided two years in succession. COCKERHAM & DEEN (1948) point out that the application of calcium arsenate dust gave satisfactory control against this pest.

The Convolvulus Hawk — Herse convolvuli L.

[Syn.: *Sphinx convolvuli* L.]
(Sphingidae, Lep.)

This moth is a frequent flier in the Near East, yet complete destruction of a sweet potato field like that which occurred in Israel in 1925 is very rare (BODENHEIMER, 1930).

Description. The moth is 45—50 mm long with a wing span of

83—100 mm. The wings, which are strong, are narrow at the base and broader at the apex. The abdomen is robust and tapers posteriorly. The colour of the forewing is grey with slightly darker patches. The hind wings are the same colour, lighter at the base, but the dark spot at the base and the three undulating dark bands which cross the wing give it a darker appearance. The thorax is grey-brown, lighter on the disc, with two posterior black spots. The abdominal segments are pink anteriorly and black posteriorly. These colours are interrupted by a longitudinal line which runs along the abdomen. The larva may be differently coloured. As a rule, the general colour is yellowish-brown, but it may be green of various shades. Oblique bands run one on each side of the segments 7 to 11; these bands are yellow and bordered above with a black line. The spiracles are black with a yellow border. The length of the larva, when mature, is 120—150 mm.

The pupa is brown, and the rostrum sheath is curled like a spring.

Distribution. *H. convolvuli* is a tropical insect, but in its migrations it reaches more northern countries. Damage is reported from Congo, Rhodesia, Uganda and the Sudan in Africa; from India, Indonesia and Israel in Asia. Records of its occurrence come also from England, the Mediterranean countries, from Russia and from Australia.

Hosts and Injury. The species feeds mainly on plants of the Convolvulaceae. Of the cultivated plants *Ipomoea batatas* is mostly damaged. In India and Java its injury is reported also on beans.

The injury consists of defoliation of the plant. In view of the fact that the larva is large and voracious, and that attack usually occurs in large numbers, defoliation of an entire field may take place within a short period.

Life History. Very little is known about its life history, except that in Java its total development lasted 45 days.

Phenology. Two years of light trapping in Israel have shown that this moth frequents the Near East particularly in the summer. June-September are the months when most of the moths were trapped, one peak occurring in June, the other in September.

Control. DDT and dieldrin dusts satisfactorily controlled an invasion of *Deilephila-lineata livornica* in vineyards. These materials will probably control also *H. convolvuli* if similar mass attacks occur.

LITERATURE

F. S. Bodenheimer, 1930. Schädlingsfauna Palästinas. 438 pp. Paul Parey — Berlin.

F. H. Chittenden, 1919. U.S.D.A. F.B. **1020,** *1-24*.

K. L. Cockerham & O. T. Deen, 1948. *J. econ. Ent.* **41,** 4, *563-565*.

H. J. Reinhard, 1923. *Texas agr. Exp. Sta. Bull.* **308,** *1-90*.

E. Rivnay, 1950. Unpublished notes.

F. C. Willcocks, 1925. Important Economic Insects and Mites of Egypt.

UMBELLIFERAE

A KEY to the Pests of Umbelliferae

A. The roots are injured

1. Cavities are gnawed in the carrot . . . . . : *Gryllotalpa* page 44
   Agrotis page 88
2. A long yellowish larva with a broad thorax bores in the carrot . . . . : *Phytoecia* page 342

B. The stem is injured

1. A long yellowish larva bores in the stem . . : *Phytoecia* page 342

C. The foliage is injured

I. By caterpillars

1. On the leaves there are dark grey-black spotted caterpillars . . . . : *Prodenia* page 102
2. On the leaves there are bright green caterpillars with black and orange markings . . : *Papilio* page 343

II. By sucking insects

1. Globular soft-bodied plant lice infest the foliage . . . . : Aphididae page 60
2. Flat green hardly noticeable insects cling to the plant; the adults are active spindle-shaped green or black aphid-like insects . . . . . : *Trioza* page 340

III. By mining insects

1. Mining larva is a Lepidopterous caterpillar with legs . . . . : *Epermenia* page 344
2. Mining larva is a legless Dipterous maggot . : *Phytomyza* page 344

The Carrot Psyllids — Trioza spp.

(Psyllidae, Homop.)

Several Homoptera infest carrots. In addition to the various aphids, including *Semiaphis dauci*, *Trioza* sp. is quite frequent. In northern countries the damage caused by these insects is more conspicuous than in the Near East. The species involved is the green species, *Trioza viridula* ZETT. and to a lesser extent *T. nigricornis* FORST. In Israel the latter, *T. nigricornis* is the common species, and not *T. viridula* as was formerly believed (RIVNAY, 1960). It is quite possible that both species are present in the Near East, at least in the northern countries of the region, therefore both species will be discussed.

Description of the genus. The following characteristics distinguish *Trioza* from other genera of the family: The body is elongate, tapering posteriorly. The head is slightly narrower than the prothorax, and constricted behind the eyes. The clypeus is bilobed, each lobe cone-shaped. The first two antennal segments are thick, the others thin and the anterior margins of the forewings are arched.

T. viridula ZETT.: The length of the body is 3 mm, the colour yellowish-green, the eyes large and dark and small dots decorate the wings.

T. nigricornis FORST: The length of the body is 4 mm, the colour is black with yellow spots on the thorax, black eyes and hyaline wings.

The egg of the green *Trioza* is yellow, spindle-shaped, and when laid sticks directly with one end to the plant. The eggs are laid in rows.

The egg of the black *Trioza* is yellowish-red and is attached to a pedicel 0.4 mm long. The eggs are not laid in rows.

The larva of *Trioza* is flat and elliptical, with a fringe of short hair and waxy filaments.

Distribution. The green *Trioza* is distributed in Europe, Korea and Japan. Damage was recorded in Scandinavia, Finland, Estonia, Latvia, Denmark and Germany. The black *Trioza* is recorded from Central Europe, the Caucasus, Siberia and Israel.

Host and Injury. The green species is limited to carrots and other umbelliferous plants, while the black *Trioza* feeds on carrots and also on cabbage, potato, beet, and others. The green species causes leaf curl and the excretion of honey dew causes development of fumagine. As a result of these plus the feeding on the sap, the carrots are poor and meagre. The black *Trioza* does not cause leaf curl.

Life History. Both species lay their eggs on the foliage. The larvae moult five times. The green species lays about 400—700 eggs, and the incubation period lasts 12—15 days (in the summer of Northern Europe). The larva clings to the plant and is hard to distinguish as its colour merges with that of the plant. Its development lasts about 30 days.

The incubation period of the black *Trioza* lasts 12 days at 20° C, while the larval period lasts 24 days at 17° C or 32 days at 15—18° C. There is, thus, very little difference in the length of development of the two species; however, there is a greater difference in the mode of diapause and number of generations.

Number of generations. According to SCHEWKET (1931), when *Trioza viridula* matures it leaves the carrot plants and migrates to the neighbouring pine woods. Maturation of adults and the subsequent migration may last till October. According to OZOLS (1925) of Latvia, the adults hide between the needles of the pine trees.

Trioza nigricornis continues to raise further generations as long as the weather permits. In Germany it may raise three generations; this tendency was noticed also in Israel.

Control. If necessary, nicotine applications will kill these pests, as will diazinon and malathion to a greater degree.

Also the systemics, metaisosystox and phosphamidon, will kill these pests, but should not be used too close to harvest.

The Carrot Borer — Phytoecia cylindrica L.

(Cerambycidae, Col.)

Occasionally the stalk and root of the carrot is found to harbour a cream coloured, elongated larva — that of the beetle *Phytoecia cylindrica*.

Description. The adult beetle is 10 mm long and 2 mm wide; the length of the antennae is 11—22 mm. The body is black in colour and covered with hairs of the same colour.

The egg is round, bright yellow and 1.5 mm long.

The larva is cream coloured, approximately 20 mm long and devoid of legs.

The pupa is about 12 mm long, light-brown, with free wings and legs.

Distribution. According to the literature, the beetle occurs in Europe, Asia Minor, the Caucasus and Siberia. As it has been recorded in Italy and Israel, it probably exists also in other Mediterranean countries.

Hosts. FIORI (1947) states that the larva may be found both on wild and cultivated carrots. In Israel the adult feeds on soft wheat and barley grains.

Type of Injury. The larva bores into the stalk of the carrot and if the stalk is intended for seed the umbel wilts and dries. Damage to carrot results in a reduced yield; when young plants are attacked, they die. SHWEIG (1954) claims that up to 60% of the crop may be attacked by this pest.

Life History. The following description is based on observations and breeding experiments by SHWEIG in Israel and FIORI in Italy.

In Israel the beetle awakes from its winter dormancy in March, while in Italy this occurs in May. The beetles then invade wheat and barley fields and feed on grains. In April the pre-oviposition stage draws to a close and the females oviposit over a number of weeks. The eggs are deposited in a groove singly on the stalk close to the neck of the root. The hatching larva bores into the stalk and also reaches the root. Both in Israel and Italy the development of the larva is concluded by August. The larva then pupates in the root or in the upper portion of the carrot, close to the surface of the ground.

FIORI in Italy maintains that the insect passes the winter dormancy in the larval state, whereas SHWEIG states that the larva pupates in August. This stage lasts 14—18 days and the adult passes the winter in the dormant state, or, in other words, at the end of August or the beginning of September the adults emerge from the pupae and hide in the ground or among plant debris, where they remain from September until March.

Ecological Factors. SHWEIG states that eggs and larvae develop well at medium or low humidity while high humidity is lethal. This, he points out, is the reason why these beetles do not infest carrot crops growing on heavy soil, whereas in sandy soil, where the water is not retained and where the humidity near the roots is not excessive, the larva develops well.

Control. Nothing is known regarding the chemical control of this pest. The synthetic contact insecticides might conceivably destroy the adult beetles invading the carrot for the purpose of oviposition.

As for prevention by agrotechnical measures, to minimize attack by this pest it is recommended that carrots be sown in heavy soil.

The Swallow Tail — Papilio machaon L.

(Papilionidae, Lep.)

Beautiful green caterpillars, marked with black and orange, often feed on the foliage of parsley, dill or carrots; they are the larvae of the equally beautiful butterfly *Papilio machaon* L.

Description. The body of the adult is 25 mm long, with a wing-span of 65—80 mm. It is yellow with a greyish-black tapering band over it from the head to the tip of the abdomen. The forewings are yellow, the base greenish-black with three spots of the same colour along the anterior margin. Along the hind margin and parallel to it there is a black dentate band with 8 lunular spots. The hind wings are also yellow, and a broad dentate band runs along the posterior margin of this wing; in it there are six blue spots: Four or more

Fig. 113. *Papilio machaon*, after BODENHEIMER.

lunular yellow spots and a red dot are near the posterior margin. The typical wing tail is black and yellow at the base.

The larva is 30—35 mm long, the head is striped with green and black, the body is green, each segment with two black bands, the one over the anterior margin is complete, the other in the middle is interrupted by six orange dots. Additional black spots occur near the spiracles and on the abdominal legs. When disturbed, a biforked odoriferous horn protrudes from the thorax.

Distribution. This species is holarctic. In England it seems to be limited to the southern coast only.

Host and Injury. The larvae feed on plants of the families Umbelliferae and Rutaceae. The injury consists of the defoliation of the plants by the caterpillars. As a rule the infestation is not severe and the damage slight, except for certain occasions when invasions take place.

Life History. The egg is laid on the leaf and it hatches after about one week. Under favourable conditions the larva completes its development in about 30 days; adding the pupal stage, about a week or 10 days, the total development of one generation is about 6—8 weeks. During the cold winter the larva enters a facultative diapause in the larval or pupal stage; thus three generations may be raised during the summer of Iraq (WILTSHIRE, 1957) or Israel. WILTSHIRE states that the larvae are seen all year round in Iraq. Also in Israel they may be found in the spring, summer and autumn. Its occurrence in Israel is not uniform and continuous but rather sporadic and localized, which suggests that this species invades the country and enriches the residential population which may exist there.

Epermenia daucella Peys

(Elachistidae, Lep.)

According to BODENHEIMER (1930) the larvae of this moth mine between the epidermis layers of carrot leaves. Very little damage is done and control is not necessary.

Phytomyza umbelliferarum Hndl.

(Agromyzidae, Dipt.)

Another miner is frequent in the leaves of carrots — the larva of *Phytomyza.* The adult is entirely brown except for the frons, the knees, the halteres and two lines on the pleurae which are pale-yellow. The length of the body is 1.5 mm, the wingspan about 2 mm.

The puparium is shiny brown, elliptical and 1—1.5 mm long.

In Israel the larva is found particularly during February-March.

This miner also does not cause sufficient damage to warrant measures of control.

LITERATURE

F. S. BODENHEIMER, 1930. Die Schädlingsfauna Palästinas, Berlin, 438 pp.
G. FIORI, 1947. *Boll. Ist. Ent. Bologna* **16**, *291-314*.
E. OZOLS, 1925. Laciksaimniecibas parvaldes izdevums Krajuma. *Letas, 1-32*.
E. RIVNAY, 1960. *Hassadeh*, **40**, 8, *957*.
N. SCHEWKET, 1931. *Z. angew. Ent.* **XVIII**, *175-188*.
K. SHWEIG, 1954. Vegetable Pests. Hassadeh Library (in Hebrew) 150 pp.
E. P. WILTSHIRE, 1957. The Lepidoptera of Iraq. Ministry of Agriculture, Govt. of Iraq, *1-162*, pts. I-XVIII.

Pest of Onion *(Allium cepa)* and Garlic *(Allium sativum)*

In addition to the pests discussed below, onion and garlic may be infested with the onion thrips *T. tabaci* (page 74) and the leaves may be fed upon by *Laphygma exigua* (page 113), and bulbs gnawed by *Agrotis* (page 88).

Dyspessa ulula Bkl

(Cossidae, Lep.)

In the Balkan countries onions are attacked by red larvae which bore into them, feed on their parenchyma and cause them to rot. These larvae are those of *Dyspessa ulula*.

Description. The adult is 12—15 mm long, with a wing spread of 20—25 mm. Its body is hairy. The anterior wings are light-yellow in colour with bright patches, while the posterior wings are dark brown. The antennae of the male are pectinate and of the female serrate, the teeth of one side being larger than those on the other side. In the female the frenulum consists of 4 jointed hairs whereas in the male it consists of one bristle.

The egg is barrel-shaped, reddish-brown in colour and its shell is engraved with a honeycomb pattern.

The larva is cylindrical and somewhat flattened, about 23 mm long and 4 mm wide. Its dorsal surface is dark reddish in colour, its ventral surface yellowish-red. The body segments are convex, each one bearing bristles located on small tubercles. On the dorsal surface of the prothorax there is a protruding sclerite and a smaller one on the mesothorax. On the abdominal legs there are 14 hooks arranged in a semicircle. The larva produces a malodorous secretion.

The pupa is approximately 11 mm in length, brown in colour and on its abdominal segments there are rows of small teeth, one row per segment. The teeth are located on the upper surface and on either side of each segment.

Infested plants may be easily recognized. At first the lateral leaves turn yellow and finally the whole plant withers. In the bore holes excreta typical of this larva are found.

Distribution. Like other species of this genus the insect apparently originates from South West Asia. ROZSYPAL (1951) states that this pest spread to the Balkan via Asia Minor and onward to Southern Europe as far as Spain. Similarly it reached Russia and Ukraine by way of the Caspian littoral and the Caucasus, from where it spread to Central Europe. It also invaded Syria and Israel as well as Egypt and her neighbours. Today it is also encountered throughout North Africa.

The pest is reported to cause heavy losses in the Trans-Caucasus, Ukraine, Czechoslovakia and the Balkan countries.

Hosts. Dyspessa attacks only plants of the genus *Allium.* So far it has been reported from the following species: *A. cepa, A. vineale, A. montanum* and *A. flavum.*

Type of Injury. The larva bores into the bulb of the onion and feeds on its content. Having completed its feeding on one plant, it may pass on to another so that one larva may destroy several plants. Even if the onion is not eaten in its entirety, bacteria and fungi gain access through the damaged parts and cause it to rot.

Reports from Czechoslovakia state that one third of all the onion plants on a 10 dunam field were destroyed there by the pest.

In general, infestation begins at the margins of the field in patches which progressively increase in size as larvae hatch from consecutive ovipositions and attack new plants in their vicinity.

Life History. Data regarding the life history of this pest are not complete. ROZSYPAL in Czechoslovakia and BUCHELOS (1958) in Greece have done some work on *Dyspessa* and the following description is taken from their observations.

The moths emerge from the pupal cells and fly around in the evening — from 9—10 p.m. in Greece and from 8—10 p.m. in Czechoslovakia. The period of flight extends over a number of weeks reaching a peak at the end of May — beginning of June in Greece, and during the first half of June in Czechoslovakia.

The eggs are deposited beneath the outermost layer of the onion close to the surface of the ground. No information is available regarding the number of eggs laid by one female. In one group, 90 eggs were counted, so one may assume that they all originated from one female which may have laid even more eggs.

The eggs hatch after 4—7 days. The emerging larvae penetrate the onion on which they have hatched, but once they have devoured one plant they depart and attack other onions in the neighbourhood. This accounts for the patchy appearance of attacked fields, each patch representing the injury inflicted by one batch of larvae.

Each larval penetration is represented by a hole in the skin of the onion. Thus the number of holes in the skin reflects the number of larvae that have penetrated the bulb.

The larval period lasts 5—6 weeks in Czechoslovakia, larvae being encountered in the fields during the months of June, July and August. After completing its development, the larva spins in the ground an oval flattened cocoon, the walls of which are made up of silk threads and particles of earth. Sometimes this takes place among the plant remains or in the plant itself, thereby facilitating the spread of the pest. In the cocoon, the larva remains in diapause until the following spring, when it awakes, leaves its temporary cocoon and spins a new one in which it pupates. About three weeks later the moths emerge. BUCHELOS in Greece observed that at times the larva does not complete its development in summer, but enters winter dormancy and continues its development in the following summer. Such larvae may wander from one field to another in search of food.

Control. BUCHELOS in Greece suggests the following control measures:

Preventive measures include rolling the soil of affected fields, regular irrigation, clearing the fields and destruction of affected plants, and proper rotation of crops. This, of course, implies abstaining from sowing onions near a field that was affected the previous year.

Insecticides: Good results were obtained in the direct destruction of the pest by spraying with the following materials: rogor at a rate of 30 g a.i. which gave a kill of 85%, parathion at the rate of 70 g with a kill of 75% and benzene hexachloride at a rate of 120 g which killed 70%. It is advisable to spray the field soon after irrigation, when there is plenty of moisture in the soil.

The Onion Fly — Hylemyia antiqua Mg.
(Anthomyiidae, Dipt.)

In the northern countries, the onion fly is one of the obstacles to the successful cultivation of the crop whose name it bears.

In the United States, Canada, Britain and Holland the fly may damage up to 90% of plants. In the Mediterranean area the fly formerly did not constitute a serious problem since onions were grown from sets which were less susceptible to attack. It was only in recent years, when it became common practice to grow onions from seeds and when the sowing season was advanced to autumn and irrigation employed, that the fly became troublesome for the onion grower, until an effective method of control was evolved.

Description. The adult fly is 6—7 mm long and light grey in colour. In some specimens a dark stripe runs longitudinally along the middle of the dorsum. When examining the head from the side, the clypeus may be observed which is inclined slightly upwards. The colour of the genae and below the eyes is silvery. In the male the eyes are close together while in the female they are separated by a reddish stripe. The tips of the legs and the abdomen are black.

The onion fly superficially resembles the house fly but is smaller than the latter and, unlike it, its wings lie on the back with their margins parallel. Moreover, the median wing vein of the onion fly is not bent upward as it does in the house fly.

The egg is white in colour, 1.2 mm long and 0.5 mm wide and shaped like a banana, with one side somewhat flattened. On the shell there are engraved 12 tiny, longitudinal grooves.

The maggot is white and 8—10 mm long when mature. The head and thoracic segments are narrower than the abdominal segments which widen posteriorly. Close to the head there are the anterior spiracles. On the last segment, which ends abruptly, there are 16 papillae arranged along its margin and two tiny knobs in the middle which bear respiratory openings.

The pupa is 6 mm long, reddish-brown in colour and at the tip of its shrivelled body there are the papillae of the last larval segment; the two knobs carrying the posterior respiratory openings are also noticeable.

Distribution. Similar to its relative, the cabbage fly, the onion fly is an insect of the northern latitudes. It occurs from Korea to the United Kingdom in the Old World, and from the shores of the Atlantic to the Pacific coast in the New World. It is found as far north as Omsk in Siberia, in Finland and Norway and in Canada up to the same latitude. As for southern countries the fly occurs in Southern France, Italy and Israel. CHITTENDEN (1917) states that in the United States the fly is not found south of a line drawn across through the state of Pennsylvania.

Hosts. This pest infests only onions and botanically related plants, such as garlic and leek. In Italy the flies have been reported to attack lettuce, and in Czechoslovakia larvae have been collected from carnations. These reports have to be regarded with some caution as the maggots may not have been those of the onion fly but others resembling them. It is worth noting that on wild plants of the onion family no maggots of the onion fly have ever been found.

Type of Injury. The maggot penetrates the portion of the plant which is below ground, mines into the stalk or the soft head and as a result the plant is weakened and the leaves turn pale. Further damage inflicted by the maggot causes the death of the plant. If by then the larva has not yet completed its development, it attacks a neighbouring plant. In this way one maggot can attack and destroy 6—8 plants. However, when the plant is large the maggot completes its development within it or, what is more, several larvae may feed in one plant and pupate on or near it. This is the reason why there is more damage done by the generation which emerges when the onions are small (in Israel this is the autumn generation) than by the following generation, which appears at a time when the onion plants are already of larger size.

In case of a heavy attack, 90% of the onion plants may be destroyed by the pest.

Life History. After the pupa completes its diapause period, the adults emerge from the soil and, provided that there is sunlight and warmth, the insects become active in their search for food (the nectar of various flowers), mating, flight to ovipositon sites and the laying of eggs. In Europe and North America, the diapause occurs in winter and the adults awaken in the spring. In England, Holland, Finland and Canada as well as in Omsk (Siberia), the emergence of adults begins in May and continues through June. In Pennsylvania the emergence commences at the end of April. In the Near East a different pattern prevails: The diapause occurs during the summer and adults emerge in autumn. Miss YATHOM's work (1960) in Israel shows that emergence begins at the end of October and continues until December. The flies are active mainly in the morning and evening.

According to data from northern countries, the pre-oviposition period lasts from 10—12 days. In Israel active flies may oviposit as early as 4.5 or even 3 days after emergence at a temperature of 26—29° C (YATHOM).

The female lays her eggs in the soil around the plant, on the leaves of the plant or on the bulb itself, if it is exposed. The eggs are laid singly or in small clusters.

High humidity is essential for the development of the egg. According to EYER (1922) humidity determines the hatching percentage but does not affect the incubation period. EYER states that at a temperature of 15.5 ° C the incubation period extends over 5.5 days, while at a temperature of 20—21° C the incubation period is 4.5—5 days. These data differ slightly from those reported by Miss YATHOM from breeding experiments in Israel (see table LV).

The maggot hatches from the egg, makes its way to the bulb, penetrates it and feeds on its parenchyma.

Table LV.

Development (in days) of the various stages of the onion fly at various temperatures (after YATHOM (1960) and others)

Temperature in °C	Egg	Larva	Pupa	Larva (after EYER, 1922)	Larva (after MAAN, 1943)
14	5.1±1.2	31±9	—	22	30
17	3.6±1.2	24±5.8	29.0±7.5	—	26
20	2.8±0.8	15±2	16.7±2.6	19	14
23	2.2±0.6	13±2.3	11 ±1.5	—	—
27	1.9±0.7	10 ±1	7.6±1.6	14	—

A number of workers who have bred the onion fly note that the larval stage lasts about 14 days, without, however, mentioning temperature or breeding conditions. Some of them, including RUHMANN (1921) of Canada, record a larval period of 29 days. EYER states that humidity has a greater effect on the length of the larval period than temperature, because larvae that developed in onions on irrigated soil matured in 18—19 days, whereas under dry conditions they developed in 25 days. As for the effect of temperature he gives data which are completely at variance with those recorded by MAAN (1943).

In Israel YATHOM bred maggots at different temperatures. Her findings are identical with those of MAAN at low temperatures, but are totally different from those of EYER. These discrepancies are not to be regarded as based on erroneous observations — the development of the larva is dependent not only upon the temperature but also upon the food available and the moisture, as well as the environmental conditions of the parents of the flies, which served as subjects for the observations.

In the course of time the larvae leave the onion and enter the soil to pupate. The choice of the site for pupation depends upon the quality of the soil or the amount of moisture it contains. The larva chooses optimal conditions of moisture: In heavy and moist soil the site of pupation is in the upper layers of the soil while in lighter soils pupation takes place at a greater depth. ISAEV (1932), who studied the subject of light soils, noted that 54% of the larvae were found at a depth of 5 cm, 39% at a depth of 5—10 cm, and only 7% at a depth of 10—15 cm. According to the observations of many workers from the northern countries, the pupa may complete its development within 8—10 days, or it may also take 22 days (MAAN) or 29 days (RUHMANN). EYER who mentions the temperature in every instance, records a pupal period of 8 days at a temperature of 22° C, and of 12—14 days at a temperature of 18° C. These figures, too, do not tally with those obtained by YATHOM in her breeding experiments.

The following values were obtained: Threshold of development of the egg — 2° C, the constant value — 50—60 day degree °C. The same values for the larva were 8° C and 190, and for the pupa which does not enter diapause — 12.5 °C and 130. According to MAAN, the threshold of development for the egg is 10° C.

These data enable us to forecast at what temperature any of the developmental stages of the pest will cease activity at a given season.

The Adult. The length of life of the adult depends to a large extent upon the food available to the larva and itself, and the degree of moisture and temperature. But the majority of workers report data without mentioning environmental conditions. The life of the fly extends from 25 days (EYER) to almost 100 days (MAAN, KASTNER, 1929). During her life the female lays about 100 eggs (YATHOM) or, according to EYER, 40—60 eggs. KASTNER states that the female lays 40—50 eggs under field conditions and about 80 in the laboratory.

Diapause. MAAN observed that the pupae of the second generation do not all emerge the same summer; a certain percentage enter a winter diapause and emerge only in the following spring. The pupae of the third generation all pass through a winter diapause and, together with those of the previous generation, emerge in the following spring. Similar observations were made in Israel, where the fly aestivates rather than hibernates. According to YATHOM, the majority of pupae of the first generation emerge during the same winter, while a small percentage enter diapause which lasts to October-November. The pupae of the second generation almost all go into summer diapause, and only a very small proportion go on to develop during the same spring, producing what should be considered a sterile generation, as no offspring follows. The continued existence of the pest in subsequent years thus depends upon the pupae of the two previous generations.

The number of pupae which develop during a given summer and which aestivate varies in different years. When the aestivating pupae encounter a long hot, dry spell, large numbers of them die.

Phenology and Number of Generations. According to data gathered from various sources the fly produces three generations in Holland, England, Canada, Ohio and Pennsylvania. In Vienna three generations were observed with a partial fourth one (SCHREIER, 1953). In Omsk and Charkov, on the other hand, only two generations were recorded and in Finland only one. In Israel two normal generations are the rule. A third one is physiologically fertile and, in the laboratory, may produce offspring, but there is some doubt whether the latter will survive under field conditions.

In Holland adult flies are on the wing from the end of May to the beginning of October. These flies belong to three overlapping generations. The peak of flying activity of the hibernating generation is in June, of the first generation in July-August, and of the second generation in September.

In Israel, too, the generations overlap. The peak of the aestivating generation is in November-December, that of the first generation in January-February, and of the subsequent generation in April. When examining the population density over the year in the form of a graph it will be seen that the peak of the aestivating generation does not stand out to any great extent. The peak of the population occurs in April and is undoubtedly composed of flies belonging to the last two generations. During the summer months the number of flies dwindles until in July they have practically disappeared. They appear again in October-November.

Limiting Factors. As mentioned previously, the fly as well as the larvae disappear in summer. In the northern countries certain parasites attack the adult flies, but in Israel these parasites seem to be absent or at least their effect does not make itself felt. It is only on warm, rainy days that a fungus, *Tarychium hylemiae,* may cause mortality among flies on the wing. Of greater importance are ecological conditions, restricting the development of the eggs and the larvae. According to YATHOM (1960), the percentage of laying females drops during the summer months to 20% as does the average number of eggs per female (about 40). An important part is played by the low soil humidity during the summer months. The absence of rain prevents hatching of the eggs in the soil and thus no larvae develop. All these factors combined prevent the development of maggots in summer and, in consequence, the fly population reestablishes itself only in autumn when new flies emerge from aestivating pupae.

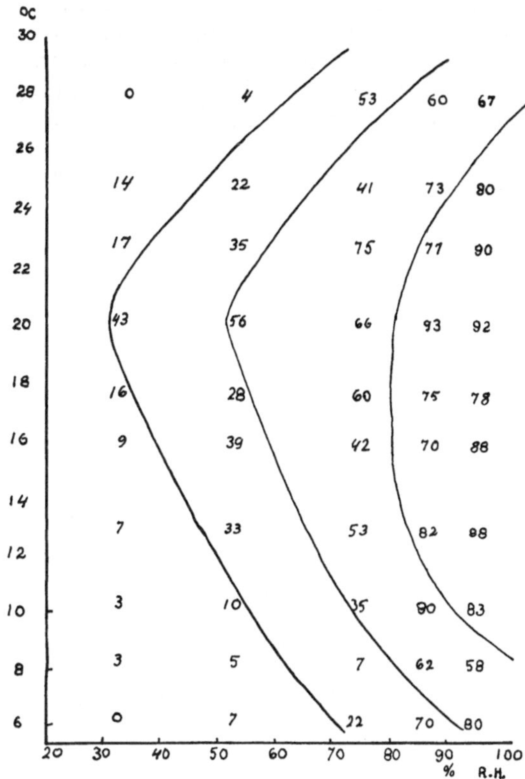

Fig. 114. Effects of temperature and humidity upon mortality of the eggs of *H. antiqua* – after YATHOM.

Control. In the northern countries the onion fly causes severe damage to the onion crop and, prior to the advent of the synthetic insecticides, control measures against the pest were expensive and not very effective. One of the measures advocated was treatment of the seeds with calomel (mercuric chloride). The same material was used for spraying the soil at a concentration of 1:1,500. These treatments were expensive and yielded only partial results.

Another attempt at control was made by sowing rows of inferior quality onions in a field to serve as bait for the flies to oviposit. In due time this reduced the damage inflicted by the pest; the bait onions were eventually removed and destroyed together with the maggots infesting them.

Another method involved the use of bait for the adult flies. Vessels were prepared containing molasses and twice the amount of water to which was added sodium arsenite at a rate of a ¼—½%. Flies were attracted to this bait and sipped from it.

Of the synthetic insecticides, DDT and benzene hexachloride gave only partial control. Satisfactory results were obtained following the use of a 0.3% spray of dieldrin W.P. containing 25% active ingredient or an emulsion containing 20% active ingredient. Both methods gave complete control.

Pests of Asparagus *(Asparagus officinalis)*

The asparagus vegetable may be attacked by *Agrotis segetum* (page 93), by the aphids *M. persicae* (page 66), *A. gossypii* (page 63), and others. The specific pest to this plant is the asparagus beetle.

The Asparagus Beetle — Crioceris asparagi L.

(Chrysomelidae, Col.)

This beetle, which originates from Europe, was unknown in Israel; it appeared there when asparagus first began to be cultivated.

Description. The adult is a small beetle, 5—6 mm long, with the sides of its body parallel. The head is blue and tapers towards its posterior end, thereby making the eyes more prominent. The antennae are filiform and consist of 11 segments. The prothorax is yellowish-red in colour and occasionally two dark spots are present on its back. The elytra are dark-blue, with red edges. On each elytrum there are three yellow spots. At times, the two anterior spots are fused and, more rarely, all three are joined together. The humeral angles of the elytra are always dark. The legs are black.

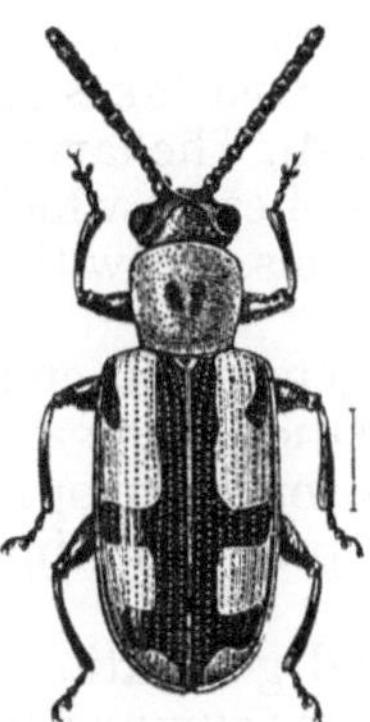

Fig. 115. *Crioceris asparagi*; after CHITTENDEN, U.S.D.A. F.B.837.

The eggs are of greyish-green colour, about 5 mm long and 0.50 mm in diameter.

The larva is 6—8 mm long, and grey, except for the head, which is shiny-black. On the back of the prothorax there are two sclerites.

The legs are black. At the sides of the abdominal segments there are light-coloured papillae covered with black dots. These papillae aid the larva in adhering to the stalk on which it lives.

The pupa is yellow and 5 cm in length. It pupates in a cell which it builds in the soil.

Distribution. The beetle occurs in Europe, all over the U.S. and in Southern Canada. In Israel the beetle arrived about 35 years ago.

Hosts and Injury. This pest attacks asparagus only. Injury is inflicted by both adults and larvae. In the beginning of spring the beetle feeds on the shoots which are intended for the market, while later they attack the large asparagus leaves. The damage to the shoots is caused by chewing and by soiling with larval excrements. In this way the market value of the shoots is reduced. The injury to the leaves weakens the tubers, thus affecting adversely the plants of the subsequent season.

Life History. The asparagus beetle awakens from its winter diapause in spring. Depending upon the geographic location and climatic conditions this may take place in late June in the vicinity of New York, in April near Washington or in March in Israel. Mating takes place a few days after the emergence of the adult from the soil, and in a short while oviposition commences. The eggs are laid on the stalk in rows, with a small space between each egg. One female may lay as many as 250 eggs. The maximum number of eggs laid in one day is about 30 while the period of incubation extends from 3 to 8 days depending upon temperature. The small larva that hatches from the egg immediately begins to chew the soft plant parenchyma on which it lives.

The development of the larva takes 15—20 days, or only 10—14 days in Israel (SHWEIG, 1954). The larva then burrows into the soil, builds a cell from particles of earth which it sticks together with its saliva and pupates inside this cell within 2 days. The pupal stage lasts 7—14 days.

The adults of the first generation emerge from the soil, spend a few days in the pre-oviposition stage and then lay their eggs, from which the second generation develops. In Israel the emergence of adults takes place in April, while in Europe and America they emerge in July.

The adults of the second generation emerge in June (SHWEIG). They do not lay eggs the same summer but enter into a summer diapause lasting until the following March. KLEIN (1954) states that these adults do oviposit but that the larvae which hatch from the eggs die immediately. This number of generations has also been reported from Europe and North America, although CHITTENDEN (1917) believes that in the southerly countries more than two generations may develop.

LITERATURE

THEODORUS K. BUCHELOS, 1958. Dept. Agric. — Phytopathological Lab. at Patra. Ph. Thesis Univ. Athens, 9-63.

F. H. CHITTENDEN, 1917. U.S.D.A. F.B. **837,** *3-13*; *J. econ. Ent.,* **9,** *57.*

J. B. EYER, 1922. *Penn. agric. Exp. Sta. Bull.* **171,** 16 pp.

S. I. ISAEV, 1932. *Bull. Leningrad Inst. Control Fin for Pests,* **2,** *13-58.*

A. KASTNER, 1929. *Z. Morph. Ökol. Tiere* **XVI,** 3, *363-422.*

A. KASTNER, 1930. *Z. Pflz. Krankh.* **XL,** 3, *124-139.*

H. Z. KLEIN, 1945. Vegetable Pests. Hassadeh Library (in Hebrew), 156 pp.

W. J. MAAN, 1943. *Tijdschr. Plantenziekt.* **49,** 4, *132-133.*

J. ROZSYPAL, 1951. *Trans. Ninth Int. Congr. Ent.* **1,** *656-659.*

M. H. RUHMANN, 1921. Report of the assistant Entomologist, 22nd Ann. Rpt. Br. Columbia Dept. Agr., *Q37-Q41.*

O. SCHREIER, 1953. *Pflanzenschutzberichte* **10,** 1-2, *4-13,* Vienna.

K. SHWEIG, 1954. Vegetable Pests. Hassadeh Library (in Hebrew), 150 pp.

S. YATHOM, 1960. *Agric. Res. Sta. Spec. Bull.* **25,** *1-65.* typescript.

VI. PESTS OF INDUSTRIAL CROPS

CHENOPODIACEAE

KEY to the Pests of Chenopodiaceae

A. The roots are injured

- I. Cavities or galleries are gnawed on the outer surface of the root or head, below the surface of the ground
  1. There are tunnels in the vicinity of the plant .: *Gryllotalpa* page 44
  2. Large white grubs are found in the ground .: *Pentodon* page 273; *Phyllopertha* page 157
  3. Curled grey-green caterpillars are found in the ground . . . .: *Agrotis* page 88
  4. Small legless white grubs cling to the cavities, or are found in the ground . . . .: *Conorrhynchus* page 365
- II. Excavations and tunnels are inside the head or root
  1. Close to the crown there is a large cavity containing white legless grubs . . . .: *Chromoderus* page 367
  2. Numerous small tunnels and galleries are in the head close to the crown, or in the petioles close to the base . . . .: *Phthorimaea* page 367
  3. Tunnels run from the crown deep into the head or root in various directions . . . . .: *Lixus* page 362

B. The petioles are injured

1. Mining purplish small caterpillars are found in the main vein or petiole . . . .: *Ephestia* page 260, 371
2. Scarred cavities in the petiole contain often yellow eggs . . . .: *Lixus* page 362
3. White legless grubs mine downwards in the petiole . . . .: *Lixus* page 362

C. The leaves are injured.

- I. By gnawing insects
  1. The margins of the leaves are bitten-off . . .: *Conorrhynchus* page 365; *Chromoderus* page 367
  2. Small or largers holes are eaten in the leaves .: *Lixus* page 362; *Chaetocnema* page 359; *Cassida* page 360; *Phyllotreta* page 293
  3. Larger or smaller sections of leaves have been gnawed off or holes made by caterpillars
     - a. Small caterpillars not over 15 mm, with purplish longitudinal stripes over the body feed on leaves and may also mine in the main vein or petiole . . . .: *Ephestia* page 260, 371; Pyralidae page 371
     - b. Green or pale green caterpillars with 3 pairs of abdominal legs feed on leaves . .: *Plusia* group page 125
     - c. When less than 15 mm larvae are green, when larger than 15 mm larvae with a purplish longitudinal line over the body which is thin and with distinct segmentation . .: *Hymenia* page 369

d. Green larvae with pale stripes, or pale green with dark-green stripes, body more or less robust, feed on leaves . . . .: *Laphygma* page 113
e. There are olive-green caterpillars, with black spots over the body . . . .: *Prodenia* page 102
f. Brown hairy caterpillars feed on the foliage: *Ocnogyna* page 371

II. The injury is caused by sucking insects
1. Small, soft-bodied, globular plant lice abound on the stem or leaves; often leaves become curled . . . .: Aphididae page 60
2. Minute, elongate, less than 1 mm long, greyish-yellow insects are on the leaves; their feeding leaves pale or silvery areas . . . .: *Thrips* page 74
3. Green, elongate Homoptera move quickly over the leaves . . . .: *Empoasca* page 53
4. The leaves become chlorotic, devoid of chlorophyll, a delicate spider web and minute red creatures may be found on the leaves . . . . .: Acarina page 80
5. Green Pentatomid bugs feed on the leaves . .: *Nezara* page 49

III. Injury is caused by mining insects
1. On the leaf there are brown patches, or pale pockets containing maggots . . . .: *Pegomyia* page 372

The Beet Flea Beetle — Chaetocnema tibialis Illig.

(Chrysomelidae, Col.)

In the Near East, this insect causes less damage than in Europe or the Western Mediterranean countries. The following discussion is based on the account of its life history in Italy, by MENOZZI (1947).

Description. The body is elliptical and very convex, 1.5—2 mm long. Its colour is dark-green with a metallic lustre. The antennae are eleven-segmented; the first is reddish and the others are brown, darker at the ends. The head and thorax are punctate and on the elytra punctate lines divide each into eleven bands. The femora are thick and serve for jumping. All legs are brown with the tibiae and tarsi light brown.

The egg is elongate, 0.6 mm long and 0.2 mm wide. The larva is long and narrow. When mature it is 1.3 mm long, white, with small dark spots. The prothoracic plate and that of the last abdominal segment are brown.

Distribution. The insect is distributed in Central and Southern Europe, Southern Russia, Turkestan, Asia Minor and Israel. Damage is reported from Austria, Italy and Spain.

Hosts. The adult may feed upon the leaves of plants belonging to the genera *Beta, Chenopodium, Atriplex, Polygonum, Amaranthus, Salicornia,* etc.

So far, the larvae have been found on the roots of beets only.

Type of Injury. Damage is caused mainly by the adults which feed on the foliage and make numerous little holes in the leaves.

Large plants can endure this injury, but when newly sprouted seedlings are attacked, they may be exterminated, necessitating a reseeding of the field.

Life History. The information available on this subject is scanty. The insect hibernates as an adult. Towards the end of March and early April it awakens and feeds on the hosts available. It feeds on the underside of the leaf, leaving the upper epidermis intact, so that window-like small holes are made. Mating and egg-laying take place about two weeks later. Eggs are laid near the plant, and after an incubation of 3—4 days the larvae hatch, enter the ground and feed upon the soft rootlets. Their development lasts about three weeks after which they pupate in cells close to the surface of the soil.

The pupal development lasts 5—8 days. Thus the whole life cycle lasts 30—37 days under Italian spring conditions.

Control. It was customary to dust the plants with calcium arsenate. Lately, synthetic insecticides have been introduced, which yield satisfactory control.

The Tortoise Beet Beetle — Cassida vittata Vill.

(Chrysomelidae, Col.)

A severe infestation by this beetle may denude the plant leaving only the veins of the leaves.

Description. The genus *Cassida* is recognized by the following characteristics: The body is elliptical, flat, strongly convex above and slightly concave below; the margins of the elytra and the thorax are extended so that they cover the entire body and the greatest part of the legs. The antennae are inserted in front close to each other.

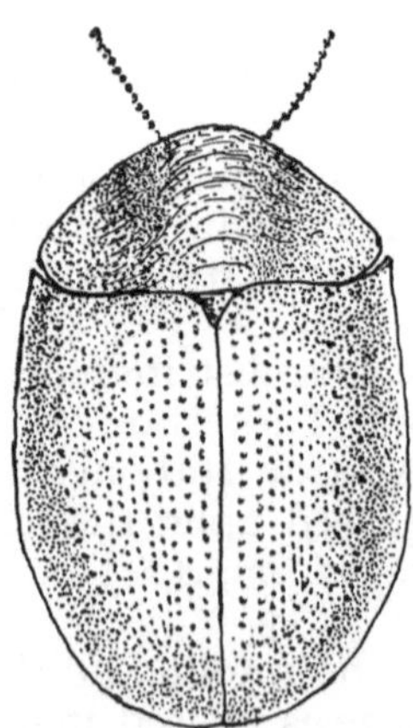

Fig. 116. *Cassida vittata.*

Cassida vittata is 5—7 mm long. Its colour is brown above and straw-yellow at the margins. In live specimens two golden-irridescent tapering stripes decorate the elytra. Ventrally the body is entirely black except for the margins of the abdomen, antennae and legs (excluding the coxae).

The egg is yellow, 1—1.5 mm long, 0.65 mm wide, and is often covered with a transparent protective mucus.

The larva is green. When mature it is 6 mm long and 4 mm wide, convex on top, and it is armed with 16 pairs of lateral ramified spines. The last segment has two long fork-like spines which can be bent over the body. They hold the moulted skins and collect the droplets of faeces. All these serve as a protective humid umbrella over the body of the larva.

The pupa is fastened to the leaf at its rear tip. The head, thorax and abdomen are quite distinct. The prothorax is like that of the adult and is armed with small spines. The margins of the abdominal segments are extended into trifurcate spines, while the last abdominal segment bears a bifurcate spine which is smaller than that of the larva. The moulted larval skin remains at the end of the abdomen.

Distribution. This species is found throughout the entire Mediterranean region. Damage has been recorded particularly from the southern countries of the region. It is found in Spain, France, Italy (Central and Southern), Malta, North Africa and Israel.

Hosts. The species is not limited to beets only. It is found on other genera of the family, such as *Amaranthus, Salicornia, Lychnis, Spinacia,* and also on the composite, *Centauria.*

Type of Injury. Both adults and larvae cause defoliation of the beet. Feeding begins with small holes on the leaf which extend gradually until only the main veins are left.

Life History. The overwintering beetles awaken as the temperature rises in the spring. According to Menozzi (1947), this takes place in Italy in May, in Morocco in February. In Israel they appear in February during mild winters or in March-April in colder years. After a week of feeding, copulation and egg-laying take place.

The eggs are laid on the lower surface of the leaf. After an incubation period of 8—11 days the larvae hatch. They feed on the soft tissue first, leaving the epidermis, but when larger they make holes in the leaf. Many larvae may devour the entire leaf, leaving the main veins only.

The larval period lasts about one month in Italy and about 18—20 days in Morocco.

When ready to pupate the larva attaches the tip of its abdomen to the leaf and moults into the pupa. The pupal stage lasts 8—15 days.

In Morocco the preoviposition period lasts about one month, while the oviposition period lasts over two months. One female may lay about 300 eggs during the lifetime at the rate of 3—12 eggs daily. Thus, the adults may live 3—4 months during the summer; overwintering females live much longer.

In Morocco there are four generations during the year, while in Italy only two generations.

Lixus junci Bohm.

(Curculionidae, Col.)

Beets are subject to attack by various weevils. Of this group, one of the outstanding in the Mediterranean countries is *Lixus junci*. If no measures of control are carried out, the damage may be great.

Description. The body is elongate, almost cylindrical, and from 7 to 12 mm long. Its colour is black, but it is covered with velvety black or golden hair; two white lateral bands pass from the head to the end of the elytra. The underside of the body is light-grey. The prothorax is wider than its length; the elytra, which taper towards the end, have eight rows of concave dots each. The male is distinguished from the female by the snout which is longer in the female than in the male; therefore, the bases of the antennae are farther from the edge of the snout than in the male.

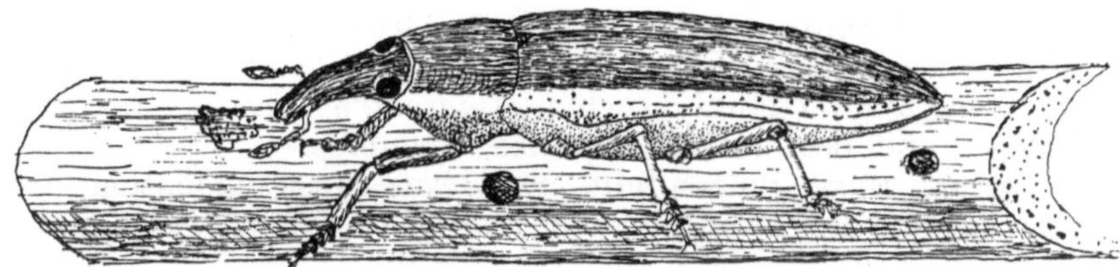

Fig. 117. *Lixus junci*. Note two oviposition holes.

The egg is elliptical, about one mm long, yellow, and delicate without any engravings. It is laid in a hole which the female excavates in the petiole. The width of this cavity is 1—2 mm, and it is covered with a mash of gnawed plant tissue and saliva.

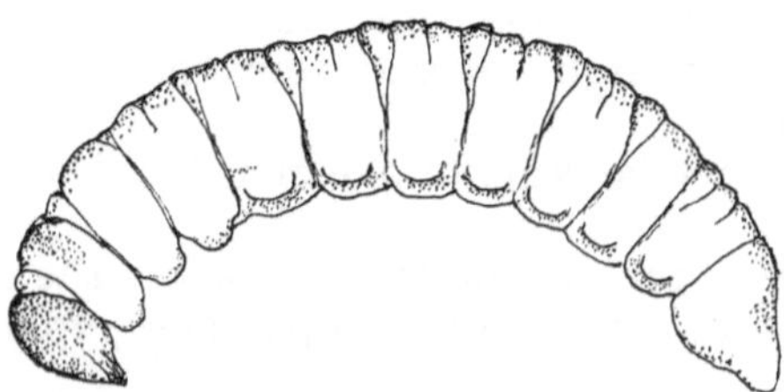

Fig. 118. *Lixus junci* – larva.

The larva is legless, its colour is yellowish-white. On the segments of the body there are transverse furrows, which divide each into several sections.

The pupa is 12 mm long. The wing pads are narrow and reach the fourth abdominal segment. The last abdominal tergite bears a wide spiny process, the other tergites bear each a row of spines which are smaller on the first segments.

Distribution. Lixus junci is found throughout the entire Mediterranean region. Its damage is recorded from Spain, France, Italy Dalmatia and from Israel. It is found also in Greece, Syria, North Africa and in Central Europe.

Hosts. The adults may be found feeding on leaves of plants other than beets. But breeding and development take place only in Chenopodiaceae. *Lixus* larvae may be found also on various wild plants of the family. Of the cultivated plants it is found in beets with broad and juicy petioles, such as fodder and sugar beets. In table beets it is less frequent.

Type of Injury. Several larvae feeding on one head perforate it. In the tunnels thus created fungi and bacteria develop which may cause partial or complete decay of the head.

Towards the end of the beet season, it is often essential to keep the beet in the fields as long as possible in order to regulate the cattle food supply, or regulate the delivery to the sugar factory. *Lixus junci* may then interfere greatly with these efforts.

Life History. The biology of this insect was studied in Italy by MENOZZI (1947) and in Israel by RIVNAY & MELAMED (1956). There are some differences in its annual cycle in both countries:

The winter is spent in diapause as an adult. The awakening depends upon the weather and the climate of the country. In Italy the insect appears at the end of March or early April. In Israel it may appear at the end of February. In warm winters it may appear as early as January and in cold winters as late as the end of March. In Italy copulation takes place at the end of April and oviposition a few days later. In Israel oviposition may begin as early as January, but the main laying period is in March.

At a temperature of 23—28° C the incubation period is 3—4 days. At the temperature of 15—16° C, two weeks pass before hatching. MENOZZI states that in Italy the egg develops in a week's time. Its calculated threshold of development is in the neighbourhood of 10° C (RIVNAY & MELAMED).

Upon hatching the larva feeds upon the inner tissue of the petiole and burrows a tunnel downwards. If it has completed its development it pupates at the base of the petiole, otherwise it continues feeding and burrowing into the head, and finally pupates there. The total development of the larvae lasted 35—40 days in Italy

(MENOZZI) and in Israel 30 days at 28° C or 50 days at a temperature of 19—20° C.

Whether in the petiole or in the head, the tunnel is widened into a cell into which the larva pupates.

The pupal period lasts 5—9 days at a temperature of 26° C, but the newly emerged adult remains in its pupal cell a few days before leaving it. Thus the total development of the insect from egg to adult lasts 8—10 weeks at 20° C, or only 5 weeks at 28° C.

In Israel the newly emerged beetles of the first generation may lay the same summer. The pre-oviposition period lasts 14—20 days at a temperature of 23° C and 9—11 days at 26° C. In Italy one female may lay from 30—40 eggs (MENOZZI). In Israel the egg-laying pattern of *Lixus* is different. Of the overwintering females, 30—45% laid 50—200 eggs; a few prolific females laid over 200 eggs. A great many of the first generation, did not lay at all the same summer, and the average number of eggs per female of those that laid was low.

The oviposition period continues until a few days before the female dies, and the daily rate of egg-laying is 3—4 eggs in Italy and 7—10 in Israel. Some females laid as many as 15 eggs per day.

Diapausing beetles lived 10—12 months. In Israel, beetles of the first generation lived shorter periods. In captivity 50% of them lived 40 days while 20% lived almost 100 days.

Number of Generations. In Italy, one may surmise from the account of MENOZZI that only one generation develops there during the year. In Israel, due to the early emergence of the overwintering adults, two generations may appear. However, as mentioned above, only a small percentage of females of the first generation lay, while all second generation females enter diapause before laying.

According to reports from Morocco, some of the females of the second generation may awaken with the onset of cooler and rainy weather, mate and lay in volunteer plants. The offspring of these develop during October and November and become adults in December. There is reason to believe that this may be the case in Israel, too.

Ecological Notes. RIVNAY & MELAMED (1956) point out that when the winter is mild, the beetles may awaken as early as January. Also, when the temperature is high, the egg-laying is accelerated and increased. Thus in 1953, when the average temperature of January and February was about 13° C, nearly 300 egg cavities were found in 100 petioles, whereas in the colder winter of 1954, when the temperature during those months averaged about 10° C, only 35 egg cavities were found in 100 petioles.

Further ecological data by these writers are contained in fig. 119. Contrary to the warm January and February of 1953 the March and April of the same year were cold, about 11° C; this retarded the development of the generation. In 1954, on the other hand, the spring months were quite warm, a factor which accelerated the development. This explains the extreme discrepancy in the time when the two peaks of adult population appeared in both years.

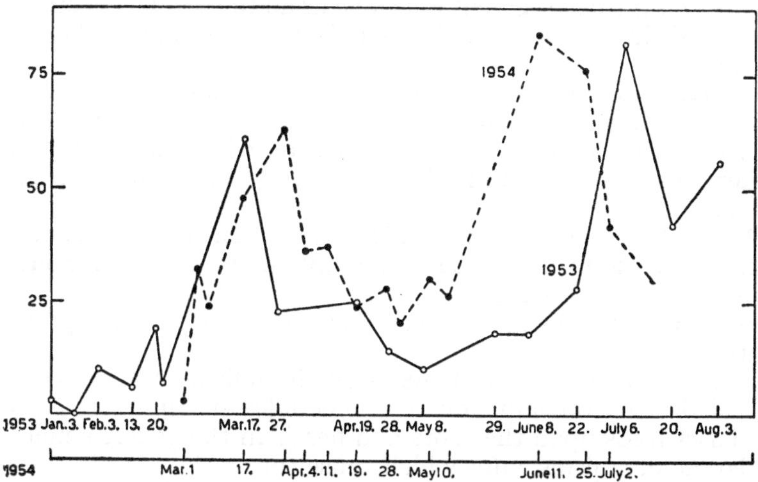

Fig. 119. Phenology of *L. junci* in Israel during 1953 and 1954. Note the difference in the time of appearance and climaxes of the population in the two years. After RIVNAY & MELAMED.

According to the same authors, the mortality of the eggs increases to about 50% when the temperature is above 28° C.

Another factor which causes mortality of the eggs is the physiological state of the plant; when the plant is in a vigorous state of growth, as many as 80% of the eggs may be crushed by the growth of the cork tissue caused by the wounds during the excavation of the cavity.

Control. Laboratory tests showed that adult beetles died 24—48 hours after contact with residues of DDT, BHC, toxaphene, and dieldrin. DDT was the quickest acting against this beetle. In the field, however, BHC gave more control than DDT, perhaps due to its penetration into the plant tissue and thereby killing the larvae. In this respect parathion was far superior to all others.

Conorrhynchus mendicus Gyll.

(Curculionidae, Col.)

In Southern Italy, Sicily and Malta, this weevil is considered an important pest of beets. In view of its proximity to the Near East, it will be discussed briefly.

Description. This weevil is 11—18 mm long, its colour is black, and it is covered with grey and white scales. The elytra are punctate and on each two parallel lines run obliquely to the margins so that the lines of both elytra make angels with the vertex pointed backwards. The rostrum is cone-shaped, the antennae clavate and the last segment is as long as all the segments combined. The tip of the abdomen of the male is concave, that of the female convex.

The egg is elliptical, 1.8 mm long and yellowish. The larva is 13—14 mm long, cream-white with a dark median line over its dorsum. The head is brown, the frons is lighter while the mandibles are darker. On the sides of the body there are two rows of rounded papillae.

Distribution. This is a Western Mediterranean species. Its northern limit in Italy is Pisa. It occurs in Sicily, Malta, Sardinia, Southern France, Spain and Algeria.

Hosts. Of the Chenopodiaceae *Beta* and *Atriplex* are preferred to other genera.

Type of Injury. The adult feeds on the foliage. Young seedlings may be destroyed completely as the petioles and stem are gnawed. The larva feeds upon the roots and head. In the galleries made by the larvae organisms enter which cause rot.

Life History. The pest hibernates as an adult. It awakens early in March and, should there be no host plants in its vicinity, it may crawl long distances in search of food plants. Occasionally it flies. After mating egg-laying begins, a period which may last about 25 days. One female may lay as many as 10—70 eggs. The egg is laid at the base of the petioles or in the ground near the plant.

After an incubation period of about 8 days the larva hatches and crawls into the ground towards the roots. It feeds on the roots and head and in doing so, unlike *Lixus* larvae, feeds on the external layers. When not feeding it rests in a cell in the ground. Its development lasts 40—50 days. Since egg-laying extends over a long period, larvae of all stages may be found at the same time in the summer. The pupal period lasts about 12—15 days. Pupae are found from the middle of August to the middle of September. The new adults remain in the soil until the following spring unless some exceptionally warm days drive them to the surface, after which they return to the ground again.

In Italy there is another species of this genus, *C. luigionii* Sol., which is similar in its habits to *C. mendicus* except that it appears later in the season when the plants are large, therefore causing less conspicuous damage.

Chromoderus fasciatus Müll.

(Curculionidae, Col.)

This species is of wider distribution than the two mentioned above.

Description. The adult is 7—11 mm long with a cone-shaped head, over which runs a median longitudinal ridge. The prothorax is narrow in front, wrinkled, and roughly punctate. The colour is brown but the scales which cover the body give it a greyish appearance except for certain bare scaleless spots as follows: A triangle on the prothorax, at the humeral angles of the elytra and one curved transverse band on the middle of each elytron.

Distribution. The species is distributed in Europe and Asia Minor.

Hosts and Injury. The larvae bore into the heads of plants belonging to the genera *Beta, Chenopodium, Atriplex* and *Erodium.*

Life History. The adults awaken in April-May and feed upon the foliage of *Atriplex.* A week later they mate and 4—5 days later egg-laying begins. One egg is laid on the plant in a cavity which the female excavates. Each female may lay from 20 to 30 eggs. The development of the generation under European summer conditions lasts 40—45 days; but only one generation develops during the year.

The Beet Moth — Phthorimaea ocellatella Boyd.

[Syn.: *Lita ocellatella* Bd.]
(Gelechiidae, Lep.)

Ph. ocellatella is a small moth hardly known to the farmer of the Near East, but much of the decay which may occur in beets is no doubt due to this pest.

Description. The length of the adult is 4—5 mm and the wing-span is 10—12 mm. The forewings are yellowish-grey with conspicuous yellow and dark spots. The fringe of hair in the forewings is longer laterally and shorter in the rear. The hind wings are yellowish with long hair in the fringe.

The neonate larva is green; as it grows it becomes purplish-green dorsally because of two purple stripes which run along the body. Each segment bears several setigerous tubercles. The head is black, the prothorax and the last abdominal segment brown. When mature it is 10—12 mm long.

Distribution. This insect is Mediterranean. Records of its occurrence in Europe come from Portugal, Spain, France, Italy, and Crimea; in Asia from Iran, Asia Minor and Israel; in Africa from Madeira, Morocco and Egypt.

Hosts. The insect feeds upon Chenopodiaceae only. If the moth occurs in the spring when no cultivated beet is available, it may lay on related wild plants. In the literature the following are mentioned as hosts: *Hyoscyamus alba, Beta maritima, Atriplex* spp., etc.

Type of Injury. By their feeding and mining into the veins and

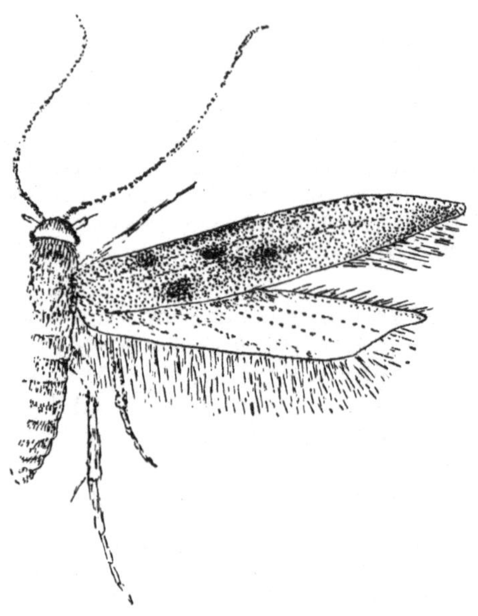

Fig. 120. *Phthorimaea ocellatella.*

stem the larvae may exterminate the entire plant when it is young. This is particularly the case with newly sprouted seedlings.

The larva also mines into the crown of the head and creates an entrance place for moulds and decay bacteria. Plants which have overcome the pest are usually small and have about 25% less sugar content.

Life History. The insect spends the winter in the larval or pupal stage. The adult emerges in the spring — in Italy in May — and after a short pre-oviposition period begins to lay. The neonate larva bores into the main veins of the leaf or into the petioles or crown of the plant. When ready to pupate the larva finds shelter between folds or between the bases of petioles and pupates in a small cocoon which it spins around itself. The total length of development from the egg to adult stage lasts 4 weeks under the conditions of Northern Italy or three weeks under the conditions of Central Italy.

Number of Generations. The moth may raise a few generations during the year. In France, Italy and Spain the second generation appears in July. In France this is the last, while in Southern Italy two more generations may appear. This is probably the case in Israel, too. In Italy the damage is caused by the latter generations.

Control. Control measures may be carried out both against adults and against neonate larvae — before they mine into the plant tissue. Parathion and endrin yield satisfactory control of both. In Israel no special applications are carried out but applications against other pests possibly keep this under control.

The Beet Webworm — Hymenia recurvalis F.

[Syn.: *H. fascialis* Z.]
(Pyralidae, Lep.)

The la' vae of this moth appear in large numbers and entirely defoliate the host plant.

Description. The length of the adult is 12 mm and the wingspan 20 mm. The colour if the body and wings is brown, two white bands cross the forewing and one crosses the hind wing. A fringe of short hair decorates the margins of both wings. White lines decorate the orbits of the eyes and the hind margins of the abdominal segments.

The egg is green, oblong, slightly flat, 0.6 mm long and 0.4 broad.

Fig. 121. *Hymenia* adult and larva. Courtesy Hassadeh.

The neonate larva is pale green and its colour changes in accordance with its food: it remains green when feeding on leaves and becomes purple when feeding on the head of the beet. When mature it is 18 mm long and 2 mm wide and the head is pale brown mottled with darker spots. The hooks on the prolegs, 24—29 in number, are arranged in two concentric, almost complete circles; in addition there is a third outer circle of vestigial hooks.

The pupa is about 10 mm long, narrow, brown, with wing pads which are dark brown and reach the fifth abdominal segment.

The pupal cocoon is 12—13 mm long, 6 mm in diameter, and is

made of compact silken threads with a coat of earth particles.

Distribution. The insect is of tropical origin, and its northern limits of distribution in the Near East are Israel, Syria and the Lebanon. In the literature, reports of damage come mostly from tropical and East Asiatic countries as follows: Bermuda, Porto Rico, Florida and Virginia in the U.S., Peru, Hawaii, Haiti, Australia, India, China and Japan. To this list Israel should be added.

Hosts. The insect may feed upon plants of various families. Chenopodiaceae are its favourites. In addition to *Beta* and *Spinacia* it has been found on *Chenopodium* and *Amaranthus*. Of other families, *Portulaca* and *Ipomoea batatas* have been attacked by the larvae.

Injury. As stated above, numerous larvae may defoliate the plants within a short period.

Life History. Our knowledge of the life history of this pest is scanty, and the papers dealing with it contain incomplete information.

The pre-oviposition period may last from two to six days. The eggs are laid in groups on the underside of the leaf. Egglaying was studied by RIVNAY (1959) and the number of eggs laid by one female is given in Table LVI. It is noticed that the maximum may be over 400 eggs while the average is 130, also that over a third of the females laid less than 50 eggs each.

Table LVI.

Oviposition by *Hymenia* (after RIVNAY, 1959)

Number of eggs	0	1—50	51—100	101—200	201—300	301—400	400+
Number of females	1	9	4	5	2	2	2

The incubation period, according to RUSSEL (1933), is 3 days at 24° C, according to JARVIS (1945) 3—5 days, in Queensland and in Israel 5—6 days. The larval period is about 10—15 days in Virginia (WALKER & ANDERSON, 1940) and 13 days in Bermuda (RUSSEL). In Israel it is 25—30 days during November-December (RIVNAY, SHWEIG, 1954). The pupal period is as long as that of the larvae. WALKER & ANDERSON state that the moth cannot endure the cold climate of the Virginian winter. Moths, larvae and pupae all die off and the renewal of the population takes place through migration from the south. They reach that state towards the end of the summer. Also in Israel they appear towards the end of the summer but never in the spring.

Control. WALKER & ANDERSON found that the best results against this pest were obtained by dusting with a 0.2% pyrethrum dust.

This should be preferred above other insecticides in view of the fact that the dusted plant parts may serve for human consumption.

Other Pyralidae

In Italy, according to MENOZZI (1947), the Pyralid *Loxostege (Phlyctaenodes) nudalis* HB. may cause damage by feeding on the beet foliage. The adult is a yellowish-white moth with amber-yellow forewings, each of which have six dark spots. It is distributed in Southern Europe, North Africa and Western Asia. In Libya it may cause more damage than in Italy but not to the extent that it necessitates control measures. Also *Pionea ferrugalis* HB. feeds among others on the foliage of beets. This is a moth with a 15—17 mm wingspan and yellow-ochre wings with undulating lines crossing the forewing. Its distribution is similar to that of *Ph. nudalis.*

In Israel, the main leaf veins or the petioles of beets may be tunnelled by larvae of an *Ephestia,* the particular species being unknown. Its tunnels are like those of the *Lixus* larvae except that they may split open. It is prevalent in the spring and is very rare in the summer. BODENHEIMER (1930) mentions *Ephestia elutella* as boring into dry petioles of beets. Both of these are of no economic importance.

Ocnogyna loewii Z.

(Arctiidae, Lep.)

Beet plants are often infested with hairy brown caterpillars, the larvae of *O. loewii*.

Description. The adult male is 12—14 mm long, with a wingspan of 30—32 mm. The forewing is brown or dark brown, with a yellow broad line running along its posterior margin, one straight line and one N-shaped line crossing the wing, dividing it thus into triangles or various differently shaped brown patches; the hind wing is cream-coloured with brownish patches. The female is 13—15 mm long, and wingless.

The neonate larva is black with silvery hair. After the third moult its dense long hair becomes brown; when mature, it is about 30 mm long.

Distribution. According to WILTSHIRE (1957) this moth is Anatolian-Iranian. There are records of its occurrence in Anatolia, Greece, Egypt, Israel, Syria and Iraq.

Hosts and Injury. This insect is polyphagous and feeds on many annual plants. From uncultivated fields it migrates often to cultivated crops including beets. The injury is made by feeding on the foliage.

Life History. The adults appear in December and after mating lay their eggs in clusters containing a few hundred each. The larvae after hatching live together in a common tent which they spin on the host plant. After the third moult they disperse and live a solitary life. During May they mature and enter the ground to pupate in a cell from which the adult emerges only the following December. Thus only one generation is raised annually.

The Beet Fly — Pegomyia hyoscyami Panz.

(Anthomyiidae, Dipt.)

On the leaves of beets pale green or brown patches may appear. Close examination reveals a pocket between the epidermis layers in which large maggots are active. These are larvae of *Pegomyia*, the most common species of which are *P. hyoscyami* and *P. betae*.

There is a diversity of opinion regarding the authenticity of the two species. Some authors believe that they are two names for the same species. On the other hand BALACHOWSKY (1936) comments: *P. hyoscyami* lives on some Solanaceae and does not oviposit on Chenopodiaceae, while *P. betae* lives on beets and does not lay on Solanaceae. In Israel larvae of *P. hyoscyami* were reared from *Beta*, *Chenopodium* and *Rumex*. The discussion which follows is based upon literature which deals with *Pegomyia* from beets only.

Description. The adult fly is 4.5—5 mm long. The thorax and the base of the abdomen are grey, the rest of the abdomen is yellow. The

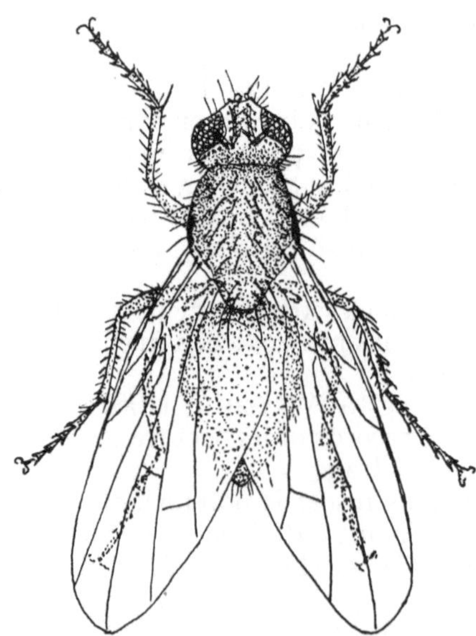

Fig. 122. *Pegomyia hyoscyami*.

occiput is grey, the genae and face silver grey, and the frons yellowish-brown. The first two segments of the antennae are yellowish-brown. The arista is not pubescent and the wings are transparent with yellowish veins. The legs are yellow except for the anterior femora which are grey and the tarsi which are black.

The egg is banana-shaped, 1 mm long an 0.3 mm in diameter; its surface is reticulate.

The larva before pupation is 6—8 mm long. The last abdominal segment is decorated with a ring of 10 papillae and in the centre are the buttons bearing the posterior spiracles.

The pupal case is 4—5 mm long and dark-red.

Distribution. This species is distributed throughout Europe from Finland to Italy. It is also found in Greenland, Canada and the U.S. In Asia it is found in Korea, Japan and the Middle East and it is prevalent also in North Africa. Not everywhere is it of equal agricultural importance. Damage by this fly is recorded from Denmark, Belgium and Austria. In the Near East it may be harmful, but not to the extent that measures of control are called for.

Hosts. The larvae feed upon Chenopodiaceae. In Israel it has been found on the plants *Beta, Emex, Rumex, Chenopodium* and *Spinacia.*

Type of Injury. When several larvae feed upon one leaf, it may become devoid of chlorophyll, the pockets coalesce as they grow larger, and young seedlings may, thus, be entirely killed off. Very often 40—50% damage is reported, and in heavy infestation the field may be totally destroyed necessitating renewed seeding.

Life History. The adult spends the diapause in the pupal stage and awakens in the spring. In Italy this occurs in April, in Austria in April-May, in Germany and Holland in May, in Estonia and Scotland in May-June. In Israel infestation was observed as early as February. After a period of feeding and mating, egg-laying begins. These are laid in groups of three to six on the underside of the leaf. According to Blunck et al. (1933), the incubation period lasts 2.5 days at 28—29° C and 13—15 days at 8° C. Moisture is essential for egg development. Blunck et al. state that R.H. above 70% is sufficient, while Bremer & Kaufmann (1931) state that at 83% R.H. all eggs died.

Upon hatching the larva bores its way from the egg directly into leaf tissue. The larva must have chlorophyll, otherwise it fails to develop. If the food supply in one leaf is exhausted, the maggot may leave it in search of another. The maggot completes its development after 9—12 days under the conditions of the European summer. Under controlled conditions the development lasted 7 days at 28° C and 17—22 days at 14.5° C.

Pupation takes place in the ground. The depth to which the larvae penetrate depends upon the moisture content of the soil; when the moisture content is average, 56% of the larvae pupated 2—4 cm deep, and 34% 1—2 cm deep; when more water was added to the same soil 61% of the larvae pupated 1—2 cm deep. The pupal development lasts 13 days and the threshold of development is 2° C (BLUNCK et al.).

The pre-oviposition period in Europe lasts 5—8 days. Adults may be active at a very low temperature. BLUNCK et al. observed egg-laying at 0° C. The optimal temperature for activity is 20—30° C; a temperature above this is detrimental to the fly. At the optimal temperature the average number of eggs laid was 50, while the maximum was 190 eggs per female.

Number of Generations. As mentioned above, the insect spends part of the year in diapause in the pupal stage. In the northern countries this occurs in the winter, and in the warmer climates in the summer. Of the first generation a very few pupae remain in diapause while in the last generation all pupae remain till the following season. In Central Europe the fly may raise 3—4 generations, in Estonia 2 and occasionally part of a third, while in North Africa 2 generations may be raised. This seems to be the case also in Israel.

Limiting Factors

1. Biotic: In all countries where the fly is prevalent it has been pointed out that parasites claim heavy toll of the population. While the first generation is little affected, 30—90% of the third generation may be parasitized.

The parasites mentioned are of the genera *Trichogramma* (egg parasites) and *Trichopria, Alysia* and *Opius* (larval parasites). All these genera are well represented in the Near East. In Israel *Entedon* sp. was also reared from *Pegomyia* larvae.

2. Climatic: Rain may easily wash the eggs off the leaves. Low relative humidity hinders the hatching of the eggs. In Turkey it was pointed out that low relative humidity is unfavourable for the pest. Also too high a temperature is unfavourable. RAMBOUSEK & NEUWIRTH (1932) pointed out that in Czechoslovakia, a hot summer, an unusually warm autumn or a premature warm spring check the pest. In the Near East these two factors apparently keep the pest in a state where control measures are not necessary.

Control. European entomologists suggest that thinning of the plants may be carried out after the egg-laying period of the first generation because, as they point out, with plenty of foliage the number of eggs per plant may be less than if the plants were thinned. Furthermore, in the process of thinning the infested plants are removed and the healthy remain.

Larvae in the fields may be killed by an application of nicotine with soap. Eighty percent of larvae have been killed in this way.

Adults may be killed by a bait of 1% sugar or molasses with 0.3—

0.4% of sodium fluosilicate. A cover spray with chlorinated hydrocarbons satisfactorily kills the fly.

General Considerations on the Control of Sugar Beet Pests

Although the list of sugar or beet pests is rather large, not all of them are so numerous as to call for control measures. With some pests, even if they are numerous, control measures are superfluous. As an example the beet fly, *Pegomyia hyoscyami*, may be considered. In northern countries it may be very harmful but in the Near East, it may attack beets as early as February, when the plants are large enough to overcome these casual assaults. In March the infestation may increase to such numbers that it causes concern as to the future of the crop. It should be remembered, however, that this generation enters diapause in the pupal stage and no further attack will take place. Therefore control measures are superfluous.

Phyllopertha larvae may attack beets when sown in a field in which infested graminaceous crops were growing the year before. The farmer should avoid cultivating beets in such fields.

Lixus may cause more concern when the winter has been very mild. One or two applications of the insecticides enumerated (see *Lixus* control) may relieve the situation. If the beets remain in the field till late in the season, further applications may be necessary at a future date. When parathion is employed fewer applications may be required.

Prodenia is one of the major pests of beets. It attacks this crop at both ends of the season. In the autumn, after peanut fields have been dried and cotton plants are no longer attractive, the increased *Prodenia* population finds the fresh beet seedlings more desirable as food. In southern Israel, the *Prodenia* attacks are then intense and continuous, thus repeated applications are often imperative. In October and November, when their activity diminishes, the injury to the plants may be far less in proportion to their numbers. The activity of the pest is resumed in force towards the end of May and June, when defoliation of the plants may be complete. The employment of dipterex is recommended at this stage.

A *Cassida* attack may easily be overcome by one or two dustings or sprayings with synthetic insecticides. BHC, DDT, dieldrin and others were satisfactory in its control.

The applications against *Lixus* will control other pests that may be on the plants, such as *Ph. ocellatella* and others.

LITERATURE

ANON, 1938. Les Lixus — Parasites de la Betterave — *L. junci* Boh. ***Défense des Végéteaux*** **53,** *1-8,* Dir. Aff. écon. Rabat.

ANON, 1938. La Casside de la Betterave *Cassida vittata* Vill. ***Défense de Végétaux*** **54,** *1-6,* Dir. Aff. écon. Rabat.

A. Balachowsky & L. Mesnil, 1936. Les insects nuisibles aux Plantes cultivées. Paris.

H. Blunck, H. Bremer & O. Kaufmann, 1933. *Arb. biol. Reichsanst.* **20,** 5, *517-585.*

F. S. Bodenheimer, 1930. Die Schädlingsfauna Palästinas. Monogr. angew. Ent., Paul Parly — Berlin, 438 pp.

H. Bremer & O. Kaufmann, 1931. *Monog. Pflanzschtz.*, **7,** *1-110.*

H. Jarvis, 1945. *Qd. agric. J.* **60,** *355-357.*

C. Menozzi, 1947. Animali e Vegetali Dannosi Alla Barbabietola da Zucchero e mezzi per combatterli, 206 pp. Societa Approvvigionamento Bietole e vendita zucchero di Genova.

F. Rambousek & F. Neuwirth, 1932. *Vest. Csl. akad. zemed* **VIII,** 193, Abst.

E. Rivnay, 1959. Unpublished Notes on *Hymenia.*

E. Rivnay & V. Melamed, 1956. *Ktavim* **7,** 1, *63-82.*

T. A. Russel, 1933. *Rep. Dept. Agric. 28-36,* Bermuda.

K. Shweig, 1954. Vegetable Pests. Hassadeh Publishing Co. *118-121* (in Hebrew).

N. G. Walker & L. D. Anderson, 1940. *Bull. Virginia Truck Exp. Sta.* **103,** *1651-1659,* Norfolk.

E. P. Wiltshire, 1957. The Lepidoptera of Iraq. Ministry of Agriculture, Govt. of Iraq, 162 pp.

K. Zolk, 1932. *Mitt. Versuchsst. angew. Ent. Univ. Tartu* **16,** 11 pp.

The Pests of the Sugar Cane *(Saccharum officinarum)*

In addition to the pests described below, sugar cane may be infested with some of the insects which infest maize and sorghum, a key to which is found on page 181. Outstanding among these is *Sesamia cretica*. Also some Orthoptera may feed upon the foliage of this plant (page 19).

In addition to the sugar cane mealy bug, other insects of this family may be found on sugar cane, but none cause as much damage as *Saccharicoccus*.

Scarabaeid larvae may feed upon and make cavities in the young shoots of sugar cane. WILLCOCKS (1925) mentions *Heteronychus* spp. and *Pentodon dispar* BANDI, while *P. corniculus* RTT. is recorded as injurious to this crop in Israel (BODENHEIMER, 1930).

The Pink Sugar Cane Mealy Bug — Saccharicoccus sacchari Ckll.

[Syn.: *Pseudococcus; Dactylopius*]
(Dactylopinae, Homopt.)

Towards the end of, and after World War I, *S. sacchari* was the major pest of sugar cane in Egypt. The discussion below is based on the paper by HALL (1922).

Description. The adult female is a large elongated coccid, 6—10 mm long, and 3—5 mm broad, oval-shaped and of a pink colour. It is sparsely covered with waxy meal and the segmentation of the body is distinct. It has neither marginal filaments nor an ovisac.

Distribution. This insect occurs in Central and South America, in Asia it is found in Formosa, the Philippine Islands, Indonesia, India and Israel, in Australia it is limited to the North Eastern littoral zone, in Africa it is found mainly in the eastern countries of the Continent — in Madagascar, Egypt and also in Madeira.

Hosts. Sugar cane seems to be the preferred host, but the insect thrives better on those varieties whose leaf sheaths adhere closely to the cane, thus offering it better protection than in the case of loose leaf sheaths (HALL).

Type of Injury. HALL describes the damage as follows:

The insects feed upon the cane by sucking the plant juices. This results in the formation of a gum on the outer surface of the cane. This gum is composed of the honey-dew secreted by the large colonies of the insect as well as by the exudate from the plant itself which is a defensive reaction induced by the feeding of the insect.

This gumming makes it very difficult to strip the cane especially in the variety of Java 105 where the sheaths are close to the cane. It was found also that the crystallizable sugar content in the gummed canes is much reduced.

Life History. The female of *S. saccharis* produces living young parthenogenetically, thus an ovisac is superfluous. Shortly after its birth, the young larva migrates to a new node where it settles down for its entire life. Males are born in small numbers and these seem to play a small role in the propagation of the species.

The insect raises a few generations annually.

Encouraging Factors. As mentioned above, this pest is of world-wide distribution and was discovered in Egypt as early as 1912. Yet it did not assume a destructive role until the end of World War I. HALL, who studied this pest in Egypt, points to the following reasons:

1. The Java 105 variety which was introduced into Egypt and which, as mentioned before, the pest favours.

2. Careless cultivation and manipulation of the crop.

a. In view of the fact that coal was unavailable, the thrash of the cane fields was used as fuel. This was conveyed long distances along with millions of insects. The dispersion of the pest was thus encouraged;

b. Poor quality "sets" were planted, which included infested parts. Thus the farmer planted colonies of the pest together with the crop;

c. In view of the high prices which the sugar fetched, cane was left in the field longer than is advisable.

Preventive Measures. HALL suggested taking the following measures in order to overcome the pest:

1. Planting non-infested "sets". In order to clean infested sets they should be dipped in an insecticide after the leaf sheaths have been removed. Parathion, malathion, lindane and endrin would probably kill the insect if sets are place in a 0.1% active ingredient solution. The fact that the insect is only sparsely covered with the waxy meal and that it has no ovisac will facilitate the insecticidal action.

In Egypt HALL recommended using a petroleum emulsion, the stock of which contained 2 gallons of petroleum, 1 gallon of water and 1 pound of soap; this stock was diluted 1 : 30 in water.

2. Cane should be grown for only two years.

3. After harvest thrash should be cleaned immediately.

4. Cane should not be grown over cane.

The Small Purple-Lined Borer (The Rice Borer) — Chilo suppressalis Wlk.

[Syn.: *Chilo simplex* BUTLER]
(Pyralidae, Lep.)

In addition to rice and other Gramineae this pest may also attack sugar cane.

Description. The moth is 10 mm long. The wingspan of the female is 22 mm, that of the male 15 mm. The body is brown and the fore-wing has a yellow-brown oblique streak extending from the apex of the wing to its disc. A row of brown spots marks the end of the wing margin, the hind wings are white.

The larva when mature is about 22 mm long, cylindrical, 3 mm in diameter, head light brown, body cream coloured with five longitudinal pink lines extending over the entire body.

The pupa is 13—15 mm long, 4 mm in diameter, the abdomen tapers gradually and ends in a bilobed tip; young specimens are dark brown, older are black; some specimens have a brown dorsal median band extending from the thorax to the abdomen.

Distribution. In Asia this insect is found in Japan, the Philippine Islands, Malaya, Java, China and Israel. In Africa it has been reported from Egypt, in Europe from Spain. Since a year ago it has become very conspicuous in the corn fields in Israel.

Hosts. Rice is apparently the most preferred host, but it also attacks maize, broom corn and cane sugar. It often affects *Andropogon halepensis* as well.

Type of Injury. This pest, according to WILLCOCKS (1925), tends to girdle the stem at the node; also the actively growing portion of the cane may be girdled. Canes thus injured are easily broken by the wind. Furthermore, injured canes cause the growth of daughter canes from the buds. Injured canes may be detected by the holes which the larvae make outside.

Life History. The life history of this pest in this part of the world was studied by GOMEZ CLEMENTE (1948, 1952), in Spain, and the details of the account have been taken from his work. The insect spends the winter as a larva in diapause. Adults emerge by the 20th April when the temperature averages 16° C. The overwintering adults live 4—8 days, and lay on the average about 200 eggs.

The incubation period lasts 5—6 days at 26—28° C and 10—12 days at 20—22° C.

The larva completes its development in 30—40 days at 27° C. The pupal stage lasts about one week. Thus the adults of the next (first) generation appear in July.The offspring of these, the larvae of the second generation, appear in August and complete their development in September. A few of these may become adult the same year, but the majority overwinter as larvae, pupate in the spring, and 10 days later emerge as adults. In the spring the pupal stage lasts 10 days at 17.6° C and 20 days at 15.4° C. The threshold of development of the insect lies thus at 14.9° C and the thermal constant is 60 day-degrees Centigrade.

Studies of the factors inducing diapause of this insect were made in Japan.

FUKAYA (1950) found that the incubation temperature is an important factor in inducing diapause. When eggs were kept at 31° C the larvae were not inclined to enter the diapause state, but when eggs were kept at 22° C and the larvae at 31—32° C, the percentage of diapausing larvae was high.

INOUYE & KAMANO (1957) point to the photoperiod as a factor inducing diapause. A photoperiod of 8—14 hours induced diapause, but 14.5—16 hours prevented it. Larvae kept in complete darkness entered the stage of diapause when kept at 20° C but not when kept at 30° C.

Limiting Factors. According to WILLCOCKS, this pest is prevented from becoming a serious pest of sugar cane by a small Chalcid egg parasite — otherwise it would be far more difficult to raise sugar cane in Egypt.

LITERATURE

F. S. BODENHEIMER, 1930. Die Schädlingsfauna Palästinas, P. Parey 438 pp.

F. GOMEZ CLEMENTE, 1948. *Bol. Pat. veg. Ent. agr.* **16,** *1-22.*

F. GOMEZ CLEMENTE, 1952. *Bol. Pat. veg. Ent. agr.* **19,** *161-188.*

M. FUKAYA, 1950. *Proc. 8th Int. Congr. Ent.* 1948-(1950), *223-225.*

W. J. HALL, 1922. *Min. Agric. Egypt, Tech. and Sci. Serv. Bull.* **26,** *1-16.*

H. INOUYE & S. KAMANO, 1957. *Jap. J. appl. Ent. Zool.* **1,** 2, *100-105.*

F. C. WILLCOCKS, 1925. Insects and related pests of Egypt. 418 pp.

Pests of Sunflower *(Helianthus annuus)* and Safflower *(Carthamus tinctorius)*

Both of these plants may be injured by *Agrotis* larvae. The foliage of *Helianthus* may be attacked by *Laphygma exigua* and by some species of *Podagria*. The leaves of *Carthamus* may be destroyed by *Heliothis peltigera* and *H. armigera*. Aphids may be feeding on the soft leaves; among them *Dactynotus carthami* H.R.L. may be found.

The larvae of *Agapanthia* bore into the stems of sunflowers, while *Lixus* feeds in the stem of safflower. Various insects feed upon the soft seeds or ovules — *Larinus* spp., Trypetidae, or some Nitidulidae.

Agapanthia cardui L.

(Cerambycidae, Col.)

Two species of this genus are known to be injurious to Compositae — *A. cardui* and *A. dahli* RICHT. The larvae of both bore into the stem. The biology of the two is more or less similar; the more familiar, *A. cardui*, will be discussed here.

Description. The beetle is 9—11 mm long, narrow and almost parallel-sided. The general colour is dark-green; a line of white pubescence runs along the inner margins of the elytra and extends as a median, bronze-coloured line on the pronotum and head. Two bronze lines run along each side of the pronotum. The underside of the body and the legs are covered with greyish-white pubescence. The basal half of each antennal segment is also clothed with white pubescence.

Distribution. These insects are distributed in the Eastern Mediterranean countries and around the Black Sea. Damage has been reported from the Caucasus and from Malta.

Hosts. The only hosts are plants of the Compositae, and the beetles prefer to lay on plants with thick and juicy stems. Injury was thus reported to artichoke in Malta *(A. cardui)* and to sunflower in the Caucasus *(A. dahli)*. In Israel both species have been found in safflower.

Type of Injury. Injury is caused by the larva which bores into the stem. In plants which have a single stem, like sunflower, this action of the larva weakens the plant and may cause breakage by the wind or when the inflorescence becomes heavy. MEGALOV (1934) also claims a lower yield from infested plants and DOBROVOLSKII (1930) relates of 40—80% of infested sunflower plants in a field. On the other hand, BYTINSKI-SALZ (1951) does not believe that the boring of the larvae causes much injury to safflower plants.

Life History. The adults begin to appear in the spring, in the Caucasus in April (DOBROVOLSKII) and in Malta in March (BORG, 1930). In Israel they appear in March too (BYTINSKI-SALZ). The adult beetle lives about two months only, but, since their emergence extends over a long period, adult beetles may be found until September. Thus they are capable of infesting safflower at a later date too, as long as the plants remain soft and succulent.

When ready to oviposit the female gnaws the cortex of the stem and with the aid of the ovipositor lays the egg deeper into the soft pith; a brown scar is left on the stem which discloses the place in which the egg was laid.

The incubation period lasts about 10 days in the Caucasus; in warmer climates this period is probably shorter. The larva bores downwards and when it reaches the root (about September) it enters a state of diapause which lasts until the next spring.

The pupation period lasts but one week.

Lixus speciosus Mill.

(Curculionidae, Col.)

The stem of the safflower may be swollen into the form of a spindle-shaped gall. In such an injured stem a boring grub may be found — the larva of *L. speciosus*.

Description of adult. The body is black, but covered with minute fine pubescence which reduces the lustre of the original colour. The length of the body may be 18 mm, that of the rostrum 6 mm. The prothorax is cone-shaped with a rugged surface.

Distribution. The beetle occurs in the Caucasus, Cyprus and Israel.

Hosts and Injury. Of the cultivated crops, safflower is infested, but no doubt other species of *Carduus* serve as hosts. The injury consists of the larva boring into the stem and weakening it. When the attack occurs on the upper parts of the plant the injury is little noticed, but when the base of the stem is attacked the number of seed bearing heads is reduced (BYTINSKI-SALZ, 1951).

Life History. The female becomes active in the spring. In order to lay, she gnaws a cavity in the stem into which she oviposits one egg. The hatching larva feeds upon the inner tissue causing the swelling of the stem. The larval development may last a few weeks. The life history of this species is possibly similar to that of *L. junci*, a pest of beets.

Larinus spp.

(Curculionidae, Col.)

The genus *Larinus* includes robust weevils of medium and small size. Over 15 species are recorded in the Near East and most of them feed upon Compositae. Four species have been found which injure safflower in Israel.

Description. The body of *Larinus* spp. is oval, the prothorax tapers sharply anteriorly to form a small round opening into which the head fits. The body is covered with pubescence or a waxy secretion.

Larinus ovaliformis CAP. is 7 mm long. The rostrum is 2 mm long and 0.75 mm thick. The colour is brown and the body is covered with bright yellow hair.

Larinus orientalis CAP. is 8 mm long. The rostrum is 3 mm long and 0.5 mm thick; the colour is dark brown, and it is covered with brown hair.

Larinus syriacus GYLL. The length of the body is 9 mm. The rostrum is 2 mm long and 1 mm thick, broader at the tip. The body is dark brown and has dense brown hair.

Larinus griseocens GYLL. The length of the body is 8.5 mm. The rostrum which is bent, is 3 mm long and 0.25 mm thick; the body is black and is covered with sparse grey hair.

The egg is elliptical, white, and twice as long as wide.

The larva is grub-like and legless. The abdominal segments are subdivided by ring-like sections.

The pupa is white, turning brown before maturation. At the end of the body there are two projections, and several setae adorn the body.

Distribution. L. ovaliformis is found in the Caucasus, Syria, Asia Minor and Israel. *L. syriacus* occurs in the eastern countries of the Mediterranean basin, Iran and Turkestan. *L. orientalis* is found in the Balkan, Asia Minor and Israel. *L. griseocens* occurs in Italy, Greece, Asia Minor and Israel.

Hosts and Injury. The four above-mentioned *Larinus* species infest safflower. According to BYTINSKI-SALZ (1951) these come from neighbouring fields in which *Carduus* and *Carthamus* spp. serve as primary hosts. The injury is caused by the larvae which feed upon the soft seeds. Several seeds may be consumed by one larva depending upon its size and the size of the kernels. BYTINSKY-SALZ found in a survey from 4% to 80% infestation in the fields of safflower grown in Israel. As many as 50% of the seeds may be destroyed, depending upon the distance from infested *Carthamus* and *Carduus*.

The adult spends the winter in the pupal cell and emerges in April. A short time after mating egg-laying begins. With the rostrum the female excavates a cavity in the inflorescence, lays an egg into it and covers it with a mash. Two or three days later the egg hatches, and the larva feeds upon the ovules or soft seeds. The development of the larva lasts several weeks. When ready to pupate the larva excavates an oblong cell, and by rubbing its body against the walls of the cell smears an exudate over them.

The pupal stage lasts about two weeks only. The adult matures in August, but remains in the pupal cell throughout the autumn and winter, and emerges the next spring.

Ecological Notes. By threshing the heads of safflowers, beetles may be killed, but the population of the beetles, in non-cultivated plants, remains to renew the infestation the next year.

Trypetidae

A few species of this fruit fly family inhabit the inflorescences of safflowers and related plants. Their maggots feed upon the soft seeds; some species cause the formation oi galls.

In Israel BYTINSKY-SALZ (1951) found five species in the heads of safflowers as enumerated below.

Description. The family Trypetidae is characterized by the wide frons in both sexes. The vein Sc in the forewing runs in a sharp angle toward the fore margin. The wings, as a rule, are striped or speckled, the middle tibiae are armed with a spur and the ovipositor is well developed.

Trypanea stellata FUESSLEY is the most common of the five species. Its body length is 2 mm and the length of the wings is 3 mm. The body colour is grey; the legs are brownish except the fore femora which are grey and brown at the tip. The pattern of the wing is illustrated in Fig. 123.

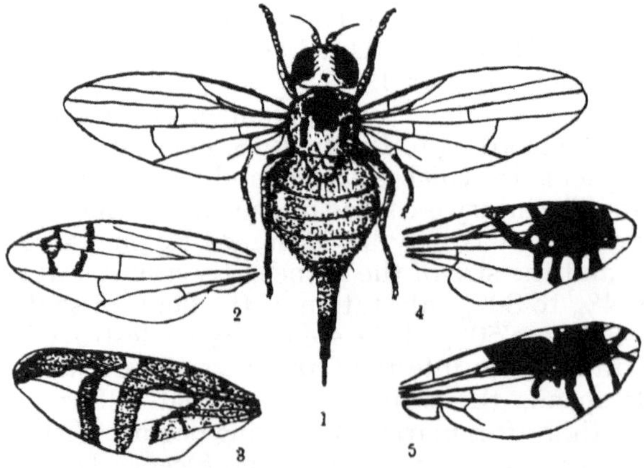

Fig. 123. (1) *Orellia colon* ab. *wenigeri* MG. and the wings of
(2) *Acanthophilus helianthi* BECK.
(3) *Chaetorellia jaceae* R.D.
(4) *Trypanea stellata* FUESS.
(5) *Trypanea augur* FRFLD.
after BYTINSKY-SALZ.

Trypanea augur FRFLD. is less common than *T. stellata*. The length of the body is 3 mm, the length of the wings 3 mm. The colour of the body is yellowish except for the notum and the last abdominal segments which are grey.

Acanthophilus helianthi BECK. The body is 4 mm long, the wing is 4 mm. The body colour is grey; the frons and clypeus are yellowish-grey; the antennae, mouthparts and legs are yellow.

Orellia colon ab. *wenigeri* MEIG. The body is 5 mm long; the length of the wing is 3.5 mm. The colour of the body is yellow-brown; the notum has grey-brown markings; the scutellum is yellow. The ovipositor is $^3/_4$ the length of the abdomen.

Chaetorellia caradjai HER. The length of the body is 4.5 mm, that of the wing 4.5 mm and the ovipositor 1.75 mm long. The body is yellowish-brown, the tergites of the abdomen are of a darker shade and the clypeus is yellow brown. On the thoracic notum there are four rows of black dots, three in each row. Three black dots also decorate the scutellum.

A fly which was reared from artichoke in large numbers is *Terellia fuscicornis* LW. Its body is 5 mm long, the wing is 5 mm and the ovipositor is 3—4 mm long. The colour of the body is greyish-brown with a brown biforked patch on the thoracic notum and brown spots on the abdominal tergites.

Trypanea stellata Fuess.

T. stellata, which is the most common species, will be discussed at more length.

Distribution. This fly has been recorded as a pest in India, Israel and the Ukraine.

Hosts and Injury. The host plants are limited to the Compositae. In addition to the genus *Carduus*, it has been found on *Calendula* in India, on *Taraxacum* in Russia and on the flowers of *Lactuca*. The injury consists of the feeding of the larvae upon the ovules or soft seeds of the host. Thus, many of the seeds of *Taraxacum* are destroyed in Russia; BYTINSKY-SALZ records 20% of the safflower heads attacked by this fly, while in India 25—30% of the *Calendula* flowers are injured and do not open because of the injury of this fly.

Life History. The female oviposits in the open flowers or in the buds before they open. From 4—20 eggs are deposited in each inflorescence. The eggs hatch within one day (in India). The larva feeds upon the ovules or soft seeds, and its development may last 3 days in India, and 12 days in the Ukrainian summer. Pupation takes place between the seeds or, in certain cases, in the ground. The pupal period lasts 11 days in Russia and 6—10 days in India.

All these flies may raise a few generations during the spring and early summer in Israel.

LITERATURE

A. BARASH, 1950. *Hateva v'Haaretz* **VIII,** *330-336.*
P. BORG, 1930. Malta Dept. Agr. ent. notes, 9 pp., (Typescript), Malta.
H. BYTINSKI-SALZ, 1951. *Trans 9th Int. Congr. of Ent., 745-750,* Amsterdam.
B. V. DOBROVOLSKII, 1930. *Bull. N. Caucas. Pl. Prot. Sta.* **V,** *75-96,* Rostov on Don.
M. S. GILYAROV & L. M. LUKYANOWITCH, 1938. Pests and Diseases of Rubber Producing Plants, *28-48,* Moscow, ONTI.
K. NIRULA KANVAL, 1942. *India J. Ent.* **IV,** 1, *90.*
A. A. MEGALOV, 1934. *Grain Prod. J.* **4,** *86-88.*

Pests of Peanuts *(Arachis hypogaea)*

Peanuts may be infested with some of the pests enumerated in the chapter on the leguminous fodder crops, a key to which is found on page 203. In Israel there are a few pests which dominate this crop. *H. cilicrura* attacks the seed and destroys it before it sprouts. *Agrotis* larvae and Tenebrionid beetles injure young seedlings. The leaves of young plants are infested with *Thrips tabaci*, and later red spider mites feed on the leaves. Finally, from July onward *Prodenia litura* becomes the major pest of peanuts in Israel. Along with *Prodenia*, larvae of *Heliothis* may be found under the plants. *Prodenia litura* feeds not only on the leaves, but it may penetrate into the ground and feed on the seed, or even on the pedicels, thus cutting off the pods.

General Consideration on Measures of Control

The extent of infestation of *H. cilicrura* is not uniform but depends upon the methods of cultivation of the field, upon its distance from other infested fields and upon the weather of the particular year. Thus in Israel, late rain and a mild temperature may increase the activity of the pest. In Israel prophylactic measures are employed. In this case soil treatment should be carried out, not seed treatment, as the insecticides injure the seeds (see page 136).

The infestation by *Thrips* occurs when the plants are very young. These are insects which invade the fresh humid fields from the uncultivated flora which has dried and withered. As the temperature rises early in the summer they disappear from the field and measures of control are superfluous. Repeated applications of insecticide against *Prodenia* aggravated the red spider situation on peanuts in Israel. Formerly this pest hardly necessitated control measures and until a few years ago, the employment of parathion against *Prodenia* controlled the mites too. Recently, the pest became resistant to phosphorous compounds. At the present time kelthane, tedion and morocide are the only miticides which control spider mites resistant to phosphorous compounds. The *Prodenia* caterpillars are hidden during the day under the bushy plant; only during the night do they move upwards to feed upon the soft foliage. Therefore it is often difficult to reach them with the insecticide, especially in low volume applications. Dusting proved superior to spraying. During the hot summer days the insecticidal residues deteriorate very fast.

Thus the insecticide may reach the *Prodenia* after the heat and sun radiation have rendered it useless. Applications at night proved far more effective than day applications.

Until a few years ago, DDT, lindane, toxaphene, sevin and parathion were very efficient in controlling *Prodenia litura* larvae. Lately, the pest has become, if not resistant, at least hard to kill. In Israel it was realized that unless a concentrated dosage of endrin and parathion is employed, no satisfactory control is obtained. 80 cc a.i. of endrin plus 150 cc a.i. parathion is often employed for each dunam of well developed peanuts.

Later trials showed dipterex, nuvan and zectran to be efficient too. For further details on measures of control of *Prodenia* see chapter on this pest page 110.

Pests of Sesame *(Sesamum indicum)*

Sesame is a dry farming crop. The fact that there is no irrigation in this field may account for the small number of pests which attack this crop. In addition to *Antigastra*, a short account of which is given below, this crop may be attacked in its early stages by crickets, *Gryllus* spp.; certain grasshoppers may occasionally feed on its leaves; various plant lice, including *Aphis gossypii* may feed on its tender shoots, and certain Homoptera, including *Empoasca*, feed on the leaves. When the field is close to crops under irrigation, *Prodenia* may migrate to the sesame, while *Laphygma* may attack it in the case of a heavy invasion. As a rule, no measures of control are necessary for sesame except in persistent attacks of *Prodenia* and *Laphygma*.

The Sesame Leaf Roller — Antigastra catalaunalis Dup.
(Pyralidae, Lep.)

This pest is active all the summer and may thus attack the plant in all its stages and on all its parts: shoots, leaves, flowers and pods.

Description. The female is 7—8 mm long and the wing span is about 20 mm. The male is much smaller. The colour of the body is yellowish brown; the legs are grey, and the wings striped. The forewings are yellow with reddish stripes along the veins; the fore and side margins of the wing are especially reddish. On the wing margins there is a fringe of white hair which is black at the base. The hind wings are entirely yellowish except for the margin which is reddish.

The egg is oval, 0.4—0.5 mm long and 0.25 mm broad. It is pale green with an iridescent shell. Eggs are laid on the plant, in corners and folds and similar places.

The larva is 11—13 mm long. Its colour is apple-green and the head is black. Between the spiracles each body segment has a row of four dark setigerous dots. Behind this row there are two more dots which are arranged so that they form with the others apices of two equilateral triangles, one on each side of the segment.

The pupa is 10 mm long, greenish-purple, tapering at both ends and the legs reach the end of the abdomen. The pupa is usually found in a loosely woven silken cocoon.

Distribution. The insect is found whenever the host is grown. Records of its occurrence come from the following countries:

Africa: South Africa, Congo, Somalia, Uganda, Tanganyika, the Sudan and Egypt. Asia: India, Burma, Israel and Cyprus. Europe: Malta and Russia.

Hosts. The various varieties of sesame serve as hosts to this species. WILLCOCKS (1925) reports that the larvae of this insect were also found on the Snapdragon *Antirrhinum.*

Type of Injury. The larvae web together on the soft leaves upon which they feed. They bore into the young shoot and when flowers appear they feed upon these as well as on the young pods. They gnaw into larger pods and feed on the seeds. In a light infestation at Beit Dagan, Israel, 10% of the pods were injured.

No record of destroyed flowers could be obtainea. A 30% infestation was reported from South Africa.

Life History. Although the insect is widely distributed and its damage is reported in various countries, still our knowledge of the pest is scanty.

Casual breedings at Rehovot, Israel, showed that the insect may complete its entire development within 3—4 weeks under the conditions of the Israeli summer.

The egg development lasts 3 days at a temperature of 28° C. The larval stage lasts 12 days while the pupal stage lasts 4 days at this temperature.

Pupation takes place in the ground in a loose cocoon.

Observations in the field showed that the insect is found throughout the summer in Israel. Similar observations were made in Africa.

Pests of Castor Beans *(Ricinus communis)*

The question may be raised as to whether a discussion of *Ricinus* pests belongs within the frame of field crop pests. It is true that *Ricinus* is a shrub or a tree but in recent years, due to the exploitation of castor oil in new lines of manufacturing, and due to the high cost of manual labour in picking the beans, efforts were made by plant breeders and geneticists to develop varieties of castor plants which grow as annual field crops, and which are harvested and threshed mechanically. These efforts were very successful. The qualities incorporated into the annual varieties were: 1. dwarf growth; 2. high yield; 3. uniform maturation; 4. capsules which do not split in the field upon maturing. In view of this, the writer thought it justifiable to treat *Ricinus* as a field crop.

For the above-mentioned reason, Scale insect pests which exist on trees only were excluded from this discussion.

KEY to the Pests of *Ricinus*

A. Seeds may not sprout
  1. Maggots are present within the seed shells . .: *H. cilicrura* page 131

B. Young seedlings droop and wither
  1. Small yellow-brown beetles gnaw into the underground part of the stem . . . .: *Hermaeophaga* page 392

C. The leaves are injured
  I. By sucking insects
    1. Colonies of small red, spindle-shaped insects abound on the leaves . . . .: *Retithrips* page 78
    2. Leaves become chlorotic; usually this process begins between the "fingers" of the leaf
       Delicate webbing contains minute spiders . .: *Tetranychus* page 80
       Eutetranychus page 83
    3. Green, elongated, lively Homoptera abound on the leaves . . . .: *Empoasca* page 53
  II. Leaves are injured by caterpillars which either gnaw their undersurface, mine into them or eat them entirely
    1. The caterpillar is a brownish semi-looper with posterior abdominal legs larger than the anterior . . . .: *Ophiusa* page 393
    2. Caterpillars are olive green with black markings over the body . . . .: *Prodenia* page 102

3. Blister-like patches are found on the leaf; a delicate green caterpillar mines into it, or a cocoon may be found on the underside of the leaf . . . .: *Acrocercops* page 394

D. Inflorescence is injured
1. Silk webbing is noticeable between the flowers or green capsules
   a. a greenish caterpillar is present . . . . .: *Phycita* page 393
   b. a reddish caterpillar is present . . . . .: *Cryptoblabes* page 394
2. Seed capsules of various sizes dry out. Green Pentatomid bugs may be found which have been feeding . . . .: *Nezara* page 49

Hermaeophaga ruficollis Lucas and Crepidodera ventralis Kl.
(Chrysomelidae, Col.)

These two insects look very much alike, and in the field it is hard to distinguish between them. The short account given below may hold true for both.

Description. The adult beetles are 2.5—3.0 mm long and 1.5 mm broad. The colour is entirely straw-yellow or reddish-yellow except for the mandibles and distal segments of the antennae which are dark brown.

Distribution and Hosts. The two insects are distributed throughout the Mediterranean and reach the Caucasus. They may both be found on various trees although they do not necessarily feed upon them.

Type of Injury. A short time after the *Ricinus* plants have sprouted, large numbers of these beetles may be found in the ground in the space between the soil and the stem. Here they hide from the sun during the day and feed upon the succulent stem of the young seedlings. Their feeding pattern may take the form of gnawing the cortex or even gnawing deeper into the soft tissue.

When many beetles concentrate upon the stem it succumbs and withers. The author witnessed whole sections of rows in which the plant had been destroyed by these beetles. Bodenheimer (1930) mentions *Crepidodera* as feeding on the foliage of this plant.

Life History. Practically nothing is known about the life history of these beetles. They belong to the group of Chrysomelidae whose larvae feed on the roots of plants. Beetles collected in the field during the spring did not lay; possibly they spend the summer in a stage of sexual diapause, yet adults have been found on various trees throughout the summer.

Control. The beetles occur particularly in heavy soil where cracks offer them shelter. Dusting the soil around the plants with BHC or aldrin gave satisfactory control. The dusting should be made only along the rows of plants where the beetles frequent.

Ophiusa algira L.

[Gen. Syn. *Grammodes, Disgonia*]
(Phalaenidae, Lep.)

The larvae of this moth are often found feeding on the leaves of the castor bean plant.

Description. The adult is 18—20 mm long with a wing span of 45 mm. Its colour is brown. The coloration of the forewing is as follows: The basic angle is brown as is the disc; between them there is a wide cream-coloured band; the side marginal area is a lighter greyish-brown, and at the apex of the wing there are two dark brown spots, one small and one larger. The line of demarcation between the dark brown and light grey-brown area is a sharply defined broken line.

According to WILLCOCKS (1925) the larvae of this species may reach the size of 55 mm. Specimens in Israel were smaller. The colour is brown or greyish-brown. Typical of this so-called semi-looper are the abdominal legs which increase in size toward the posterior end. The first pair of legs is about 1 mm long with very small hooks, while the fourth is over 2 mm long with large conspicuous hooks.

Distribution. This moth is Eurosiberian, and in the Near East it is recorded from Egypt, Israel and Iraq.

Hosts. Larvae may feed on *Punica, Rubus* and *Ricinus.*

Phycita diaphana Stgr.

(Pyralidae, Lep.)

The larvae of this species are often found in the inflorescences of *Ricinus* and may cause considerable damage.

Description. The body of the adult is brown, 10 mm long, with a wingspan of 22 mm. The forewings are grey, mottled with brown dots; the anterior margin is slightly darker. The side border of the wing is decorated with seven black dots. The hind wings are translucent; their fringe of hair is shorter and sparser than that of the forewings.

The larva is about 15—18 mm long. Ventrally it is green; dorsally it is dark olive green, lined longitudinally with about 6—8 pale green lines. The head is brown as is also the prothoracic plate. On the occiput there are two black dots encircled with a yellowish broad border. On each segment between the spiracles there are black setigerous dots which form triangles, one on each side of the heart tube.

Young larvae are pale green with a black head and prothoracic plate.

Distribution. This species was recorded from Algeria, Egypt, Israel and Iraq.

Hosts. In all the above-mentioned countries it was found feeding on *Ricinus.* WILTSHIRE (1957) mentions *Populus euphratica* and *Chrozophora verbascifolia* as host plants.

Type of Injury. The injury is caused to the inflorescences by the larvae gnawing into the stem, feeding on the male flowers and biting into the seed capsules. An infested inflorescence may be detected by the silken webbing and dry plant parts between the single flowers.

The Honeydew Moth — Cryptoblabes gnidiella Mill.

(Pyralidae, Lep.)

This moth is polyphagous, occurring on cotton amongst many other plants. The moth is attracted to the honey dew secreted by bugs and to plant exudate. The female lays her eggs where she feeds and the larvae develop on the plant. This moth is considered a pest of carob and citrus, but it is of little economic importance to cotton and *Ricinus.*

Acrocercops conflua Meyr.

(Gracillariidae, Lep.)

On the *Ricinus* leaves there may be small, pale patches which are the result of a small mining caterpillar, *Acrocercops.*

Description. The adult is a minute moth, 3 mm long with a wing-span of about 9 mm. Its antennae have brown and white alternating rings. The forewings are pale brown mottled with markings of darker and lighter shades of brown; in the fore margin the darker and in the hind margin the lighter spots dominate. The ventral part of the abdomen is light grey and four pairs of oblique lines run from the pleurites to the sternites. The legs are light grey with oblique maculae, which give them, esspecially the tibia, an external ringed appearance. The hind wings are grey.

The larva is a glistening pale green, about 6 mm long when mature, with very distinct segmentation. The head is brown and the first thoracic segment is yellowish with two dark markings.

The cocoon is loose, about 8 mm long and is usually stuck to the underside of the leaf.

The pupa is bright green with antennae which project beyond the tip of the abdomen.

Distribution. This insect has been recorded from Somalia, Egypt, Israel and Iraq.

Host and Injury. Apparently *Ricinus* is the only host of this species, and the damage consists of its mining into the leaves.

According to WILLCOCKS (1925), the mines may be so numerous that they cause injury to the plant, especially when the plant is small and has only a few leaves.

Life History. No data are available about the life history of this species except that it lays its eggs on the underside of the leaf, and the newly hatched larva bores into the leaf and feeds on the parenchyma. The mine is narrow and sinuous at first and then it becomes enlarged into a blister-like pocket. The spinning of the cocoon usually takes place on the underside of the leaf. The insect may be found throughout the year but becomes more numerous towards the end of the summer.

Control. As a rule no control is needed except when infestation is very severe at which time an application of BHC emulsion, as in the case of the wheat miner, will give satisfactory results.

LITERATURE

ANON., 1926. Farming in South Africa **I**, 7, *223*.
F. S. BODENHEIMER, 1930. Die Schädlingsfauna Palästinas. P. Parey, Berlin. 438 pp.
E. RIVNAY, 1961. Unpublished Notes on *Antigastra*.
F. C. WILLCOCKS, 1925. Insects and related pests of Egypt. 418 pp.
E. P. WILTSHIRE, 1957. The Lepidoptera of Iraq. Ministry of Agriculture, Govt. of Iraq, 162 pp.

The Pests of Flax *(Linum usitatissimum)*

In the Near East flax is attacked by a few insects only. *H. cilicrura* (page 131) may destroy the seed before it sprouts. *Agrotis* caterpillars (page 88) and *Gryllotalpa* (page 44) may gnaw at the stem close to the ground. Young plants may be destroyed by *Aphthona euphorbiae*, a short account of which is given below, and by *Longitarsus parvulus* PAYK. which is similar to it in its habits. *Laphygma* larvae may feed upon the foliage, while *Heliothis* larvae may feed upon the soft parts of the plant.

Aphthona euphorbiae Schr. *

(Halticinae, Chrysomelidae, Col.)

This beetle together with other species of this family may attack flax in its early stages and exterminate it.

Description. This pest is a small Halticid, 1.5—2 mm long, greenish-black, except for the anterior and middle femora which are yellowish. The first segment of the antennae is distinctly shorter than the distance between the antennae. The pronotum is punctate and without a transverse furrow; the elytra are punctate, without lines.

Distribution. This pest is found throughout Europe, Siberia and the Mediterranean region. There are records of its occurrence from Syria, Cyprus and North Africa.

Hosts. The insect is polyphagous, and has been recorded from plants of the family Euphorbiaceae, from beets, mustard and *Linum.*

Type of Injury. Young *Linum* plants are attacked by the adults which may destroy them entirely, necessitating a reseeding of the field.

Life History. The emergence of the adults from winter quarters and their mating takes place during May. The female enters the ground near the plant and lays her eggs 1—2 cm deep. The maximum number of eggs observed was over 250 per female.

The incubation period lasts 11—22 days. The young larva burrows into the rootlets of the host plants and feeds upon them. The larval development lasts about 30 days after which it enters the ground to pupate in a little cell which it builds. About 9 days later,

* Based on the paper by KURDJUMOV-quoted from BALACHOWSKY & MESNIL (1936).

sometime in the middle of July, the adult emerges and feeds upon the blossom of the *Linum* plants.

In August the beetles enter the state of diapause before raising another generation that year.

Control. As a prophylactic measure early planting is recommended. By the time the beetles emerge the plants are mature and strong enough to withstand attack.

The dusting of the plants with a stomach insecticide will yield satisfactory control.

The pests of Cotton *(Gossypium)*

KEY to the Insects injurious to Cotton

In addition to the pests enumerated below *Cryptoblabes gnidiella* and *Plutella maculipennis* were found on cotton in Israel — but these are only casual visitors.

A. Seeds do not sprout
   1. Maggots feed upon the seeds and may be found within the shells . . . . : *H. cilicrura* page 131

B. Young plants are injured

   I. A short time after sprouting, seedlings may droop and wither
   1. Maggots are seen boring in the stem . . . . : *H. cilicrura* page 131
   2. Caterpillars are found in the soil . . . . . . : *Agrotis* page 88

   II. Leaves are injured
   1. The growth of the plant is stunted, leaves are curled and folded and minute and elongated insects are present between the folds and in the depressions of the leaves . . . . : *Thrips* page 74
   2. Small soft-bodied green insects are found on the leaves . . . . : Aphids page 60
   3. Leaves become chlorotic; on the underside of the leaves delicate webbing may be detected with minute red spiders . . . . : *Tetranychus* page 80

   III. Leaves are gnawed by caterpillars
   1. Green striped caterpillars feed on the foliage . : *Laphygma* page 113
   2. Olive green caterpillars with dark brown spots feed on the foliage . . . . : *Prodenia* page 102

   IV. The stem is injured, the top of the plant droops and withers. Traces of boring larvae and their faeces may be found between leaflets
   1. Boring larva is white with a pink thorax . . : *Platyedra* sp. page 413
   2. Boring larva is brown with yellowish spots; soft spines cover its body . . . . : *Earias* page 402

C. Fruiting plants are attacked

   I. The leaves are eaten
   1. by green striped caterpillars . . . . : *Laphygma* page 113
   2. by olive green larvae with black spots on the body . . . . : *Prodenia* page 102

   II. The leaves are injured by sucking pests
   1. Lively green elongated insects move over the leaves . . . . : *Empoasca* page 53
   2. Small soft-bodied insects feed on the leaves, secreting honey dew; plants are sticky . . . : Aphids page 60
   3. Minute white flies hover over the plant when disturbed; their larvae are flat, greenish, scale-like creatures . . . . : *Bemisia* page 56

4. Leaves become chlorotic, usually the infestation begins between the "fingers" of the leaf and spreads gradually . . . .: *Tetranychus* page 50
5. Large green bugs feed on the plant . . . .: *Nezara* page 49
6. Small greyish bugs may be found around flowers or soft fruit . . . .: *Oxycarenus* page 399

III. Fruiting bodies are injured

1. The bud or soft boll is entirely or partly consumed, older bolls have a hole in which a caterpillar may be found
   a. The caterpillar is striped in two shades of green . . . .: *Heliothis* page 119
   b. The caterpillar is olive green with dark spots on the body . . . .: *Prodenia* page 102
   c. The caterpillar is brown with yellowish spots. Soft spines cover the body . . . .: *Earias* page 402
2. On the surface of the boll there are pink dots or a hole 2 mm wide with the epidermis intact; inside the seed a pink worm is found . . . .: *Platyedra* page 413
3. The boll is open, small greyish bugs may be found in the lint or around the seeds . . . .: *Oxycarenus* page 399

The Cotton Seed Bug — Oxycarenus hyalinipennis Costa
(Lygaeidae, Hem.)

There is a dispute as to the economic importance of *O. hyalinipennis* to cotton. Some writers claim that it causes a good deal of damage, while others think of it as "scarcely a pest". The discussion below is based on the paper by KIRKPATRICK (1923).

Description. The male is 3.8 mm long, the female 4.5 mm. The head is blackish, punctate and clothed with white hair. Of the antennae the first joint is black with a yellow spot at the end, the three

Fig. 124. *Oxycarenus hyalinipennis* – adult. After SILVESTRI.

remaining joints are black, only the second is light brown in the middle. The prothorax and scutellum are punctate, black and the prothorax is slightly brown at both ends. The legs are shiny brown except for the femora and part of the tibia which are black. The fore wings are hyaline, and the triangular base slightly brownish with a small brown spot at its apex. The underside of the body shows many colours of which brown, black and white are dominant.

The egg is oval, pointed at the posterior end, 0.95 mm long and 0.30 mm wide. The shell has a longitudinal corrugation which gives it a striated appearance. It is a pale straw colour, which later becomes orange. Upon hatching the shell splits.

Distribution. This species of *Oxycarenus* is chiefly African. Records of its distribution come from Egypt, the Sudan, Somaliland, Uganda, Tanganyika, Nyasaland and Angola. It is distributed also in Israel, probably Jordan too, and it has spread also to Brazil (KIRKPATRICK, 1923).

Hosts. Although this insect may be found on various plants, it feeds only on the seeds of Malvaceous hosts. Species of the following genera serve as food: *Gossypium, Hibiscus, Malva, Althaea, Pavonia, Sida, Abutilon* and others.

Type of Injury. Although the bug may be seen feeding on green plant parts, KIRKPATRICK (1923) claims that this is a mere casual intake of moisture and that real feeding takes place on the ripe seed only. The bugs do not damage the cotton lint except that during the ginning process some may be crushed and thus stain the lint. The damage to seed is caused by the lowering of the germination power and by the reduction of the seed weight. A reduction of as much as 15% is recorded.

Life History. During the spring, after feeding on the seeds, copulation takes place and egg laying begins. The eggs are laid as a rule in the lint of the open boll, or in crevices and holes in the fruit. A female may complete its oviposition in one day, and the average number of eggs per female is about 20 (the maximum number of eggs laid was 26). The incubation period lasts 4 days at a temperature of 30—35° C, 12 days at 21° C, and 43 at 14.6° C.

The larva moults five times before reaching adulthood. The length of each moult at two temperatures is shown in Table LVII.

It is noticeable that the last stage is the longest at both temperatures, while the fourth stage is shortest at the high temperature and the second to the longest at the low temperature. In addition it is evident from this table that the total larval development lasts 19 and 31 days at the higher and lower temperatures, respectively; this means that the summer generation lasts about three weeks and the autumn generation about 6 weeks (from egg to egg). According to KIRKPATRICK, the bug raises 3—4 generations and then enters a sexual diapause.

Table LVII.

Length of development of various stages of *Oxycarenus hyalinipennis* (after KIRKPATRICK, 1923).

Temp. in C° \ Larval stages	1	2	3	4	5	Total
25—32	5	3	3	2	6	19 days
23	5	4	4	8	10	31 days

Diapause. The nature of this diapause is somewhat different from that of other insects discussed in this book. As a rule it begins in the autumn, but some individuals may enter diapause as early as August. The breaking of the diapause usually occurs in April, when the first malvaceous seeds ripen, but some bugs may remain until late summer before they begin to feed on seeds and reproduce. During the winter, in warm days, the bugs may move about and feed on moisture and nectar. Diapause may be broken in the middle when temperature and humidity are raised and seed is available as food.

Control

Preventive Measures. Picking the bolls soon after they open may greatly reduce the infestation.

Chemical Control. There is no doubt now that the various contact insecticides which are employed in the control of Lepidopterous larvae keep this pest under control, too.

The Hibiscus Mealy Bug — Maconellicoccus hirsutus Green

[Syn.: *Phenacoccus hirsutus* GREEN]
(Coccidae, Rhynchota)

In Egypt, when cotton is grown not far from infested plants, it may be subjected to attack by the hibiscus mealy bug. In India this species, together with other mealy bugs, damages cotton. The following discussion is based on the paper by HALL (1921).

Description. The adult female is 2—3 mm long and 0.9—2 mm broad, ovate, and sparsely covered with mealy wax through which the reddish colour of the body shows. Marginal appendages are absent, while caudal appendages are either very minute or absent. The antennae are long and nine-jointed. The legs are longer than the antennae, with simple tarsi and a single claw.

Distribution. This insect is apparently of South-eastern Asian origin. It is recorded from India and Pakistan, Burma, Malaya,

Indonesia, the Philippine Islands, Formosa and New Guinea. In Africa it has been recorded from Zanzibar, Sudan and Egypt.

Hosts. This species is polyphagous; many shrubs and trees serve as hosts. As the primary host, plants of Malvaceae, notably *Hibiscus* and *Gossypium*, rank first. Several Leguminous trees are attacked, including species of the genera *Acacia, Robinia, Cajanus, Ceratonia, Erythrina* and others.

Type of Injury. The hibiscus mealy bug usually attacks young shoots and causes leaf curl or, in the case of heavy attack, stunts the growth of the plant.

Life History. The length of life of the female from hatching to the adult stage is five weeks. When the female has matured and is ready to reproduce, she looks for a suitable place to settle and lay; usually this is in the company of other laying females.

The eggs are laid within an ovisac, which increases in size and content as the female decreases and finally dies. The egg laying lasts about a week, and the number of eggs laid varies from 150—300 eggs per ovisac. The incubation period lasts 6—9 days under conditions of the Egyptian summer. Early in the spring the number of males is very small; it increases towards the end of the summer but even then they are far less numerous than the females. During the winter the egg stage lasts a long time and hatching takes place towards March.

Control. HALL points out that the infestation of cotton takes place only where infested plants are nearby. Such sources of infection should be removed. The phosphorus insecticides which are employed against other cotton pests will probably kill this mealy bug as well.

The Spiny Bollworm — Earias insulana Boisd.

(Phalaenidae (Noctuidae), Lep.)

Until recently cotton was not grown in Israel although conditions of soil and climate were favourable. The main limiting factor was the spiny boll worm *E. insulana*. The ravages of this pest prevented the Arabs from growing cotton, and it also undermined the recent efforts early in this century to establish this crop in this country. Only when adequate methods of control were found, was it possible to raise cotton in Israel.

E. insulana is one of a group of species belonging to this genus which attacks cotton in various countries. Thus *E. fabia* is known to damage cotton in India, while *E. biplaga* is found in South Africa. *E. insulana* is found in many countries including the Mediterranean countries, so this discussion will mainly involve this species.

Description. The length of the moth is about 12 mm and the wing span from 20—22 mm. The scales on the body and wings are small and dense. Their colour is as follows: The abdomen and hind wings greyish-white or grey, the thorax and forewings applegreen. In the winter the moths may be pale-green or even straw-yellow. Three undulating lines cross the forewings. Often a dark-blue or brown-blue spot is located on the disc of the forewings. This spot varies very much with the individual and is not influenced by external factors while the general colour — green or yellow — may be due to the season and temperature (YATHOM, 1956).

The egg is spherical and slightly flat, 0.6 mm in diameter. The colour is pale bluish-green and small grooves and ridges run from the top centre to the periphery. The ridges extend slightly beyond the surface forming a minute crown-like structure; the whole egg looks like a tiny pomegranate.

The larva is 15—18 mm long, broad in the middle and tapering slightly in both directions. On each body segment, except the prothoracic, there are two pairs of setigerous papillae. The colour of the body is brown or greyish-brown, while the first, third and fourth, sixth and seventh abdominal segments are yellowish on the top. There are yellowish spots at the bases of the papillae and these

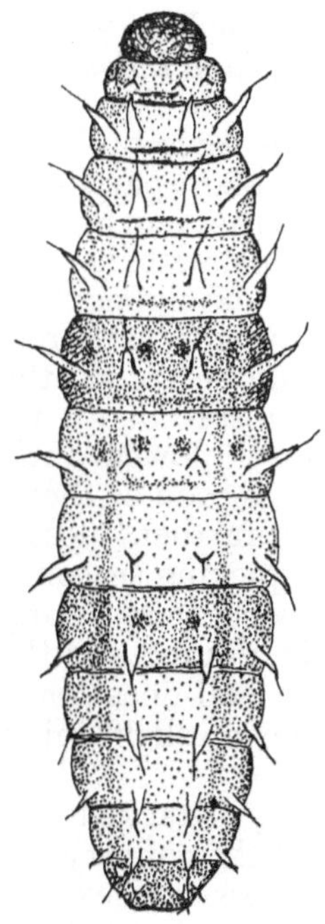

Fig. 125. *Earias insulana* – larva.

are more conspicuous on the thorax. In addition to the setigerous papillae there are other setae scattered on the body.

The pupa is chocolate coloured and both ends are rounded. The body bears neither setae nor papillae except for three processes at the last abdominal segments. The length is 13 mm. It is found in a cocoon of dirty white colour.

Distribution. The genus *Earias* is distributed in Africa, Asia, Australia and Southern Europe. *E. insulana* is the most widely distributed. It occurs in Africa, both North and South, and East and West and it penetrates into the Mediterranean Islands, the Balkans and Spain; in Asia it is recorded from Turkey, Syria, Jordan, Israel, Iraq, Iran and Arabia, in the west, in India and Indochina in the south, and the Philippines and Japan in the east.

However, damage is not equal in every country. It depends upon various factors as will be discussed below.

Hosts. As a rule, Malvaceous plants serve as hosts to *Earias insulana*. There are some records of its occurrence on other kinds of plants. Thus MASON (1915) records mulberry infested with *Earias insulana* in Nyassaland. VAYSSIERE & MIMEUR (1925) state that they found larvae on carob and maize in West Africa. In Israel, too, *Earias* was found on maize, but this was an exceptional case.

The Malvaceous genera, which serve as food to *Earias insulana* are *Althaea*, *Abutilon*, *Gossypium*, *Hibiscus*, *Malva*, *Malvastrum*, and *Sidon*.

Type of Injury. The larva feeds by boring into the plant tissue. As a rule soft and succulent plant parts are chosen. When the plant is young the larva bores into the stem from the top, and consequently the upper part of the plant withers. BALLARD (1921) does not consider this damage serious; he claims that the plant becomes more ramified than otherwise, and the yield remains normal in spite of this injury. Naturally it may be argued that this early population, if left uncontrolled, may result in an increase in the following generations which later attack flowers and bolls.

The flowers are attacked as soon as they appear. As a result they drop, and this stimulates further blossoming which ultimately causes an abnormally long picking season.

Bolls may be attacked when very young. They may be entirely eaten by a larva, or may be injured and thus drop prematurely. At a more advanced stage the boll does not drop, but it may dry or remain closed as a result of even slight injury, or the contents of the boll may be destroyed.

Not every field is equally attacked. The extent of infestation varies with the locality and the country.

It is customary to express the extent of infestation by the percentage of injured susceptible parts of plants such as buds, flowers, and bolls.

In Egypt it was found that the infestation fluctuates between 1—6%; at the end of the season the percentage may be higher. In the Sudan the percentage of infestation is even lower (1—5%) (PEARSON, 1958). A higher infestation is recorded in India. In that country the percentage is low at the beginning, in August, and higher at the end of the season, in October, when 40—60% of susceptible parts are infested. In Israel the infestation is low at the beginning of the season, in June; in July and early August it is even lower. Toward the end of August there is a sudden rise (RIVNAY & YATHOM, 1958). Before this rise the infestation may fluctuate from 4 to 12% but afterwards the infestation may range from 60 to 80% and often, when no control measures are taken, the entire crop may be exterminated.

PEARSON criticizes the customary method of expressing the extent of infestation. He claims that an impression is formed that the infestation is higher at the beginning of the season and at its end. He believes the reason to be a misjudgement of the real situation. Early in the season the dry plant tops and later in the season the infested large bolls are quite conspicuous and noticeable, while in the middle of the season the susceptible parts are very numerous, and many of the flowers and young bolls drop unnoticed and unrecorded. In Israel, too, for the purpose of a quick assessment, in order to determine the need of control applications, this method is used; but for studies of fluctuations of populations a method was introduced where the number of caterpillars is counted in a given area or on a given number of plants (RIVNAY & YATHOM, 1958). This method eliminates the erroneous picture whereby the percentage of infestation went up at the end of the season because the number of susceptible points is reduced.

In fig. 126 the fluctuations in the population of *Earias* larvae in Israeli fields is shown by both methods. It is distinctly evident that the difference is greatest at the end of the season when there are few available parts to count. This figure also shows the depression in the population in the middle of the season.

Life History. The adults are active at night. During the day they hide under the leaves and remain motionless until dusk. Eggs are laid singly. As a rule folds and cracks are chosen for this purpose, e.g. between the soft leaves, between petioles and between fruit and sepals. Usually one or two eggs are found in one site, but when the flowers are large more eggs are laid in one place.

The incubation period lasts about 3 days under the conditions of the Israeli summer. Similar records were obtained by VAYSSIERE & MIMEUR, (1925) in West Africa, and by HUSAIN & LAL (1920—'23) in India where eggs of *E. fabia* were under observation. At the end of the summer the incubation period in West Africa lasts 19—22 days and 6 days in Israel at a temperature of 19° C (YATHOM,

1956). Upon hatching the neonate larva may bore directly into the tissue where the egg was laid, or it may wander about before beginning to feed and bore. As mentioned before, if no flowers or

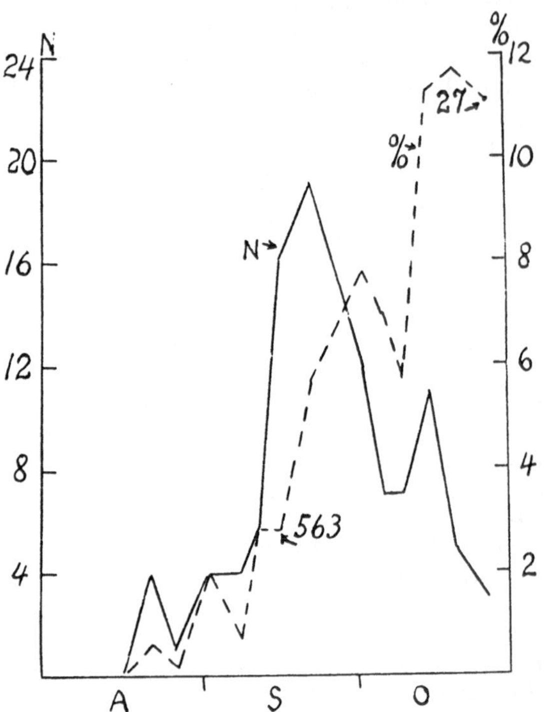

Fig. 126. *Earias* infestation during Aug.-Oct. in a cotton field at Beit Dagan during 1959, showing two methods of assessing infestation. Scale and respective line N indicate the number of *Earias* larvae found on 30 plants picked at random. Broken line shows the same numbers, expressed in percentage of infested plant parts susceptible to attack (buds, flowers, bolls etc.). Figures indicate the number of such parts on the respective dates. Note the ascent of the percentage line as the number of parts diminishes while the number of larvae is low.
After RIVNAY & YATHOM.

fruit are available it bores into the stem from above. When it finishes one square or boll it may crawl out of it, in search of another. If it drops to the ground with the empty fruit, it crawls back to the plant to attack new bolls. In this case it is usual that nearly mature bolls may be attacked; in such bolls the larva reaches and feeds on the seeds. One larva may thus feed and destroy several bolls.

The larval period is quite short. In India this stage of *E. fabia* lasts 10—16 days during the summer or 50—60 days in the winter. That of *E. insulana* in West Africa lasts 15—28 days. The records obtained by YATHOM in Israel are given in Table LVIII.

Table LVIII.

The length of development (in days) of the spiny boll worm at various degrees of temperature.

Average and range (in brackets) After YATHOM (1956)

Temp. in °C \ Stage	Egg	Larva	Pupa	Preoviposition period
29	3	—	—	—
28	—	9.1 (6—13)	8.7 (6—12)	2.7 (1— 3)
27	—	—	—	4.6 (4— 6)
26	3	10.7 (7—18)	10.2 (7—16)	8.6 (7—12)
25	4			
22	5			
21	—	18.8 (13—28)	19.8 (12—29)	
19	6			

According to these figures the threshold of development of the larva is about 14.5, which is about the same as given by HUSAIN & LAL (1920—'23) for *E. fabia.* The larva moults five times during its development and these changes take place in the tunnel or the cavity where it feeds. For pupation the caterpillar leaves its feeding place, finds a sheltered spot, such as between leaves, in corners or on the ground, and spins a cocoon, The cocoon is densely woven and its texture is like dirty white felt.

The pupal period for *E. fabia* in India lasts 4—9 days in the summer and 23—37 days in the winter. That for *E. insulana* in West Africa is 10—52 days (VAYSSIERE & MIMEUR). This period in Israel as recorded by YATHOM, is given in Table LVIII. From these records the threshold of development was found to be 15.5° C. The preoviposition period at 28° C is 1—3 days, at 27° C 4—6 days, while at 26.5° 7—12 days. However, it is not unusual that over two weeks may pass before laying begins at a temperature of 27° C.

As with other insects, the number of eggs laid by one female depends upon the temperature and upon the food of the individual during its larval period. In Table LIX the records of oviposition by several females are given at various degrees of temperature, as recorded in Israel. These data show that at a temperature below 20° C the percentage of laying females is quite small and the average number of eggs laid quite low. At 26—28° C the percentage of laying females is the largest and the average number of eggs also the highest. As the temperature rises beyond this point the number of laying females diminishes and average egg laying decreases. Observations close to these were made in India (AHMAD & GHULAMULLAH, 1941). The females from which the records in Table LIX were obtained were reared on hibiscus flowers. The highest egg production by one female was 291. A comparative study on the effects of food upon the egg production was made in Egypt (DUDGEON, 1916). Moths which were reared in their larval period on cotton stems laid 88 eggs. Those reared on cotton buds laid 196 eggs, those reared on cotton bolls 342 eggs while those reared on okra pods laid 399—451 eggs.

As a rule most of the eggs are laid within the first week of the life of the female; weak females have a prolonged laying period.

The length of life of *E. fabia* in India is seven days in the summer, that of *E. insulana* is 28 days in the winter. In Israel, at a temperature of 28—29° C, 50% of the moths died within 10 days; at 26° C 50% died within 19 days, while at 14° C 50% died within 39 days. The longest life span at this temperature was 70 days. (YATHOM).

Table LIX.

Fertility of the Spiny Boll Worm at various degrees of Temperature. After YATHOM (1956).

Date of rearing	Average monthly temp. °C	Number of females	Percentage of laying females	Average number of eggs
8. 1952	29	15	16.5	53
9. 1952	28	8	50	109
8. 1953	28	45	51.0	136
10. 1952	26	12	50	137
9. 1953	26	35	51.3	144
11. 1953	19	40	7.5	39
12. 1952	20	45	0	0
1. 1954	14	8	0	0

Factors affecting density of population. In considering the population density of this insect which has such a wide geographic distribution, a distinction must be made between the typical population of a certain geographic area and the seasonal fluctuations of the population in a given area.

As stated earlier, although *E. insulana* is widely distributed, its damage may be negligible in one area but very severe in another area of similar climate. PEARSON (1958), who collected the available data on cotton pests in Central Africa, enumerates five causes which determine the population of this pest in a given area. They are as follows: *a.* The length of the life cycle; *b.* The situation of host plants other than cotton; *c.* Period of cotton growing; *d.* Rainfall; *e.* Natural enemies.

a. The outstanding characteristic of the life history of *E. insulana* is that it has no true diapause. In other words the insect may continue to develop as long as food is available and temperature permits. In countries where the winter is cold (but not cold enough to kill the pest) the development rate is only reduced. At favourable temperatures the insect completes its development in 25 days and can raise a few generations during the cotton period, or even possibly over a dozen generations during the whole year as in the Sudan.

b. Hosts. Since cotton is limited to a certain period of the year, the continual existence of this pest depends upon other food plants. The bridging over from one cotton season to the other depends upon the availability and abundance of host plants during the so called "cotton free" period. An abundance of such plants in this interim period will carry over a strong population into the cotton season. But should these host plants be also abundant during the cotton period, they may compete with cotton for the pest. Plants such as *Hibiscus* and *Abutilon* attract *Earias* moths more than *Gossypium,* and thus may relieve the cotton fields of severe attacks.

c. The Cotton Season. The cotton sowing season is an important factor in controlling the population of the pest in the cotton field. Should the cotton be sown before the end of the intermediate host period, then the invading moths are more numerous than when a longer host free period is established. If cotton is sown during the winter the chances of the pest to establish a strong population are less than if the cotton period is in summer.

An earlier variety of cotton will offer less opportunity for a population build-up than a variety with a prolonged growing and harvest period. Of primary importance is volunteer cotton. The role of this crop is similar to that of the intermediate hosts in offering food to the pest during a host-free period.

PEARSON compared the cotton growing season in five countries and the number of generations which the pest raises in each. Three of these will be given as examples.

In Egypt the cotton season is from May to the end of September. During these five months *E. insulana* may raise six generations: three more generations can develop on hosts other than cotton during the intermediate period.

In Uganda the cotton period is from August to the end of January. During these six months the pest may raise five generations; three additional generations may develop on hosts other than cotton during the cotton-free period. In both of these countries the spiny boll worm is injurious.

In the Sudan (Gezireh) on the other hand, cotton is grown from October to the end of January. During these four months the moth may raise five generations; eight more generations could develop during the intermediate season but hosts are not available (there is a law prohibiting growing of host plants during this period). In this part of the Sudan the spiny boll worm is not a serious pest.

d. Rain. This factor may influence the *Earias* population in two ways. Rains encourage the growth of hosts in the intermediate season, or prolong the hosts' season. On the other hand, it seems that *E. insulana* thrives best in more arid places and seasons, and rain may directly injure it.

e. Natural Enemies. From the references on this subject an impression is formed that parasites play a small role in regulating the spiny boll worm population. In Egypt *Bracon brevicornis* parasitizes *Earias*, and in India *B. lefroyi*. Students in India point out that the climatic changes in India prevent the parasites from building up a denser population.

Still, it is evident that natural enemies at large apparently have some influence in regulating the population of the pest. In places where insecticidal applications were made against other pests, the *Earias* problem has become aggravated. It has been suggested that bugs, spiders, beetles, lacewings, etc. apparently play a greater role in keeping the pest in balance than it was originally thought. At any rate further studies on this subject are essential.

PEARSON tries further to point out how the various factors enumerated above may be responsible for the *Earias* conditions in the different countries.

In Egypt, as mentioned above, the cotton season begins in May. Formerly long season varieties were grown and the season extended till December. It was also the practice to leave ratoon crops, and volunteer plants were given little attention. Two factors created a problematic situation as regards *Earias*, namely: A long cotton season and intermediate hosts. When short season varieties replaced the old, ratoon crop practice was eliminated, and volunteer plants cleaned, the situation improved a great deal.

Similar conditions exist in the Northern Sudan, Nigeria, Chad and Tanganyika where hosts other than cotton bridge the intermediate period carrying over populations from one season to the next.

The situation in the Gezireh district in the Sudan is entirely different. There, cotton is grown in the winter when conditions are unfavourable for the pest. During the favourable intermediate months, when the pest could successfully develop, there are no hosts available. Wild plants do not grow and cultivated plants are prohibited. Thus the pest cannot build up a dense population. Wild *Abutilon* plants are sparsely scattered over the country; they are infested with *Earias* and other larvae, a factor which has also helped to build up a parasite population. Also in the savannah, north and south of the Equator, *Earias* is no serious problem in the cotton industry. Cotton is grown, there in the rainy season when other hosts are also available. In the prolonged, dry season when there are no hosts at all, conditions prevail which do not favour the development of the pest.

In Uganda there are both limiting and favourable factors. On the one hand the cotton season is very prolonged, 9 to 10 months; this encourages the build-up of a pest population and also creates a short intermediate period in which the pest passes from one cotton field to the other without the need of other hosts to bridge the two seasons.

On the other hand, many other hosts grow during the same period; some of these are species of *Hibiscus* and *Abutilon* which are more attractive to *Earias* than is *Gossypium*. On these plants there are also other Lepidopterous larvae which, as mentioned before, encourage parasitism.

Conditions in Israel

In Israel, as in Egypt, the cotton season is in the summer, from May to October, 6 months, when climatic conditions favour the occurrence of *Earias*. The conditions during the intermediate period vary from year to year and accordingly they effect the situation of the pest in this season as follows: A very cold and prolonged winter reduces the *Earias* population regardless of the availability of host plants, while a warm spring-like winter stimulates the development and reproduction of the pest. However, the fate of the cotton fields depends more upon conditions at the end of the winter. Rains usually cease by the end of March and early April. If dry desert winds occur which dry the flora earlier than usual, then there is a smaller chance of carrying over a heavy population to the cotton fields. The dry desert winds have also a direct detrimental effect upon the insect. On the other hand, if rains continue to a late date they prolong the occurrence of intermediate hosts upon which a rich spring generation develops to invade the new cotton fields with greater force.

From the study by Miss YATHOM (1956) it is evident that when larvae were reared at 29° C the percentage of laying females decreased while the number of eggs laid by one female was also far less than these reared at 26° C. At a temperature higher than 29° C the ill effects would surely be more severe.

Therefore, when the average temperature rises to these levels, the mortality of the adults increases, their reproduction diminishes and the population dwindles down. This factor may account for the depression in the population which occurs so frequently in Israel during the months July—August. Thus, if we follow the course of the density of the population in Israel, we find it has two peaks and two depressions. The spring peak, end May—June, and the autumn peak – September early October; the winter depression November—April and the summer depression during July—August. There are six generations in Israel as follows: the first spring generation, on intermediate hosts, from February to April, 4 summer generations on cotton, and the last, winter generations, on intermediate hosts from October to February.

Translocation. It has been pointed out on several occasions that *Earias* moths move from one host in one season to another host in another season. No mention is ever made of the distance between fields which permits invasion by moths from another host. The impression is that the movements of moths may take place over long distances, especially when changes in climate may also stimulate such displacements.

Methods of Control

Agrotechnical Methods. In order to control *Earias* it is essential to reduce as far as possible the factors which encourage its reproduction. It was stated that (in some places) the moth may raise a few generations during the cotton free period if other hosts are available. Therefore, the destruction of secondary hosts, where this is possible, is of primary importance. In Egypt, ratoon cotton was abolished as a measure against the pink boll worm, and thereby the spiny boll worm situation was also improved. In Iraq the cotton farmers are forced to clean their fields of cotton remains as well as volunteer crops. HUSAIN (1925) in India also recommends such measures. In the Gezireh region of the Sudan, the growing of *Hibiscus esculentus* and other hosts is prohibited during the intermediate host-free period. Short season varieties of cotton give the pest less chance to build up dense populations.

Trap Crops. Some writers suggest the planting of trap crops between the cotton rows in order to attract the spiny boll worm moth. KING (1918), for instance, proposes *Abutilon* which attracts *Earias* more than cotton. ZANON (1921) mentions *Hibiscus esculentus* as a good trap crop, while FLETCHER (1921) believes that *Hibiscus abelmaschus* is still better. Naturally, in the case of *H. esculentus*, the edible fruit is an additional asset to the farmer.

Chemical Measures. In various countries it was, and still is customary to employ stomach insecticides, arsenates or silicate compounds, against the spiny boll worm. Of the synthetic insecticides DDT, toxaphene and BHC, were, and still are, the most popular substances against this pest. Some of the insecticides, for instance BHC and DDT are mixed together with sulphur to form the so called "cotton dusts". However, as PEARSON mentions, the application of these insecticides encourages the increase of spider mites. In one large scale field experiment PEARSON even mentions an increase of *Earias* as a result of DDT applications. In places of heavy infestation of the spiny boll worm the above mentioned insecticides gave no control of the pest. In Iraq and in Israel not one of these three proved to be of any use. WALKER (1954) in Iraq obtained very satisfactory control with endrin and to a lesser extent with dieldrin. In comparative tests in the laboratory RIVNAY & YATHOM (1956) showed that of the eight insecticides used, DDT and perthane were the least effective against the adult of *Earias* while endrin and parathion were the most efficient. Lindane and diazinone were also effective but the duration of their residual action was too short. In field trials the same authors (1957) showed that endrin at various rates of application was more effective than toxaphene or cryocide. Later trials by these writers showed that gusathion and sevin are also effective. A comparison of the effectiveness of these latter against *Earias* is shown in fig. 127. Other suggestions are given in the chapter on general considerations on control measures.

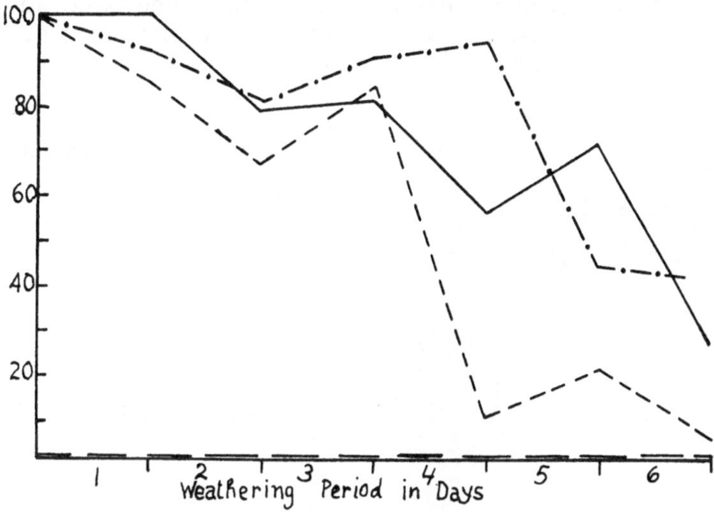

Fig. 127. Comparative effects of three insecticides and the duration of their action on the spiny bollworm. The plants were treated outdoors with a 0.2% emulsion or suspension; after various periods of weathering they were brought into the laboratory and fed to the caterpillars. The curves indicate percentage of their survival 24 hours after contact with treated food.
Straight line = zectran;* broken line = gusathion; line and dot = sevin; level line = control.
After RIVNAY & MEISNER.

The Pink Boll Worm — Platyedra gossypiella Saund

[Syn.: *Pectinophora gossypiella* SAUND]
(Gelechiidae, Lep.)

The pink boll worm is one of the most notorious pests of cotton. The size of the moth, its frail structure and obscure colours are in complete contradiction to its role as a pest. In many cotton growing countries it is the menace of the cotton industry. Due to its habit of diapausing in the cotton seed, it has spread and reached countries far remote from its origins. Many quarantine regulations have been

* Trademark of the Dow Chem. Co.

promulgated against it and yet it is found wherever cotton is grown and the climate favours its development.

Description. The adult is a moth 8—9 mm long with a wing span of 15—20 mm. The colour is greyish-brown; two blackish spots decorate each forewing. The apex of the forewing is darkened and a black line runs along its margin. The hind wings are silvery, broader than the forewings and are extended sidewards into a typical finger-like projection. The labium is long and bent upward like a scythe. It is brown and two black spots decorate the second segment of the palpus; the last segment has two black rings, one basal and one medial. The antennae are serrate, and on the first segment there are five large setae directed forwards.

Fig. 128a. *Platyedra* – adult after HUNTER, U.S.D.A. Bull. 723.

Fig. 128b. *Platyedra* larva. after HUNTER, U.S.D.A. Bull. 723.

The egg is elliptical, 0.6 mm long, 0.4 mm broad, with delicate folds over its surface. Its colour is iridescent green, and before hatching it changes to pink.

The mature larva is 11—13 mm long. It is cylindrical, white below and purple above. The head is almost spherical, broader than long, reddish-brown, with blackish-brown mandibles. The prothoracic plate is small, brown and divided. The plate on the last abdominal segment is also brown. The body is covered with small setigerous papillae which are pink and yellowish-brown in the middle. The abdominal legs bear 15—17 hooks.

The pupa is 8—10 mm long and reddish-brown. The end of the abdomen tapers into a short spine bent upward. The body is velvety with neither setae nor hair except for the last segment. The tips of the antennae are directed sidewards and the tips of the legs reach the fifth abdominal segment.

Distribution. The pink boll worm was first described from India in 1843, and there is reason to believe that this country is its original homeland. HOLDAWAY (1926), on the other hand, believes that the pest originates from Australia because other species of the genus *Platyedra* occur there, and because this insect infests plants other than cotton, namely *Notoxylon australis*. Whichever the country of origin, the pest has spread and invaded many cotton growing countries. In Asia the spread of the pest was more eastward and south-eastward than westward. The reasons, according to PEARSON (1958), are the hosts. *Gossypium herbaceum*, the more resistant cotton variety, spread to the countries west of India, while *Gossypium arboreum*, the susceptible variety, spread eastward carrying with it the pest. Thus the pink boll worm is quite prevalent in Burma, Malaya and the Philippine Islands but is not found in Transcaucasia, Iraq and Turkestan. Its penetration into southern Iran and Israel is probably of a more recent date and took place through a different route.

The penetration of the pest into Africa was probably through Madagascar and Tanganyika. From here it spread in two directions; Northward through Kenya to Somalia (1910) and southward to Mozambique and Nyasaland (1939). The penetration into Egypt was independent from this line of march; it was introduced there with infested seeds directly from India early in this centruy. The pest was first discovered there in 1910. From Egypt the pest spread to the Sudan and other countries in Central Africa. Today it is found in most of the cotton growing countries in Africa. It has not yet been recorded in Rhodesia and the Union of South Africa.

In Australia the pest is an old resident. It reached the U.S., Mexico and Brazil in 1916 through seeds imported from Egypt.

Hosts. Plants of the genus *Gossypium* are the most attractive to the pink boll worm but not all species are equally preferred. As mentioned above *G. arboreum* is more attractive than *G. herbaceum*. In the U.S., *G. thurberi* attracts the moth while other non-cultivated species are not attacked at all. The genus *Hibiscus* is second in preference. In the Sudan as well as in India and the U.S. *H. esculentus* (okra) is a preferred host, especially when cotton is not available. In West Africa *H. canabinus* and in Uganda *H. marcanthus* are attacked. Other species are less attractive to the boll worm. Other genera whose species attract the moth are *Thespesia, Th. lampas* in Java, and *Notoxylon, N. australis* in Australia. The genera *Althaea, Sida* and *Abutilon* are seldom attacked.

Type of Injury. Injury may be caused to buds, squares, flowers and bolls. Loss is incurred through the fact that the plant produces buds which never mature into bolls. When very soft bolls are attacked they drop prematurely. In more mature bolls, the larvae feed on the seeds causing disturbances in the development of the boll.

The cotton is then of an inferior quality; the presence of larvae during the ginning time may mar its appearance and lower its marketing value. The situation is accentuated at the end of the season when the pest activity has increased and the potentialities of the plant — to replace injured fruit — have decreased.

In Egypt, before measures were taken against the pest, the losses caused by the pink boll worm were estimated at 50%. Today, in spite of measures of control, the loss to the cotton industry is estimated at 17%.

Life History. After the hibernation period, the adults begin to emerge early in the spring and continue until the end of August. These data are based on observations made both in Egypt and India. The moths are active only at night; if a moth happens to emerge during the day it hides and becomes active at dusk. According to BUSCK (1917) in Hawaii the height of activity is from 6.5—8 in the evening.

LOFTIN (1921) observed that oviposition may begin as early as April. The eggs are laid singly or in groups on all parts of the plant. They are usually laid in folds, along the leaf veins, in grooves, or between the leaflets of a bud. However, most of the eggs are laid on squares and bolls. LOFTIN in Mexico found 51% of the total oviposition on bolls. Many eggs may be laid on one boll; BUSCK counted as many as 20 eggs per boll.

The incubation period lasts three or more days depending upon the temperature. Periods of incubation according to various sources are given in Table LX.

Table LX.

Length of Development in days of the Various Stages of the Pink Boll Worm in Various Countries

Source / Stage	According to LOFTIN (1921) Mexico	According to BUSCK (1917) Hawaii	Data collected by PEARSON (1958)		
			Kenya, Congo, Somalia, Uganda	Egypt & India Summer	Egypt & India Winter
Incubation Period	3—12	4—12	4— 6	3— 4	7.5
Larval Period	8—16	20—30	12—20	9—14	19
Pupal Period	6—20	10—20	9—13	8—10	13

Upon hatching, the larva begins to bore into the tissue of the boll. Not all the larvae succeed in penetrating and not all the larvae which enter the boll survive to maturity. Larvae which hatch from distant eggs crawl about until they find a boll; yet OHLENDORF (1926) found that the population of larvae in bolls does not generally increase due to invasion of larvae from sites outside the boll.

The entrance hole is very small and is difficult to detect. A reddish spot surrounds the entrance hole, but such spots may be due to other causes as well. Thus estimations of the rate of attack are made only after bolls are opened. The entrance site of the pink boll worm is usually on the exposed part of the fruit and not under the sepals as is the case with other pests. Out of 5790 entrance holes OHLENDORF found only 113 under the sepals.

The larva bores directly into the center where the seeds are located; often the young caterpillar may wander about in the peel before it bores directly to the seeds. Young soft bolls are consumed in their entirety, in older bolls the seeds alone are consumed. Mature seeds cannot be penetrated except by larger larvae. Thus the chances of seed infestation depend upon both the age of the boll and the age of the caterpillar. As a rule the larva completes its development in one square or one boll unless these parts drop prematurely. In such cases the larva dies if it is too young; otherwise it continues to feed on non-living organic matter, an entirely different situation than is the case with *Earias*.

When a bud is attacked, the larva feeds upon the pollen and ovules. A bud attacked in this way develops into an abnormal flower with petals only partly open.

Upon completing its development, the larva makes a 2 mm hole in the cortex of the boll, leaves it and drops to the ground where it pupates under rubbish or dried leaves. This habit is characteristic in the summer of warmer countries. In a cold weather the procedure is different; an exit tunnel is bored to the cortex, but a layer of epidermis is left over the future exit hole. The caterpillar returns to the seed and pupates in it; the maturing adult thus finds a ready made exit way. Often the boll opens before pupation, larvae may be then seen crawling in the cotton and pupating therein.

Under certain conditions the development of the larva is arrested, and it enters a state of diapause. The factors affecting this and the length of the diapause period will be discussed separately below. The discussion here will be limited to non-diapausing larvae; Table LX contains records of development collected by PEARSON, by BUSCK of Hawaii and LOFTIN of Mexico. It should be added that the length of development also depends upon the food of the larva. Bollfed larvae develop 30% faster than flower-fed individuals. When the larva reaches the pupation stage it spins a cocoon. Anteriorly this cocoon is loose in order to enable the moth to emerge. The cocoon is spun in the ground or above it under rubbish.

The diapausing larva spins a more compact cocoon. The larva remains inside curled head to tail throughout the diapause period. The larva which fed upon the seeds diapauses in the seed shell, and the cocoon is spun within it. When the seeds are too small, two or even more seeds are attached together to form a large capsule in which the larva hibernates. As a rule the larva does not leave the diapausing site unless the environmental conditions change. Thus, if infested seeds are found in moist soil which causes the awakening of the larva, this latter makes its way to the surface of the soil, spins a silken lining in the tunnel and pupates inside it. The pupal period of both types, whether diapausing or not, lasts a week or two depending upon the temperature (see Table LX).

Most of the eggs are laid during the early period of the life of the female. Often most of the eggs are laid during the first day. The number of eggs per female varies according to the larval food and the climate. Thus BUSCK claims that females in his rearings laid, on the average, 100 eggs, while WILLCOCKS (1925) in Egypt gives an average of 400—500 eggs per female. The adult may lay her eggs without feeding, but when sugar syrup is supplied, the length of life is doubled from 7 to 14 days. These records of LOFTIN are identical with those obtained from Africa or India.

Diapause. It was stated in an earlier paragraph that under certain conditions the larva may enter a state of diapause for a definite period. Special attention was given to this subject by various entomologists in several countries. Efforts were made to study the causes of this diapause period, its time of occurrence and the time of its termination. An extensive literature has accumulated on this subject as a result of observations and statistics. Due to the distribution of the pest in various countries of varied climates there are apparent discrepancies and contradictions. After reviewing and analyzing this literature PEARSON's discussion on this subject will be followed. Also FIFE (1949) points out the accepted view that not one but several factors are responsible for the diapause in the pink boll worm. Each one will be discussed separately.

Food. Reports from various countries point out that diapausing larvae are found particularly in mature bolls. When an analysis is made of both kinds of bolls it is found that soft immature bolls contain 80% water while mature bolls contain only 20%. The former are rich in sugar while the latter are rich in fats and protein. Counts by SQUIRE (1939) showed that in bolls of low water content 62% of the larvae enter a state of diapause, while only 5% do so in bolls rich in water. OWEN & CALHOUN (1932), on the other hand, point out that a great many larvae enter a state of diapause although their food consists of soft buds and flowers.

Those in favour of the view that food is a factor inducing diapause in the pink boll worm point out that the highest percentage of diapausing larvae occurs at the end of the cotton season when most of the bolls are mature. In opposition, it is pointed out that even at the end of the season some larvae remain active and do not enter the state of diapause.

In summing up this discussion it may be stated that food may act as a factor inducing diapause, but it is not the only factor.

Humidity. FIFE of Porto Rico showed by tables based on statistical observations in the field, that from 20—100% of larvae diapaused when precipitation was less than five inches (12.7 cm); when the rains were more plentiful, above 6 inches (15.2 cm), the number of diapausing larvae was only 8%. A similar situation exists also in the Sudan, and in certain islands of the West Indies, such as Barbados and St. Vincent. On the other hand, observations in other islands of the West Indies, in Antigua and Monsaret showed that the high percentage of diapausing larvae occurs in August-November at the height of the rainy season. Furthermore, a survey in arid countries shows that diapause does not occur during the dry season, while in rainy countries it takes place in the dry season.

Temperature. Low temperature may retard the development of an insect even to a point of standstill without inducing the true diapause. With some insects low temperature may induce true diapause. In India it is stated that most of the larvae enter a state of diapause when the temperature drops to 15° C. In other countries, on the other hand, the temperature never drops to that level, yet diapause takes place nevertheless. It is evident therefore, that low temperature is not the primary cause which induces diapause.

Photoperiodicity. Certain facts point out that photoperiodicity is an important factor in inducing diapause in the pink boll worm.

1. In the Equatorial Zone between latitude 7.5° N and 7.5° S, where no distinct difference exists between the length of day and night, pink boll worm larvae do not enter the state of diapause except for rare exceptions; (PEARSON).

2. In countries distant from the Equatorial Zone, where there is a distinct difference between the length of day and night, diapause occurs during the short day period. It is true that some larvae may enter diapause as early as August, but the percentage of such larvae rises suddenly after the autumn equinox;

3. Data collected by PEARSON show that the farther a country is from the Equator, and the larger the difference between the length of day and night, the earlier the larvae enter the state of diapause, and vice-versa. In countries close to the Equator, diapause may be as late as January.

It may be assumed that the pink boll worm needs a short day of at least 11.5 hours.

The photoperiodicity factor explains why larvae may enter diapause in spite of high humidity, high temperature, and in spite of food rich in water. This factors also explains why the time of diapause may be different in islands which have the same climate and agricultural conditions but which are located in different latitudes.

The tables given by FIFE, in which he shows that precipitation below 5 inches (12.7 cm) induce diapause, were analysed by PEARSON who claimed that the high percentage of diapausing larvae coincides also with the short day season. (Fig. 129).

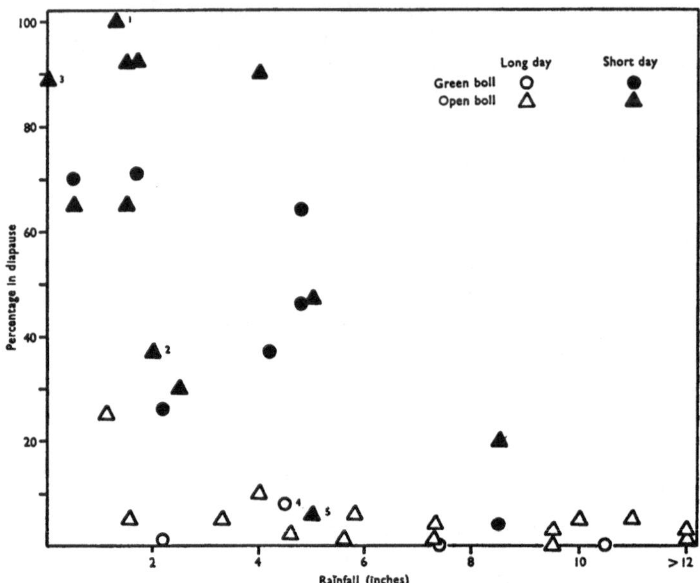

Fig. 129. Relationship between Diapause in *Platyedra gossypiella* and Photoperiod and Rainfall during Development.
Each point represents the proportion in diapause amongst a batch of larvae (usually several hundred) collected in green or open bolls in the field in Puerto Rico (approx. 18° 30N.), plotted against the rainfall in the 30 days preceding collection. Larvae collected between 21st March and 21st September treated as having experienced long-day conditions, with the exception that batches collected on 26th March (3), 17th April (1) or in early April (2) have been treated as short-day, and on 30th September (4) as long-day cases. Batch (5) was collected on 21st November. (Data derived from FIFE 1949 as follows: table 2 (22 values), fig. 3 (12 values), table 5 (3 values not included in table 2) and table 7 (single mean value)).
From PEARSON – Courtesy of the Cmwlth Inst. Ent.

But even with photoperiodicity a definite stand should be taken only after the confirmation by experiments.

Time of Diapause. In Egypt, cotton matures and is picked during the autumn. Diapause begins in September and a sharp rise in the percentage of diapausing larvae occurs in November-December. Here, ripe fruit and photoperiodicity are dominant factors. For the same reasons the rate of diapausing larvae in India rises from 3% early in October to 81% toward the end of the month. Similar conditions exist in Texas with 2—3% diapausing larvae in September, and 73% in October. In Nigeria, at 11° NL, where cotton is sown in September, the diapause occurs by the end of October and continues till January. A similar situation is found in the Gezireh of the Sudan at 15° NL and in Central India where cotton matures in January-December.

The Time of Awakening. The time of awakening from the diapause does not necessarily depend upon the length of the diapause period. The moths of larvae which entered the diapause period early may

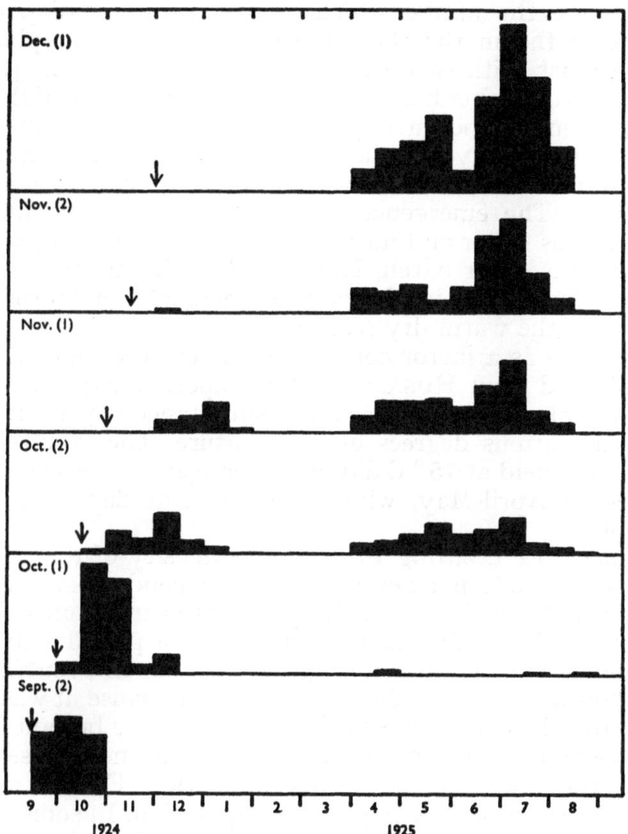

Fig. 130. Patterns of Emergence of adult *Platyedra gossypiella* derived from Larvae formed at different Stages of the Cotton Crop.
Histograms based on data obtained in the Punjab, 1924, for total moths emerging each fortnight from weekly collections of green bolls. For convenience of illustration, the collections and the emergences have been grouped by half-months. This may introduce a slight error, since the numbers of bolls in successive weekly collections differed. Arrows indicate start of each two-week collecting period. All emergences on same scale.
(Data from Bindra 1928, Table XII).
From Pearson – Courtesy of the Cmwlth Inst. Ent.

awake together with those that entered diapause later. The emergence period of moths may last a few months, but there is a climax which extends over a shorter period, and this climax is typical for each country. Organized observations were made by BINDRA (1928) in India (quoted from PEARSON). According to these, larvae were collected from the field every two weeks throughout the September-December period. The emergence dates of the moth are given in Fig. 130. From this it is evident that the later the larvae were collected the greater was the number which entered a state of diapause. The emergence of the moths the following summer took place from April to August, with two peaks, one in May and one in July.

Certain factors either hasten or retard the emergence of the moths. Humidity seems to be an important factor. FIFE collected larvae from the field. Thirty-two days later he placed one batch of the larvae in humid environments while another batch he placed in dry surroundings. The emergence of the moths from the humid environments was faster and more uniform than the emergence of the moths from the other batch. However, high humidity is not absolutely necessary since in Egypt the emergence of moths usually occurs during the warm dry season.

Temperature as a factor controlling the emergence of the moths may be studied from HUSAIN'S (1931) experiments: Larvae which entered the state of diapause were held simultaneously, from January onwards, in various degrees of temperature. The results were as follows: Those held at 15° C did not emerge at all, those held at 20° C emerged in April-May, while those held at 25° C emerged in March-April.

Favourable and Limiting Factors. The density of the population of the pink boll worm in a new cotton field depends upon the number of moths which survive from the previous year's population. The availability of hosts during the intermediate period plays a role only in the case of non-diapausing individuals. In countries where the pink boll worm does not enter a state of diapause, it was pointed out that after the cotton is picked, volunteer crops become infested. Thus the cleanliness of the fields during the intermediate season will determine the size of the surviving population. Food is a decisive factor also in the case of a diapausing population. In some countries dried cotton plants are collected from the fields and preserved as fuel. These remain throughout the winter, and in the spring swarms of moths leave the dry bolls that were preserved along with plants. In order to reduce the diapausing population it is advisable to bury plant remains in the ground by plouging soon after harvest.

In the cotton field, the build up of the pink boll worm population depends upon the following factors: 1. temperature; 2. the length of the cotton season; 3. the initial date of infestation.

1. To a certain limit the higher the temperature, the faster the

development of the generation; thus in colder climates, less generations develop during the season and vice versa. However, in a country where the average temperature rises to 30° C and above the population decreases.

2. The term "cotton season" means the period from sprouting to maturation. Certain varieties of cotton mature faster than others; in this case the pest raises a smaller number of generations than when the variety stays in the field for a long period.

3. The term "initial infestation" refers to the date when this pest first establishes itself in the field. An example of this may be taken from the situation in Egypt. There the moths begin to emerge in May and continue to do so till October (See fig. 131). In the case of annual cotton no bolls are available until June. Thus the early moths find no hosts upon which to lay. In the case of ratoon cotton, however, there are flowers and buds already in May and thus an additional generation may develop in that field.

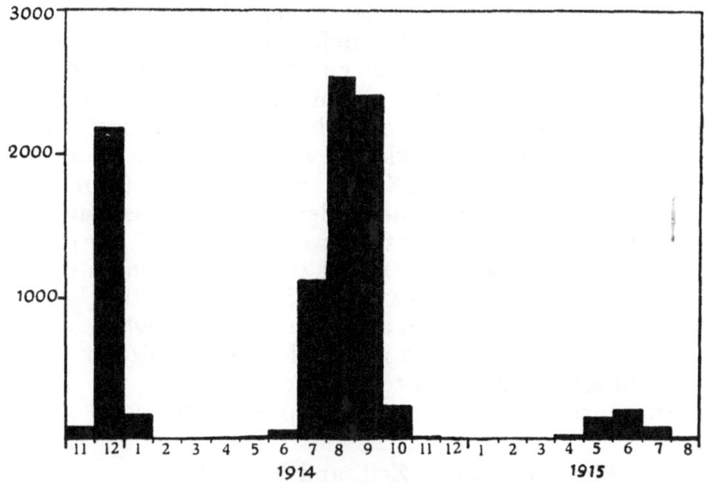

Fig. 131. Emergence of adult *Platyedra gossypiella* in Egypt
Numbers of moths emerging per month from 50,000 bolls collected in early November, 1913, at Bahtin, Egypt, and kept under conditions characteristic of those in heaps of cotton sticks after harvest
(Data from Willcocks 1916)
From Pearson – Courtesy of the Cmwlth Inst. Ent.

Observations indicate that moths may fly and migrate to another field after the supply of food becomes exhausted in the field. This is the only explanation as to how a new field becomes infested when the source of infestation is several hundred metres away. Moths were also caught in aeroplanes in altitudes of 1000 metres.

Control

a. Preventive Measures. In order to maintain the lowest level of the moth population, it is essential to prevent the favourable factors discussed above.

The field should be ploughed and plant remains buried as early as possible after picking. Ratoon cotton should be avoided; volunteer cotton should be destroyed before blossoming. All these measures, when strictly carried out, will help keep the population at low levels. A short season variety of cotton greatly reduced moth damage after it was introduced in Egypt. The practice of gathering dry cotton plants for fuel should be abolished.

b. Seed Treatment. In order to prevent the carry-over of the pest in the seed, these should be treated. If destined for export the treatment should be undertaken before the emergence of the moths. Seed for sowing should be heated for 5—10 minutes at a temperature of 55—65° C. This treatment will killl larvae and pupae of the pest and will not injure the seed. Seed may be also fumigated with hydrocyanic gas, carbon bisulphide or methyl bromide.

c. Treatment in the Field. Before the synthetic insecticides were adopted in agriculture it was customary to spray or dust the cotton plants with compounds of arsenic and flour. These treatments did not yield satisfactory control. When DDT was first applied it gave far better control. In Mexico it was customary to use a dust containing DDT 10%, BHC 2% and sulphur 40% (Rude, 1954). The latter was aimed against the spider mites. It was claimed that DDT alone or mixed with BHC at the rate of 200 g.a.i. per dunam — 5—6 applications — reduced 99% of the infestation as compared with non-treated plots. Rainwater (1956) showed that gusathion 25—100 g a.i. per dunam, gave a satisfactory kill, when applied at weekly intervals it reduced the infestation from 53% to 1.3% and the number of larvae from 1.4 to 0.2 per boll.

For further insecticides — see the chapter on "General consideration".

Platyedra vilella Zell. and Platyedra sp.

During the spring, cotton plants in Israel are attacked by larvae which bore into the soft stem from the top and cause damage similar to that caused by the spiny boll worm early in the season.

Their colour is white with the thorax pink; in the one the prothorax only, in the other also the mesothorax. One of the species is *P. vilella* Zell., which was recorded as a cotton pest from Iran, Iraq, Turkestan, Transcaucasia and Morocco.

As a rule their infestation is low, not above 8—10%, and their population dwindles further as the summer advances, so that they cause little concern. Measures against other pests control it too.

Apparently they invade cotton fields from wild Malvaceae.

General Consideration in the Control of Cotton Pests in Israel

As mentioned above cotton in Israel is a recently re-established crop. There is reason to believe that it was not grown before because of the pests which destroyed the blossoms and bolls. Today, not only have the pests been overcome but the yields in this country are of the highest in this geographic zone. Successful pest control is not the only reason for high yields. Intensive cultivation, proper timing of irrigation and lavish fertilization are also responsible. However, without adequate control of the pests there would be no cotton to pick from the healthy stands of cotton plants, as was the case in Israel in 1953 and 1956. In view of this it is advisable to follow pest problems and their control in Israel where conditions are similar to those of other cotton growing countries in the Near East.

The cotton plant is a source of food for insect pests from the time of seeding until it is destroyed, for it is attacked before the seeds sprout and pests inhabit it even after the cotton is picked.

It is wise here to follow the plant chronologically and along with it the pests which attack it.

The seed of cotton may be exterminated before it sprouts by larvae of *H. cilicrura*. In this case soil treatment is recommended since the treatment of the seed may interfere with the mechanized seeding. However, this prophylactic measure is not necessary unless rains are late and the temperature during March and April is very mild; otherwise the seed escapes damage by this pest.

Cutworms, particularly *Agrotis ypsilon*, may injure young seedlings. At the same time the leaves of the young plant may be infested with *Thrips tabaci*, with *Aphis gossypii* and with *Tetranychus telarius*. Occasionally *Laphygma* larvae may also be found feeding on the foliage. As the plants grow the stem may be bored by larvae of *Earias* and by the so called "white larva" *Platyedra* sp. If a few of these infest the plants, an application of endrin + phosphamidon, 100 cc endrin + 80 cc of phosphamidon a.i./dunam, will control any of the pests mentioned above, except the mite.

For each pest alone it is not always advisable to employ insecticides. *Tetranychus* infests young seedlings of cotton particularly when the field is adjacent to a beet field infested with this pest. Observations have shown that as the temperature rises in June the population of this pest in the cotton fields dwindles and the plant overcomes it. The population may increase again in August-September, but this does not always occur. Both field and laboratory experiments have shown that kelthane, tedion and morocide are the best acaricides.

The infestation by *Thrips* is also temporary. As with peanuts, fresh cotton fields attract the thrips from surrounding uncultivated fields in which the flora has withered and dried. With the rise of the temperature the thrips population also dwindles, so that if it is the only pest in the field, insecticide application is superfluous. The deformed leaves may be a cause for concern, but in irrigated fields the plants overcome this. In non-irrigated fields the plants may succumb to this injury and control measures should be carried out in due time.

The activity of *Earias* larvae and of *Platyedra* sp. in the young stem of the plant is not very injurious, very often the plants overcome it as described by BALLARD (1921). Furthermore, observations carried out since 1953 point to the fact that the *Earias* population diminishes and remains at low levels for a few weeks during the period of late June-early August (See fig. 132). The *Platyedra* species also disappears. The only serious pest during these high summer months in the Israeli cotton fields may be the leaf worm *Prodenia litura*.

During August, early or late, depending upon the weather of that year, the cotton fields are reinhabited by numerous pests both in

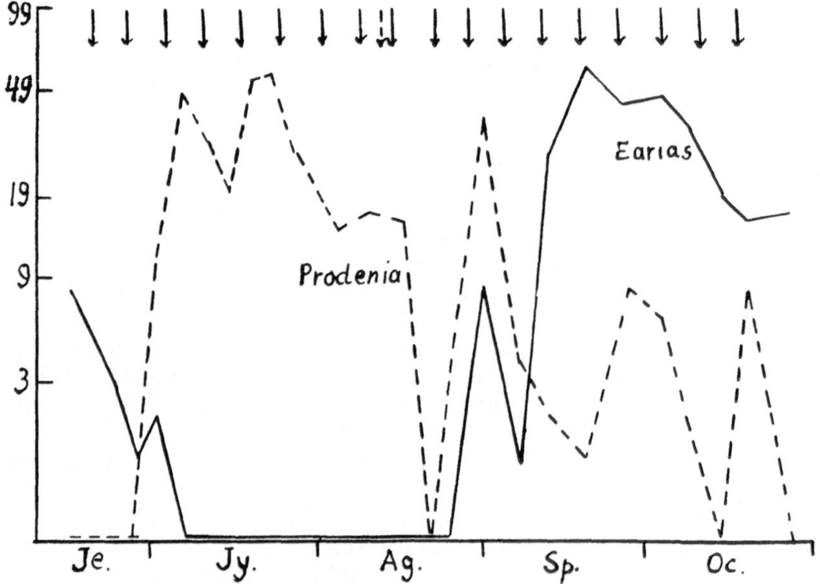

Fig. 132. Fluctuations of *Earias* and *Prodenia* populations in a cotton field at Beit Dagan in 1960.
Note the *Earias*-free-period during July and August, the rise of *Prodenia* during this period, and that of *Earias* late in the season.
The scale (logarithmic) indicates the number of larvae found on each 60 plants picked at random. The field was treated with sevin dusts at weekly intervals (arrows).
After RIVNAY & YATHOM.

numbers and in species. By then the pink boll worm has had a chance to build up a noticeable population: *Heliothis armigera* may be more numerous than when the plants were small; a new strong generation of this pest may rise in the cotton fields; and finally the spiny boll worm may sharply increase in numbers of individuals per area. So much for lepidopterous caterpillars. Of the sucking insects the white fly, *Bemisia tabaci*, may rise in numbers, *Aphis gossypii* may cause the plants to become sticky from honey dew excretions, *Empoasca* may appear in the cotton fields; and spider mites may render the foliage chlorotic.

The extermination of a few or all of the pests in one stroke is preferable to controlling each pest separately.

A few principles should be considered in order to achieve successful control of cotton pests.

a. No pesticide should be applied unless absolutely necessary. Insecticides may upset favourable biological balance and cause more harm than good. As pointed out above, application against thrips is superfluous. Early in the season, applications against *Earias* may also be unjustified, if the infestation is not very severe.

b. If possible, an insecticide should be chosen which will kill the pests in all their stages.

c. Insecticides in a combination of two or three should be selected which kill as many of the pests as possible. Thus, DDT, which was the least efficient against *Earias*, proved best against *Heliothis armigera*, or gusathion which was one of the best against *Earias* larvae was less efficient against *Prodenia*. Adding a systemic insecticide will help to destroy sucking pests if present.

d. A thorough cover of the plant is essential. Many pests attack the new soft foliage and fruiting parts which are, as a rule, on the top of the plant. Thus the cover of the upper parts of the plant is essential. However, the spiny boll worm, when it drops to the ground within an injured boll or bud, crawls back to the plant and attacks the lower nearly ripe bolls. If these are unprotected they may be destroyed. Applications of insecticides should therefore cover the entire plant. In case of dense foliage, shields should be built in front of the nozzles to clear the foliage and allow a thorough spray.

e. Night applications are advisable as many of the cotton pests are night feeders and it often happens that insecticides, due to the heat and sun radiation, deteriorate too quickly; by the time the larvae come in contact with their residues they are no longer effective. Night applications proved far superior to those carried out during the day.

f. The frequency of application should be established according to conditions in the cotton field. Early in the season, for instance, before fruiting, more tolerance may be shown to pest infestation than later or when the plant is laden with flowers and fruit. During this critical period a threshold of infestation should be established according to which is decided the need of control measures. In Israel, for instance, 2—4% of *Earias* infestation calls for applications: thus early in the season weeks may pass without application while later they may be repeated at five days intervals.

The effectiveness of a pesticide against a certain pest is not permanent and stable. As in other places so in Israel changes in the efficiency of certain insecticides were witnessed. For instance, until two or three years ago *Tetranychus* was satisfactorily controlled by parathion but this is no longer true today. Thus pest control recommendations change frequently. New substances may appear which prove superior to those which were in practice. It is worthwhile, then, to survey the findings and practices regarding cotton pest control in Israel since 1953.

In 1953 small plots of cotton were treated with DDT and BHC dust against *Earias insulana*. The pest was not controlled and the yield was entirely exterminated. Endrin was introduced and during 1954 and 1955 it proved to be the most effective substance against the spiny boll worm in the field. Laboratory tests during that period showed this substance to be an effective contact insecticide against the adults, and that the residues remained effective for 4—5 days. Second to this substance in efficiency were parathion and malathion, while DDT was not all effective (Rivnay & Yathom, 1956). Field experiments during these years showed that endrin, 50—70 g a.i./dunam, was more effective than toxaphene, 300 g a.i./dunam, and that doubling the dosage of endrin increased its control to a very small extent, while cryocide hardly controlled this pest. (Rivnay & Yathom, 1957).

In 1956 the *Earias* population was so dense and infestation so severe that the customary application of endrin was ineffective and the entire yield was exterminated. The experimental plots were also destroyed.

Field trials in four localities in Israel in 1957 (Rivnay & Yathom, 1958) showed the following:

1. Gusathion and endrin, when applied in the same quantities per area, gave the same results in the yield of cotton.
2. When *Prodenia litura* was the major pest, as was the case in Gilat, endrin was superior to gusathion.
3. When malathion or dipterex were added to the gusathion, at the rate of half active ingredient of each, and compared with gusathion alone no significant difference was obtained in the yield of cotton; the major pest was *Earias*, and the infestation was not over 10%.

In 1958 *Prodenia* was the major pest in all the experimental localities. At Gilat and the Jordan Valley, when applied in the same quantities of act. ing. per area, endrin was superior to gusathion;

when dipterex was added to gusathion, giving thus a double dosage of act. ing., endrin remained superior (RIVNAY & YATHOM, "Ktavim" — in press) Laboratory experiments showed that gusathion was effective against the larvae of *Earias* but had very little effect against the adults.

The field experiments in 1959 showed that sevin, whether as a wettable powder or dust, at the rate of 200—400 g a.i. per dunam, resulted in the highest yield of cotton when compared with endrin, gusathion or dipterex. Again *Prodenia* infestations were very severe.

In the laboratory, sevin and gusathion were compared against the larvae of the spiny boll worm. Sevin residues remained active longer (see fig. 127).

During these past three years in general practice, parathion was used in addition to the above mentioned insecticides. In 1960 again, *Prodenia* infestations were very severe. In order to strengthen cottonion — a local brand of gusathion — DDT or toxaphene were added against *Prodenia*. Other substances used were sevin in two formulations, endrin, and endrin together with phosphamidon. Judging from the yields of cotton sevin in both formulations gave the highest yield; next was endrin, especially when enforced with phosphamidon. (RIVNAY & YATHOM, 1960). Also laboratory tests showed that the combination of endrin and phosphamidon was more

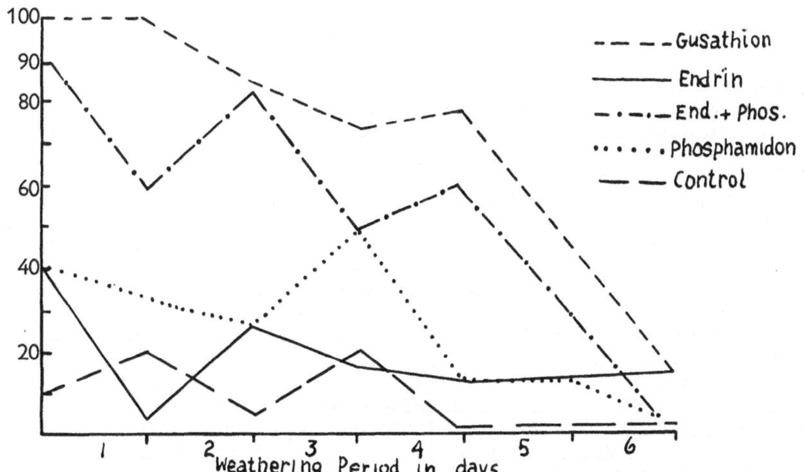

Fig. 133. The effects of endrin and phosphamidon used separately or in combination, on the spiny boll worm – compared with the effects of gusathion.
The concentrations of the emulsions in percentage of active ingredients were: gus. 0.2; end. 0.2; phos. 0.22; end. + phos. 0.11 + 0.11. Further explanations as in fig. 127.
After RIVNAY & MEISNER.

effective against *Earias* than when each of them was employed separately at the same concentration (Fig. 133). These tests showed also that zectran* is effective against the larvae of this pest. (Fig. 127) (RIVNAY & MEISNER, 1962).

That summer, mite infestation was severe in some localities. The employment of kelthane and tedion gave adequate relief from this pest. In general practice it was discovered that even high dosages of parathion did not kill the red spider mite, and in the laboratory ASHER & CWILLICH (1960) showed that resistance to phosphorous compounds had apparently developed in this pest in some localities.

From practice it was discovered that neither endrin nor parathion, nor sevin satisfactorily killed large mature *Prodenia* larvae. The only effective means of extermination was brought about by the application of 150 cc a.i. parathion together with 100 cc a.i. endrin. This practice was used in many localities, and farmers claimed that it was the only way to save the yield of cotton from *Prodenia* larvae.

In Egypt five applications, at 14 days intervals of gusathion, DDT and endrin, at the rates of approximately 70, 100 and 35 g ac. ing. per dunam, reduced the infestation of *Earias* and *Platyedra* by 72% (KAMEL et al., 1958). The same authors showed in 1960 that lebaycid was more effective than gusathion against *Prodenia, Earias* and *Platyedra* (KAMEL & SHOEB, 1960).

LITERATURE

T. AHMAD & GHULAMULLAH, 1941. *Indian J. Ent.* **3,** *254-284.*

S. ASHER & R. CWILLICH, 1960. *Ktavim* **10,** *159-163.*

A. BALACHOWSKY & L. MESNIL, 1936. Les insectes nuisibles aux Plantes cultivées. Paris.

E. BALLARD, 1921. Proc. 4th ent. Mtg. Pusa. *70-83.*

S. S. BINDRA, 1928. *Mem. Dept. Agric. Indian Ent.* **10,** *167-216.*

A. BUSCK, 1917. *J. agric Res.* **9,** *343-370.*

I. BISHARA, 1936. *Ministry of Agric. Egypt. Tech. and Sci. Serv. Bull.* **163,** *1-23.*

G. C. DUDGEON, 1912. *Agric. J. Egypt* **5,** *45-48.*

G. C. DUDGEON, 1916. Trans. 3rd Mt. Cong. Trop. Agr. 1914, London, 36 pp.

L. C. FIFE, 1949. *U.S. Dept. Agric. Washington, D.C., Tech. Bull.* **977,** *1-26.*

T. B. FLETCHER, 1921. Sci. Repts. Agric. Res. Inst. 1919-20 Calcutta, *68-94.*

W. J. HALL, 1921. *Ministry of Agric. Egypt. Tech. and Sci. Serv. Bull.* **17,** *1-28.*

F. G. HOLDAWAY, 1926. *Bull. ent. Res.* **17,** *67-83.*

M. A. HUSAIN, 1925. *Ann. Rept. Ent.* 1923-24. Rep. Dept. Agr. Punjab.

M. A. HUSAIN, 1931. *Ann. Rept. Ent.* 1930. Rep. Dept. Agr. Punjab.

M. A. HUSAIN & M. M. LAL, 1920-23. *Repts. Dept. Agr. Punjab,* Pt. 2, 1918, *153-157;* 1919, *173-184;* 1920, *52-67.*

A. A. M. KAMEL, A. SHOEB, A. HANNA & S. SOLIMAN, 1958. *Agr. Res. Rev.* **36,** 1.

A. A. M. KAMEL & A. SHOEB, 1960. *Agr. Res. Rev. (Cairo)* **38,** *27-46.*

H. H. KING, 1918. *Wellcome Trop. Res. Lab. Khartoum, Ent. Bull.* **8,** *4.*

T. W. KIRKPATRICK, 1923a. *Ministry of Agric. Egypt. Bull.* **33,** *1-15.*

---

* Trademark of the Dow Chem. Co.

T. W. KIRKPATRICK, 1923b. *Ministry of Agric. Egypt. Tech. and Sci. Serv. Bull.* **35,** *1-106.*
U. C. LOFTIN, 1921. *U.S. Dept. Agric. Bull.* **918,** *1-56.*
C. MASON, 1915. Report of the Entomologist for the year 1915. — Dept. Agric. Nyasaland.
W. OHLENDORF, 1926. *U.S. Dept. Agric. Bull.* **1374,** *1-64.*
W. L. OWEN & S. L. CALHOUN, 1932. *J. econ. Ent.* **25,** *746-751.*
E. O. PEARSON, 1958. The insect Pests of Cotton in Tropical Africa, London, 339 pp.
C. F. RAINWATER, 1956. *Agr. Chem.* **11,** 2, *32-33.*
E. RIVNAY & Y. MEISNER, 1962. Lab. Tests against Earias larvae. Pro. Rep. No **372,** Agr. Res. St. Rehovoth.
E. RIVNAY & S. YATHOM, 1956. *Ktavim,* **7,** *59-62,* 8.
E. RIVNAY & S. YATHOM, 1957. *Ktavim* **8,** *57-63.*
E. RIVNAY & S. YATHOM, 1958. *Ktavim,* **9** *3-17.*
E. RIVNAY & S. YATHOM, 1960. Field Trials against Noctuids in Cotton in 1960. Pro. Rep. No. 318. Agr. Res. St. Rehovot.
C. S. RUDE, 1954. *J. econ. Ent.* **46,** *1038-1041.*
F. A. SQUIRE, 1939. *Emp. Cott. Gr. Rev.* **16,** *194-196.*
T. H. C. TAYLOR, 1936. *Rept. Dept. Agr. Uganda,* 1935-36, Pt2, *19-39.*
P. VAYSSIERE & J. MIMEUR, 1925. *Agr. colon.* **85,** *6-14.* Paris.
R. L. WALKER, 1954. *J. econ. Ent.* **47,** *367.*
F. C. WILLCOCKS, 1916. Insects and related Pests in Egypt. pt 1, 393 pp.
F. C. WILLCOCKS, 1925. Insects and related Pests in Egypt. p. 418 pt. 2.
S. YATHOM, 1956. *Ktavim* **7,** *43-57.*
V. ZANON, 1921. *Riv. Agric. Parma,* **XXXVI,** 1-2, 5, *23-24.*

AUTHOR INDEX

ARTHROPOD INDEX

Bold type: page where a lengthier description is given; Italics behind the names of families: A pest belonging to that family is described there at some length

PLANT INDEX

Bold type: pages containing the keys to the pests.

SUBJECT INDEX

In some chapters the reader will find lengthier discussions on certain subjects of general entomological interest as follows: